Environmental History

Volume 8

Series editor

Mauro Agnoletti, Florence, Italy

More information about this series at http://www.springer.com/series/10168

Ana Isabel Queiroz · Simon Pooley
Editors

Histories of Bioinvasions in the Mediterranean

Springer

Editors
Ana Isabel Queiroz
IHC—Institute of Contemporary History, NOVA-FCSH
Lisbon
Portugal

Simon Pooley
Department of Geography
Birkbeck, University of London
London
UK

ISSN 2211-9019 ISSN 2211-9027 (electronic)
Environmental History
ISBN 978-3-030-09120-0 ISBN 978-3-319-74986-0 (eBook)
https://doi.org/10.1007/978-3-319-74986-0

Printed on acid-free paper

This Springer imprint is published by the registered company Springer International Publishing AG part of Springer Nature
The registered company address is: Gewerbestrasse 11, 6330 Cham, Switzerland

Acknowledgements

The editors are very grateful to the reviewers who contributed their time and expertise to peer review the chapters included in this book. We would like to thank: Janine Adams, Leo Füreder, Frederick J. Kruger, John Measey, Francisco Moreira, Stephen J. Pyne, Patrick Schembri, Paulo Sá Sousa, Andrew V. Suarez and Ioannis Vogiatzakis.

Contents

1 **Introduction: Historical Perspectives on Bioinvasions in the Mediterranean Region** 1
Simon Pooley and Ana Isabel Queiroz

2 ***Mare Nostrum, Mare Quod Invaditur*—The History of Bioinvasions in the Mediterranean Sea** 21
Bella S. Galil, Agnese Marchini and Anna Occhipinti-Ambrogi

3 **The Apparently Relentless Spread of the Major Decapod Alien Species in the Mediterranean Basin and European Inland Waters** 51
Chiara Manfrin, Catherine Souty-Grosset, Pedro Anastácio, Julian Reynolds and Piero G. Giulianini

4 **Scourge in the Mediterranean. Policies Against the Argentine Ant in the First Half of the 20th Century** 87
Ana Isabel Queiroz

5 **Herpetological History of the Balearic Islands: When Aliens Conquered These Islands and What to Do Next** 105
Iolanda Silva-Rocha, Elba Montes, Daniele Salvi, Neftalí Sillero, José A. Mateo, Enrique Ayllón, Juan M. Pleguezuelos and Miguel A. Carretero

6 **The *Xenopus laevis* Invasion in Portugal: An Improbable Connection of Science, Mediterranean Climate and River Neglect** 133
Mónica Sousa, Angela Maurício and Rui Rebelo

7 **Ecologically Benign Invasions: The Invasion and Adaptation of Common Waxbills (*Estrilda astrild*) in Iberia** 149
Gonçalo C. Cardoso and Luís Reino

8 Cordgrass Invasions in Mediterranean Marshes: Past, Present and Future 171
Bernardo Duarte, Enrique Mateos-Naranjo, Susana Redondo Goméz, João Carlos Marques and Isabel Caçador

9 The Forest on the Peninsula: Impacts, Uses and Perceptions of a Colonial Legacy in Cyprus 195
O. L. Pescott, S. E. Harris, J. M. Peyton, M. Onete, A. F. Martinou and J. O. Mountford

10 The Long and Entangled History of Humans and Invasive Introduced Plants on South Africa's Cape Peninsula 219
Simon Pooley

Index 253

Contributors

Pedro Anastácio MARE—Marine and Environmental Sciences Centre, Departamento de Paisagem, Ambiente e Ordenamento, Escola de Ciências e Tecnologia, Universidade de Évora, Évora, Portugal

Enrique Ayllón Asociacion Herpetologica Espanola, Espanola, Spain

Gonçalo C. Cardoso CIBIO/InBIO—Research Centre in Biodiversity and Genetic Resources, University of Porto, Porto, Portugal; Behavioural Ecology Group, Department of Biology, University of Copenhagen, Copenhagen, Denmark

Miguel A. Carretero CIBIO/InBIO—Research Centre in Biodiversity and Genetic Resources, Universidade do Porto, Porto, Portugal

Isabel Caçador MARE—Marine and Environmental Sciences Centre, Faculty of Sciences, University of Lisbon, Campo Grande, Lisbon, Portugal

Bernardo Duarte MARE—Marine and Environmental Sciences Centre, Faculty of Sciences, University of Lisbon, Campo Grande, Lisbon, Portugal

Bella S. Galil The Steinhardt Museum of Natural History, Tel Aviv University, Tel Aviv, Israel

Piero G. Giulianini Department of Life Sciences, University of Trieste, Trieste, Italy

Susana Redondo Goméz Departamento de Biología Vegetal y Ecología, Facultad de Biología, Universidad de Sevilla, Sevilla, Spain

S. E. Harris Eastern Michigan University, Ypsilanti, MI, USA

Chiara Manfrin Department of Life Sciences, University of Trieste, Trieste, Italy

Agnese Marchini Department of Earth and Environmental Sciences, University of Pavia, Pavia, Italy

João Carlos Marques MARE—Marine and Environmental Sciences Centre, c/o Department of Zoology, Faculty of Sciences and Technology, University of Coimbra, Coimbra, Portugal

A. F. Martinou Joint Services Health Unit (JSHU), Akrotiri, Cyprus

José A. Mateo Black Market, Palma de Mallorca, Spain

Enrique Mateos-Naranjo Departamento de Biología Vegetal y Ecología, Facultad de Biología, Universidad de Sevilla, Sevilla, Spain

Angela Maurício Departamento de Ambiente e Equipamento, Município de Oeiras, Oeiras, Portugal

Elba Montes Department of Zoology, Faculty of Biological Sciencies, Valencia University, Valencia, Spain; Environment Department, Ibiza Island Council, Ibiza, Spain

J. O. Mountford NERC Centre for Ecology & Hydrology (CEH) Wallingford, Wallingford, Oxfordshire, UK

Anna Occhipinti-Ambrogi Department of Earth and Environmental Sciences, University of Pavia, Pavia, Italy

M. Onete Institute of Biology Bucharest, Romanian Academy, Bucharest, Romania

O. L. Pescott NERC Centre for Ecology & Hydrology (CEH) Wallingford, Wallingford, Oxfordshire, UK

J. M. Peyton NERC Centre for Ecology & Hydrology (CEH) Wallingford, Wallingford, Oxfordshire, UK

Juan M. Pleguezuelos Department of Zoology, Faculty of Sciences, Granada University, Granada, Spain

Simon Pooley Department of Geography, Birkbeck, University of London, London, UK

Ana Isabel Queiroz IHC—Institute of Contemporary History, NOVA-FCSH, Lisbon, Portugal

Rui Rebelo Departamento de Biologia Animal & Centre for Ecology, Evolution and Environmental Changes, Faculdade de Ciências da Universidade de Lisboa, Lisbon, Portugal

Luís Reino CIBIO/InBIO—Research Centre in Biodiversity and Genetic Resources, University of Porto, Porto, Portugal

Julian Reynolds Emeritus, Department of Zoology, Trinity College, University of Dublin, Dublin 2, Ireland

Daniele Salvi Department of Health, Life and Environmental Sciences, University of L'Aquila, Via Vetoio, Coppito, L'Aquila, Italy

Neftalí Sillero CICGE: Centro de Investigação em Ciências Geo-Espaciais, Faculdade de Ciências da Universidade do Porto, Vila Nova de Gaia, Portugal

Iolanda Silva-Rocha CIBIO/InBIO—Research Centre in Biodiversity and Genetic Resources, Universidade do Porto, Porto, Portugal; FCUP, Faculdade de Ciências da Universidade do Porto, Porto, Portugal

Mónica Sousa Departamento de Recursos Naturais e Conservação da Natureza, Instituto da Conservação da Natureza e das Florestas, IP, Lisbon, Portugal

Catherine Souty-Grosset Laboratoire Ecologie & Biologie des Interactions-UMR CNRS 7267, Equipe Ecologie Evolution Symbiose, Université de Poitiers, Poitiers Cedex 9, France

Chapter 1
Introduction: Historical Perspectives on Bioinvasions in the Mediterranean Region

Simon Pooley and Ana Isabel Queiroz

Abstract This chapter introduces the main themes of the book and states its aims and scope. It defines and provides an overview of the study region, discussing the biophysical and historiographical dimensions of defining a 'Mediterranean region'. Key polemics arising in debates over how to think about bioinvasions are considered, and the editors' positions clarified. An overview of the book's ten chapters follows, informing a concluding discussion on directions for future study of bioinvasions in the Mediterranean region.

Keywords Bioinvasions · Mediterranean region · Environmental history
Natural sciences

1.1 Bioinvasions in Interdisciplinary Context

Since the 1980s, bioinvasions[1] have come to be regarded as one of the major threats to global biodiversity, contributing to the ecological homogenisation of the earth. In this view, the distribution of biota around the planet will no longer be determined primarily by the traditional biogeographical regions, but also by our changing climate and by the socioeconomic relationships which mediate the spread of alien species around the planet (Capinha et al. 2015). Bioinvasions are, in short, a global environmental problem of anthropogenic origin.

---

[1]The term is increasingly used in scientific literature with the same meaning as "biological invasions"; see Valery et al. (2008) for a review of existing definitions.

S. Pooley (✉)
Department of Geography, Birkbeck, University of London, London, UK
e-mail: s.pooley@bbk.ac.uk

A. I. Queiroz
IHC—Institute of Contemporary History, NOVA-FCSH, Lisbon, Portugal
e-mail: ai_queiroz@fcsh.unl.pt

A. I. Queiroz and S. Pooley (eds.), *Histories of Bioinvasions in the Mediterranean*, Environmental History 8, https://doi.org/10.1007/978-3-319-74986-0_1

The focus of studies of bioinvasions has overwhelmingly been on the harmful consequences of anthropogenic transfers of living organisms, from their places of origin to other regions where previously they did not occur. While bioinvasions are not new phenomena—well known examples include the chestnut blight in North America in the early 1900s, and phylloxera which attacked European grapevines from the 1860s to 1890s—their frequency and distribution seem to be intensifying with accelerating globalisation (Simberloff 2015). Along with accelerating climate change, they have taken on considerable symbolic valence as vectors of humanity's destructive influence over nature in the Anthropocene.[2]

Bioinvasions have ecological, socio-economic and cultural effects, and may be the least predictable of all the major forms of environmental disruption (Daehler and Gordon 1997). *Which* organisms will become successful invaders? *Where* and *when* will invasions occur? *What* kinds of impacts will they have? (see Bright 1998, pp. 25–28). Despite important advances in modelling the vulnerability of ecosystems and performing risk identification assessments (e.g. Andreu and Vilá 2010; Vicente et al. 2013; Almeida et al. 2013), there are still obvious limits of predictability (Hulme 2012) that mean that bioinvasions remain challenging for science, policy and management.

Bioinvasions have been studied mainly by the natural sciences, and the field of invasion biology or invasion ecology has become a well-established subdiscipline in the last three decades. Scientists focus on the biological characteristics and mechanisms that allow certain species, at certain times and in certain places, to become invasive. They also define the means and methods for managing them in the field (containment, control or eradication, according to particular circumstances and needs). These topics illuminate the natural phenomena, but alone do not address the complexity of the problem.

It has now been widely acknowledged that it is also vital to understand and address the human dimensions of bioinvasions (e.g. McNeely 2001; Rotherham and Lambert 2011; Frawley and McCalman 2014; Keller et al. 2015). There are many facets to the study of the anthropogenic dimensions of bioinvasions, for example: social science studies of their social and economic impacts; cultural and social variations in attitudes to and uses of invasive species; the politics of biocontrol and individual rights to import species; and philosophical debate over the ethics of defining species as 'alien' and how to eradicate them (also an animal rights issue).

Bioinvasions may take generations to take hold, at a pace too slow to offer the striking narratives necessary to engage the public imagination (at least until it is too late), and offering enough time for humans to develop strong cultural and economic relations with invading species (Essl et al. 2011; Lidström et al. 2015). On the other hand, as Lidström et al. (2015, p. 1) argue, narratives about "invasive alien species"—and particularly those developed to advocate control measures—can

[2]See Chew (2015) on the history of the narrative about bioinvasions as 'the second greatest threat' to species extinctions, and the wider resonance of this narrative with dismay over humanity's impacts on nature.

'[reduce] complex webs of ecological, biological, economic, and cultural relations to a simple "good" versus "bad" battle'. The challenge for environmental histories of bioinvasions is to devise compelling narratives which nevertheless capture this complexity and reveal how these invasions have unfolded.

We have chosen the Mediterranean region (discussed below) as our geographical focus, because of the great antiquity of human and ecological relations in the region, with a long history of internal maritime communications and trade, and from the late fifteenth century, an ever widening network of maritime communications with the rest of the planet. Historians have paid much attention to the impacts of European trade, and the suite of invasive plants, animals and diseases exported to the rest of the planet (Crosby 1972, 1986; McNeill 2003). They have paid less attention to the reverse flow (Beinart and Middleton 2004).[3] The introductions and circulations of some well known naturalised (and non-invasive) plants such as tomatoes, potatoes, tobacco and citrus are better known, and there is interesting work (mostly not by historians) on the colonisation of islands and the impacts of the introduction of domesticated fauna on the indigenous biota (e.g. Massetti 1998, 2007).

1.2 Aims and Scope

This book has three major aims. First, we aim to provoke natural scientists working on bioinvasions in the Mediterranean region to think more historically (environmental history, not historical ecology). Second, we aim to convince environmental historians to engage properly with the science of bioinvasions. Third, through sharing the research presented in the selected chapters, we aim to convince both historians and scientists of the richness of the available research materials and themes, the importance of the issues, and through doing so to stimulate interdisciplinary collaborations to explore the novel and unanswered questions which an environmental historical approach to bioinvasions brings to the fore.

The book brings together environmental historians and natural scientists to share their studies and experiences, producing narratives which draw on a bricolage of biophysical data and ecological methods, and historical sources and methods. Central to our combined endeavours is the recognition that chronological and spatial scales are crucial variables in all our narratives seeking to explain the movement of invasive species across ecosystems and landscapes. Recognizing the connections between the expansion of species and historical phenomena, an environmental history perspective suggests we focus on:

[3]But see an interesting collection from 2003 in a themed issue of *Landscape Research* 28(1) edited by Marcus Hall, notably papers by Hall, Donald Hughes and John McNeill.

- the vectors and the reasons for intentional or accidental introductions in the past;
- the perceptions and attitudes of individuals, societies, sectors and administrators regarding the transfer of species;
- and the responses of society to the damage and disruption caused by invasions.

Transfers of biota been a feature of the human past for millennia. People have brought with them, or introduced, exotic species for food, for providing natural resources like timber or rubber, as companions, and for moulding the landscapes they inhabit—for utilitarian ends (e.g. stabilising driftsands, draining swamps or pest control), for recreational ends (e.g. hunting or fishing, or providing shade and shelter), and for aesthetic ends (e.g. beautifying harsh landscapes and creating gardens).

Although introduced species may be benign (see Chap. 7) or even have benefits (noted in Chap. 2), the focus of most of this book is on invasive species which have been introduced and then proliferated in a self-sustaining way, causing damage to local biodiversity, the economy and welfare of the human population. We argue for the centrality of a historical understanding of these phenomena. Many historical species introductions continue to have ecological, socio-economic and cultural consequences in contemporary landscapes, and intervene materially and immaterially in the relationship between humans and nature. What is more, for all taxonomic groups, the increase in numbers of introduced species does not show any sign of saturation (Seebens et al. 2017).

This appeal to include historical approaches, and integrate anthropogenic influences into our explanations, does not amount to a call for a constructionist view which privileges culture or human agency over all else. We do not aspire to "reframe, downplay or deny the role of IAS in global change" (Russell and Blackburn 2017, p. 4). Neither do we advocate any generic position "against" or "in favor" of introduced species. Instead, we present a set of histories from the Mediterranean region (or the Mediterranean-type ecosystems, in Chap. 10), ranging from the ancient past to current challenges, in order to raise questions about the ways in which people and societies, local landscapes and seascapes, and indigenous and introduced species have interacted in the course of bioinvasions across the region.

1.3 The Study Region

The Mediterranean region is an imprecise concept, broadly understood but fuzzy around the edges, and it can be defined differently depending on your choice of biophysical, political and cultural variables and definitions (Hughes 2005). In this book we start from the basis that it includes the entirety of the Mediterranean Sea, and terrestrially, more or less the Mediterranean-type climate (MTC) region surrounding that sea (also referred to as the Mediterranean biogeographic region). A limitation we struggled in vain to overcome is that all of the histories of

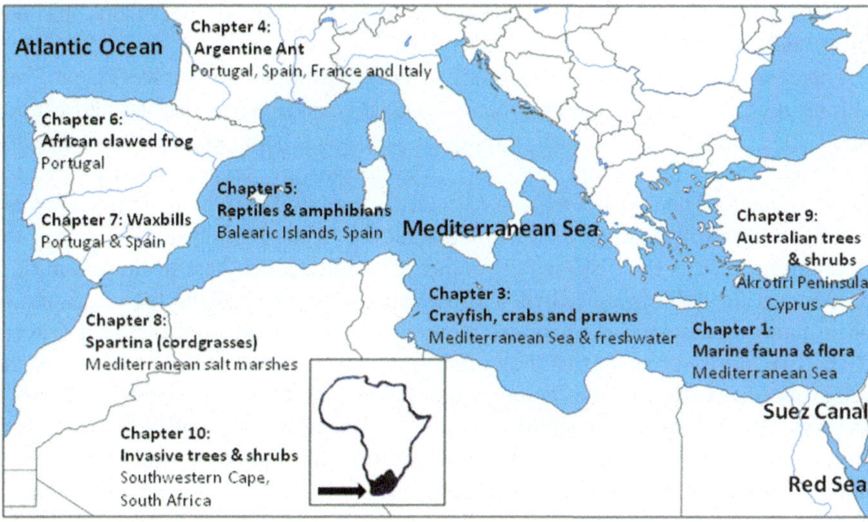

Fig. 1.1 Study region showing approximate distribution of the chapters

bioinvasions in this volume focus primarily on the northern and eastern region of the terrestrial territories in the region (mainland and islands) (Fig. 1.1).

The MTC region is characterised by summer drought and winter rains. Proximity to the coast moderates winter temperatures, so suitable growing temperatures coincide with rainfall. Primary productivity is high for such semi-arid regions and the vegetation forms dense thickets which dry out over the long hot, windy summers. Fires are frequent and most plants are adapted to fires, within certain ecological parameters (see panel in Chap. 10). These established fire regimes may be altered by the introduction of fire adapted plants which (for example) regenerate more quickly after fires, result in more intense fires, or develop fire-resistant stands. There are variations in average temperatures, and the seasonality of rainfall, across the region.

This description of the biophysical template should not obscure the fact that this region has been transformed by human actions—from agriculture to herding, burning to hunting and deforestation. This has occurred in so many ways and so gradually over millennia that it represents perhaps the archetypal landscape in which humans and their physical and biotic environment can be said to have coevolved. Like the Braudelian cultural conception of the Mediterranean, not defined only by “its geographic spaces and determined by its center and periphery, thereby prescribing and proscribing its identity/ies” (Goldwin and Silverman 2016, p. 4), “Mediterranean nature” is not simply comprised of ‘typical’ natural landscapes and assemblages of biota which have evolved over geological time. In addition, it has evolved from a long history of interactions between human and natural agents, of tides and currents in cultural and ecological histories, including intentional and accidental introductions and transfers of exotic organisms into and across the region.

Therefore, to this primarily ecological conception of the Mediterranean, and without rehearsing all of the arguments admirably outlined in Horden and Purcell (2000), we offer a brief historiographical comment here. We acknowledge the difference between history *of* the Mediterranean versus history *in* the Mediterranean. This book includes historical case studies which collectively are intended to contribute towards our understanding of the bioinvasion history *of* the Mediterranean. It is the histories of marine invasions which best fit this aspiration, in taking in the full sweep (albeit with regional emphases) of the region. Some of the remaining chapters are perhaps more histories *in* than *of* the region (except Chap. 10), but we contend that they offer illuminating perspectives on histories and processes of invasion in the region which have important commonalities across a diversity of ecological and cultural contexts. We hope that collectively through addressing a common theme they contribute towards outlining what a history *of* bioinvasions in the Mediterranean might look like.

Acknowledging the fragmentations and variations that characterise Mediterranean landscapes and history, there are nevertheless some key longer term processes and events which have influenced biological introductions and invasions in the region. In addition to climatic and topographical factors, and longer term processes like island creation and other geological shifts, these include: seaborne exploration, colonisation and networks of communication and trade; the terrestrial impacts of agriculture, forestry and pastoralism; and the convergent evolution of the flora in response to the aforementioned human interventions and the use of fire to manage landscapes. The individual chapters in this book offer histories of introductions, radiations, connectivities and interactions occurring across a range of specific local or regional biophysical and cultural contexts.

Contra Braudel (or, one possible reading of his magisteral work), these histories of bioinvasions undermine notions of timeless continuities or geographical or environmental determinism. They feature a bewildering array of introduced species from far flung seas and shores, following diverse pathways of invasion facilitated by unexpected vectors, imported for numerous purposes (or accidentally). Just possibly, in aggregate—and alongside climate change and transformations and accelerations of anthropogenic impacts including urbanisation, tourism and transport—bioinvasions portend a terminal disruption of the ecological coherence still perceived by some as the best unifying framework for defining a 'Mediterranean basin' region. In this sense, this book offers both an argument for the utility of thinking about a Mediterranean region and its history, and at the same time, a warning of its fragility, both physically and conceptually.

1.4 Natural and Cultural Dimension of Bioinvasions

The interest in bioinvasions shown by scholars from various disciplines, and the assumption that greater cooperation is desirable when addressing global environmental problems, have not yet resulted in many joint projects or consensual perspectives on bioinvasions. In this as in other subjects, it sometimes appears as if "historians are from Venus and ecologists are from Mars" (Pooley 2013). We believe that historical analysis is revealing with regard to the cultural matrix of the problem, which underlies many of the difficulties encountered in recent years in relation to the prevention of new introductions and the management of bioinvasions. An important theme of many of the cultural debates about bioinvasions has been differences in how disparate social and cultural groups value invasive species. Underlying that are more fundamental questions about the nature of 'nature' and 'wildness,' and of indigeneity and alienness, which have been well rehearsed elsewhere (Cronon 1995; Woods and Moriarty 2001; Coates 2007; Ritvo 2012).

Polemics have occurred within and between disciplinary fields. In general, conservationists have worked to prevent invasions as threats to local faunal and floral assemblages, and disruptors of natural (often 'indigenous') ecosystems (Richardson 2011). Some social scientists on the other hand (e.g. Comaroff and Comaroff 2001) have maintained that such narratives verge on xenophobia. Certainly, in some regions, polarising debates over native versus alien biota has resulted in counterproductive tensions between competing land management sectors and interests (see Chap. 10). More recently, some ecologists have argued in favour of 'hybrid' or 'novel' ecosystems, claiming that anthropogenic influences are pervasive and we live in a post-natural world. They have made a case for the contributions of introduced species to ecosystem services (e.g. Maris 2011). Davis et al. (2011) argued that in such a radically transforming world, environmental impact rather than place of origin should be the main criterion for conservation decision making. Hoffmann and Courchamp (2016) argue that while distinguishing between natural colonisations and invasions is useful for policy makers and managers, these value-laden distinctions are not appropriate for scientists, who should rather focus on processes and mechanisms.[4]

We argue that such debates over how to value species are ethical rather than ecological debates—and this is precisely what makes dealing with bioinvasions a wicked problem.[5] We are not convinced by Davis et al.'s (2011, p. 153) claim that taking cognisance of cultural and historical associations of humans with indigenous and introduced species is not relevant as it '[makes] little ecological or economic

---

[4]The concept of 'biocomplexity,' where 'coupled natural and human systems' are studied in interrelation, tries to address such diversity in biological and cultural relationships, and remind us that biodiversity is not a simple, objective concept (Callicott et al. 2007; Pfeiffer and Vouks 2008).

[5]These are problems where there are disagreements about how the problem should be defined, information is lacking about causes and possible effects of chosen policies, and complete solutions may not be possible.

sense'. Attempts to disarticulate 'differing interpretations of the evidence' from 'underlying values' (Russell and Blackburn 2017) and polarise disagreements over the impacts of invasive species into science versus 'invasive species denialism' are in our view profoundly unhelpful.[6] At the point at which ecologists, or historians for that matter, become 'experts' and advise on policy and management, it becomes essential to be reflexive about our individual values and how they shape our research and our analyses. The social impacts of invasive species management should be factored into policy making (Crowley et al. 2016), and this includes taking into account relationships with species and ecosystems built up over many years or even generations.

It is important to think about how *particular* human and non-human individuals and communities have interacted in specific places at specific times. It is unhelpful to think of humans as an abstract, generalised force, as has too often been the case in narratives of the Anthropocene (Lövbrand et al. 2015). Doing so implies normative assumptions about the value or non-value (and kinds of value that are important) of the invasive species, and also about the relative responsibilities and vulnerabilities of communities affecting and being affected by bioinvasions.

For some of those who support the existence of a culture/nature continuum, Nature is a construct of the human species; of humans' accumulated experience of life on Earth. In this view, transfers of species and all phenomena of transformation that succeeded them (including bioinvasions) are part of Nature, as it is today. For advocates of a 'post-nature' position, all such anthropogenic transformations are acceptable as long as change is controlled and engineered to support human life and the environmental systems and 'services' it depends upon in perpetuity (Lynas 2011). For others, this position overestimates humanity's capacity to control natural environments and processes, and is unacceptably anthropocentric. While acknowledging that it is most realistic to regard 'humans' as 'part of nature', environmental historian Smout (2003, p. 18) argues that humans' power to transform their environment 'gives them a unique position of responsibility and demands of them a sense of reverence for life other than their own'.

In our view the assumption that humans are just one more species on Earth, and that there is an inextricable link between human history and natural history, should not legitimize a discourse that naturalises anthropogenic biotic transfers, equating humans with other animals and plants when it comes to migration and colonization of other territories other than those of their origin.

[6]And see responses to Russell and Blackburn in *Trends in Ecology & Evolution*, April 2017, 32(4).

1.5 Overview of the Chapters

Chapters 2 and 3 focus on marine bioinvasions. Chapter 2, by Bella Galil, Agnese Marchini and Anna Occhipinti-Ambrogi, retraces pathways of introduction for different groups of species (marine plants and animals) into the Mediterranean Sea, and the consequences of these processes. Although the Mediterranean Sea attained something like its modern form with the re-opening of the Straits of Gibraltar c. 5 Ma BP, it is believed that significant introductions of nonindigenous marine species dates to as recently as the late fifteenth century, as European navigators began to travel to the Americas and first entered the Indian Ocean. The impact of possible introductions via the excursions of earlier seafarers entering the Atlantic is unknown. Subsequent introductions were mostly accidental (especially through fouling of hulls), but some like the Asian oyster *Crassostrea angulata* (possibly introduced in the 1500s) were intentional. As a result of limited knowledge of the indigenous marine biota, c.1800 is used as a baseline for assessing whether species are indigenous or not, and to date, c.736 species are believed to have been introduced since then.

Most marine non-indigenous species have arrived into the Mediterranean through shipping, the opening of the Suez Canal, aquaculture and the aquarium trade. Since the late 1400s there have been opportunities for the introduction of species from the Atlantic Ocean into the Western Mediterranean (first noted in the 1800s), notably from the Americas. The opening of the Suez Canal in 1869 impacted particularly on the biota of the eastern Mediterranean, which was first noted in the 1920s. Another key factor was the development of industrial aquaculture in the 1970s, introducing species from the Indian and Pacific oceans.

As Galil et al. remind us, awareness of marine bioinvasions as an environmental problem impacting on the indigenous biodiversity of the region (and the first important legislation to address this) only came to the fore in the 1980s, following the invasion of sea grass meadows by the 'killer alga' *Caulerpa taxifolia* at Monaco. There was little protection for Mediterranean marine species until the 1970s, and none for invertebrates until the 1990s. Guidelines for controlling introductions through ships ballast water and sediment discharges were introduced in 1991. Scientific research into the problem only began to take off in the early 2000s, and the first international action plan for reducing introductions of invasive species (focused on data gathering and dissemination) was adopted in 2003, and updated in 2016. An international convention for controlling ships ballast water and sediments came into force in 2017.

Chapter 3, by Chiara Manfrin, Catherine Souty-Grosset, Pedro Anastácio, Julian Reynolds and Piero Giulianini, focuses on the second most invasive group of organisms in the Mediterranean Sea, the decapods, including crayfish, crabs and prawns. These species have entered the Mediterranean from two main directions: from the Americas via the Atlantic Ocean, and from the Indian Ocean via the Suez Canal. They have been introduced for commercial purposes, and accidentally as bycatch or discarded ships' ballast. The chapter features an extensive table listing the first records for the occurrence of key invasive species in the Mediterranean basin, focussing on the 86 'worst' species' (those identified as having particularly negative impacts). Besides marine species, it also lists 17 introduced freshwater decapods and discusses the invasion histories of the Chinese mitten crab (*Eriocheir sinensis*) and the red swamp crayfish (*Procambarus clarkii*). The chapter discusses the protection of native species and policies to control bioinvasions, noting that billions of euros are spent in the region each year for that purpose. The economic impacts, health and cultural effects of these bioinvasions by decapods are noted. The authors warn that warming sea temperatures are likely to favour the spread of introduced species westward across the Mediterranean Sea.

Of the terrestrial invertebrates introduced to Europe which have proved invasive, ants comprise an important group with at least 32 species known to have become established. Chapter 4, by Ana Isabel Queiroz, focuses on one of the best known and longest established of these, the Argentine ant (*Linepithema humilis*). The first records of occurrence on islands and mainland Europe from c. 1850 to 1921 have been collated and are shown on a map. The Argentine ant had become recognised as a pest by the 1920s in the Mediterranean climatic region. They were regarded as a problem for locals and the tourism industry on the Côte d'Azur, and for farmers in Italy, southern France, Portugal and Spain. France, Italy and Portugal formulated policies and took direct management actions against the Argentine ant over the next 30 years, with Spain focussing instead on several pest species including this ant. The chapter investigates the different approaches to legislation and management designed to eliminate the ants across these four countries which suffered invasions, over the first half of the twentieth century. It compares the resulting legal regulations, investigating how they: state the main goal (confinement, control or eradication); define the territorial scope (regional or national); and prescribe implementation (which entities are responsible to take actions). The chapter relates policy against this invader to the political context and the ideology of the regimes in force. In the long run, all attempts to eliminate the ant have failed. Management actions are no longer mandatory, though they are applied in particular high priority areas, particularly areas of importance for biodiversity.

The book then moves on to consider examples of terrestrial vertebrate invasions in the region. The Balearic herpetofauna are the focus of Chap. 5, by Iolanda Silva-Rocha, Elba Montes, Daniele Salvi, Neftalí Sillero, José Mateo, Enrique Ayllón, Juan Pleguezuelos and Miguel Carretero. They investigate the introductions of frogs, toads, lizards, snakes, turtles and tortoises going back to ancient times. The authors discuss the original fauna of the islands and the particular vulnerability of island ecologies to bioinvasions, as well as the long history of anthropogenic introductions. Some of these anthropogenic introductions are believed to have been accidental, for example of gecko species. Other species were intentionally introduced, whether for utilitarian reasons for example tortoises as food sources or frogs to control insects, or for cultural reasons, such as snakes which were believed to bring good luck.

Since the 1800s, ten herpetological species have been introduced to the Balearic Islands, largely due to increased tourism and trade. These include introductions from mainland Spain, North America, and Algeria. Some species were introduced by the pet trade, with the remainder thought to have been introduced as a result of the trade in live ornamental plants, particularly in olive trees from Spain since the 1980s. The authors discuss the management of invasive introduced reptiles in the archipelago, focussing on the outcomes of two contrasting responses to invasive reptiles: the passive response to Italian wall lizards on Menorca, and a much more active response to invasive snakes on Ibiza.

A recent introduction to Europe is the African clawed frog (*Xenopus laevis*), whose curious trajectory from the Cape Province of South Africa to waterways near Lisbon in Portugal is the subject of Chap. 6, by Mónica Sousa, Angela Maurício and Rui Rebelo. Widely exported from South Africa since the 1940s, for scientific purposes, a population was shipped to the UK in 1968. In the mid-1970s, individuals were brought to Portugal by a researcher, and are likely to have escaped captivity during a flood not long thereafter.

These highly invasive frogs (this known from other invasions around the world) survived undetected in the wild more than 30 years. They were found as a result of efforts to conserve riverine ecosystems initiated in the 1980s, and reinforced by Directive 2000/60/EC of the European Parliament. The Water Framework Directive introduced a more broadly ecological perspective to water management and this included stream restoration efforts including clearing invasive vegetation and planting indigenous riparian vegetation, with biological surveys used to assess surface water quality. The frogs were identified during a fish survey for this purpose in 2006, and the first field surveys specifically to establish their abundance and distribution were conducted in 2007 and 2008. A control programme was initiated in 2010. This is a paradigmatic example of invasions resulting from escapes from research institutions, and prompted more stringent biosafety measures in Portugal.

In contrast to the other invasions analysed in this book, Chap. 7 by Gonçalo Cardoso and Luís Reino discusses a benign invasion, the spread of the common waxbill (*Estrilda astrild*) into Portugal and adjacent areas of Spain. A pet trade in the species seems to have been established by the 1800s, with the birds most likely originating from Portugal's colonies in Angola and Mozambique. It is notable that the first invasions in Portugal, from escaped or released birds, date to the 1960s, when independence wars broke out in these colonies and Portuguese settlers began returning to Portugal.

The chapter showcases the innovative ways in which historical observational data and mathematical modelling approaches are being combined to teach us new things about how bioinvasions progress. There is surely a history of science oriented narrative to be developed here. Further, an implication of this research is that there appear to be exciting possibilities for interdisciplinary research on personality and behaviour changes in biological populations of colonising species, in relation to environmental histories of the landscapes they colonise. It appears that waxbills forage in grasslands in agricultural landscapes in Iberia in summer, while native granivorous birds forage on forbs in less transformed areas. The authors suggest that agriculture establishes exotic non-crop plants which are less used by native faunas, especially plants with small seeds which are ideal for small birds like waxbills to feed on.

The book also examines plant invasions, beginning with Chap. 8, by Bernardo Duarte, Enrique Mateos-Naranjo, Susana Redondo Goméz, João Carlos Marques and Isabel Caçador. This chapter investigates the spread of two species of cord grass (*Spartina versicolor* and *Spartina densiflora*) from the Americas. They were first observed in the Mediterranean region in the 1840s. Nowadays, these species are invading the salt marshes fringing the Mediterranean's shores. Salt marshes are habitats of great ecological importance for the estuarine systems, as they are characterised by high productivity, provide important habitats for wildlife, constitute good carbon sinks, and act as shoreline stabilisers.

Spartina versicolor is thought to have been introduced as packing material in crates offloaded at various ports around the Mediterranean Sea, at first in France, Italy, Algeria and Portugal. *Spartina densiflora* may have been introduced to the southwestern Iberian Peninsula as a result of the maritime timber trade between South America and Spain. However, as the authors note, there is currently insufficient information to establish an accurate chronology of the appearance and spread of these species across the region.

Chapter 9, by Oliver Pescott, Sarah Harris, Jodey Peyton, Marilena Onete, Angeliki Martinou and J. Owen Mountford, focuses on the impacts of invasive Australian trees and shrubs on the Akrotiri Peninsula, on the island of Cyprus. The authors situate this in the wider narrative of the 'ruined landscapes' of the Mediterranean which importantly influenced British forestry policy on the island. The chapter begins with an account of the period of British administration on the island (1878–1960), before bringing the story of the management of this area up to the present.

In 2003, the complex of seasonal and permanent saline water bodies, salt and freshwater marshes and sand flats at Akrotiri were designated as a Ramsar wetland site of international importance. However, when the British took control in the 1870s, the prevailing narrative held that anthropogenic deforestation of the region's mountains was resulting in flooding and soil erosion, and creating malarial marshes in the lowlands downstream. Eucalypts were introduced to try to drain the wetlands, and pines, Australian acacias (wattles) and casuarinas were introduced to stabilise driftsands. The invasiveness of some of the wattles and casuarinas seems only to have been noted in the mid-1990s, with the first fieldwork attempting to quantify this being undertaken in 2003.

Fire, in conjunction with introduced invasive plants, has had a major impact on Mediterranean type regions worldwide. This important aspect of bioinvasions is addressed in the final chapter of the book, which includes a panel on fires in Mediterranean regions. Chapter 10, by Simon Pooley, provides an environmental history of plant introductions (and fire) in one of the world's other Mediterranean regions, the southwestern Cape region of South Africa. It investigates which plants were introduced to the Cape (mainly from America, Europe, and as is the case in Chap. 9, especially Australia), when and why. It reveals the evolution of attitudes towards, and scientific knowledge of, their ensuing impacts on local landscapes and ecosystems services. The main contentious areas were debates over the impacts of these introduced plants on water supplies, and fire incidence and severity.

The chapter examines cultural and political dimensions of attitudes to introduced plants, and the development of a nationalist narrative about indigenous plants and local landscapes. It takes a detailed look at the development of policies and management initiatives over the twentieth century, arguing that these were shaped by complex interactions of ecological, social, political and economic factors and events, and not simply scientific research and the physical impacts of the plants. The chapter concludes with a discussion of recent controversies over invasive introduced plants between local invasion biologists and commercial foresters. It does so in the context of the very successful narrative developed by invasion biologists internationally about fighting the 'McDonaldisation of nature,'[7] considering how this has been deployed in the region, and some policy implications of such a polarising narrative. The chapter makes a case for the inclusion of the kind of environmental history it offers into the comparative studies called for by invasion biologists, notably studies of the Mediterranean basin with its long history of social-ecological interactions.

In conclusion, this book features a range of case studies of bioinvasions from the plant and animal kingdoms in the Mediterranean region, encompassing marine, freshwater and terrestrial environments. These range from introductions and invasions of mainland territories by Argentine ants and waxbills, through invasions of islands by reptiles and amphibians and Australian plants, to invasions of salt marshes along the coasts by cord grasses, and of freshwater waterways by African clawed frogs.

The numerous invasive species discussed have been transported, deliberately or accidentally, from around the globe or internally across the Mediterranean region. Introductions from outside of the Mediterranean basin region include reptiles and amphibians from South Africa and North and central America, birds from sub-Saharan Africa, ants from South America, trees from Australia, cord grasses from the Atlantic coasts of South and North America, crayfish and crabs from North America and the Gulf of Mexico, and numerous molluscs and crustaceans from Indo-Pacific regions. The important role of fire coupled with invasive introduced plants is addressed in the Mediterranean basin and South Africa's southwestern Cape.

The chronological range of this collection is extensive and variable, ranging from the Neolithic and the Bronze Ages through Classical times, the ages of European maritime discovery, to the present. Most of the chapters focus on the more recent past, however, beginning their discussions in the nineteenth century, partly because the baseline for scientific studies in the region is c.1800. Developments in shipping and trade (including the opening of the Suez Canal),

---

[7]The concept of McDonaldisation was first introduced by Ritzer (1993), and has been widely adopted by invasion biologists to describe the homogenisation of the planet's biodiversity which they fear biological invasions are bringing about (for an example of this usage, see Richardson 2002, p. 123).

within and beyond the Mediterranean Sea, have proved deeply influential, but there exists a surprisingly imprecise knowledge of the chronology of introductions and spread of species prior to the twentieth century. What is more, awareness of most of these invasions seems to have come surprisingly late, with many of the discoveries and interventions described here originating in the late twentieth or early twenty-first century. Histories of the science, and of the institutions guiding research, policy-making and management of bioinvasions in the Mediterranean region, would make a useful contribution to our overall understanding of the problem across the Mediterranean basin.

This book brings new and fascinating material into the purview of environmental historians, and the editors are grateful for the patience and enthusiasm of the scientists among our authors in working with us through an editorial process involving many exchanges and versions to steer their work towards more historical writing. As a result of our approach to chapter selection, the material reads unevenly as history, but we feel the gains are substantial.

1.6 Directions for Future Study

Drawing on the case-studies included herein, we highlight a few research opportunities where environmental historians could make a substantial contribution to studies of bioinvasions. From a historical perspective, it seems very little is known about some of the major pathways of invasion into Mediterranean Europe. Further historical research could inform less conjectural accounts of the pathways of invasive species, for example of invertebrates and plants from South and North America, or of birds from former African colonies to Portugal. As suggested by Chaps. 7 and 8 for instance, there is work to be done by historians here, on the history of imports and trade, and indeed more broadly on the environmental impacts of colonial ventures and their aftermath in the colonising countries.

Historians have also paid scant attention to the histories of invasions from species moved around within the region. A handful of interdisciplinary studies suggest there is much interesting work to be done on such translocations, for instance collaborations between historians, taxonomists and geneticists (e.g. Clavero et al. 2015). Much groundwork for such studies has been done, in the shape of histories of landscape use and change in the region since ancient times (Braudel 1972; McNeill 1992; Hughes 2003, 2005, 2014; Horden and Purcell 2000; Grove and Rackham 2001). For instance, in his history of the mountains of the region, McNeill (1992) charted the importance of other major introductions into the region: of pastoral peoples and kingdoms; of irrigation techniques; of malaria carrying mosquitoes (diseases are oddly missing from most overviews of bioinvasions); and of American food crops. McNeill (2003) also offered a general framework for biological exchanges in European history more broadly, though he did not consider bioinvasions, weeds or wild animals explicitly, focussing instead on crops, diseases and domesticated animals.

Work on invasive plant pests in the region has focussed mainly on national or global consequences,[8] although more recent studies do not ignore the local scale, and recognise the social and biophysical dimensions of these invasions, and the consequences for particular landscapes and land use practises (e.g. Oestreicher 2005; Macedo 2011). Historians could investigate these previously neglected dimensions, researching for example how such bioinvasions have prompted the introduction of new crop species, or of new varieties of existing crop plants, or the abandonment of certain crops and plantations.

Comparative histories of the kinds of management efforts discussed by Duarte et al. in Chap. 8 could provide important contextual information for attempts to control bioinvasions in the Mediterranean. It would be interesting to know to what extent such projects have influenced subsequent management thinking—or failed to, through having been forgotten, ignored, or simply remaining unknown to managers in different parts of the region. The authors suggest that disputes and developments in taxonomy, changes in conservation philosophies, and public environmental awareness, have all influenced efforts to control introduced cord grass invasions in the region.

The global dispersion and impacts of invasive Australian trees and shrubs since the 1850s, understood within larger networks of colonial botanical and forestry expertise, and material exchanges, and fostered by acclimatisation societies, are being actively investigated by environmental historians (e.g. Kull and Rangan 2008; Bennett 2010; Carruthers 2011). Some of these histories inform Chaps. 9 and 10 in this collection. It is notable that these Australian invasive plants have had somewhat less success in overwhelming the indigenous vegetation in the Mediterranean basin, than they have in other Mediterranean type climate regions. Keeley et al. (2012) argue that the very long history of human disturbance in Mediterranean basin ecosystems has resulted in plants well adapted to coping with clearing for crop cultivation, livestock grazing and the use of fire as a land management tool. The shrubs of the region survive or recover from such disturbances well, and on the whole resist invasions by plants introduced from the other Mediterranean climate regions. An exception is parts of the western Iberian region, characterised by acidic soils and higher rainfall, and it would be very interesting to have more detailed environmental histories of the introduction of Australian trees and shrubs to this region, their resulting impacts, and attempts to manage them. These plants are widespread in parts of Portugal, and together with the abandonment of rural areas, they have contributed to extensive destructive fires in recent decades.

There is also scope for developing more comprehensive histories through integrating the methods of environmental history and historical ecology—and of the humanities, social and natural sciences more generally—to produce comparative

[8]Phylloxera invasions were often analysed in the framework of national histories of wine in the nineteenth and twentieth centuries, and their consequences for agriculture, industry and trade (e.g. Lachiver 1988; Martins 1991; Pan-Montojo 1994).

histories of bioinvasions in the Mediterranean type regions worldwide. Chapters 4 and 10 in particular suggest ways in which environmental histories of bioinvasions could enrich studies of this environmental problem across the Mediterranean basin and the other Mediterranean-like regions, which include coastal regions of California and Chile, the south-western Cape in South Africa, and south-western Australia.

Indeed, while there are significant differences in the geographies and histories of these regions, there are also some striking similarities, as described in Chap. 10. These include the introduction of Australian and other overseas trees and shrubs by colonial experts and land managers at roughly the same time as occurred in territories around the Mediterranean basin, notably in France, Italy, Portugal, southern Spain and parts of North Africa, to improve allegedly degraded local landscapes. Many of the same scientific explanations (at first entirely positive) were used to justify these introductions, including an assumed positive role in encouraging rainfall and preventing erosion, somewhat contradictorily alongside (for eucalypts) a role in drying out malarial marshlands. Plantations of introduced trees were also introduced to modernise rural landscapes and support industrialisation. These transformations had unintended consequences, impacting on local fire regimes, and invading indigenous ecosystems beyond the bounds of controlled plantations.

The integration of histories of land management sectors notably agriculture and forestry in the region (imperial, colonial and postcolonial), could add much to our understanding of why and how plants and animals where introduced, and how their subsequent invasions have been managed. For example, the history of introductions and invasions on the Akrotiri Peninsula could be further enriched through integrating the fire history of Cyprus including the period of Ottoman rule preceding British occupation. Virtual guerilla warfare over burning and grazing practises during the period of British rule explain much about foresters' attitude to the land and local land use (Pyne 1997). Drawing together studies of fire histories and policies in other parts of the Mediterranean (e.g. Pereira et al. 2016) with studies of bioinvasions could be similarly informative. Such historical studies of land management sectors in colonial and postcolonial contexts, given focus through overarching themes like bioinvasions and fire management, could also help to overcome the north-south divide that tends towards a focus on Europe and a neglect of Africa and the Levant in studies of the Mediterranean.

Finally, there are fascinating opportunities for drawing together regional-scale historical studies of the colonisation of the mainland and islands of the region by the terrestrial fauna, before and after humans arrived on the scene. As historians have noted (e.g. Hall 2003; Pooley 2014), in many cases decisions about which biota are 'indigenous' and which are 'introduced' depend on the timescale of the analysis. Most of the surviving megafauna of the region have been introduced by humans, and there are vigorous debates over the taxonomic status of species of goat and mouflon (for instance) and whether these should be protected as unique species (having been present in their current locations for centuries) or controlled as anthropochorous (or essentially feral) animals. The distinction between domesticated and wild animals is not as clear-cut as it may seem, with some animals being introduced and allowed to roam free as a resource for hunting as and when necessary (Gippoliti and Amori 2004; Massetti 2006).

The first international initiative to study bioinvasions in the Mediterranean-type ecosystems was instigated by conversations between researchers from all of the Mediterranean type regions, at the third meeting of MEDECOS, an international group of ecologists assembled to study them comparatively, in Stellenbosch, South Africa, in 1980 (Pooley 2014). This book, in a certain measure, is a sequel to these pioneering initiatives and many subsequent held on this topic, and there is a certain symmetry to ending with a chapter on bioinvasions in South Africa's fynbos. This book seeks to widen the range of disciplines involved in investigating bioinvasions, encourage historians to further investigate this important social-ecological phenomenon, and proposes that for these purposes the concept of a Mediterranean region remains a productive one.

References

Almeida D, Ribeiro F, Leunda PM, Vilizz L, Copp GH (2013) Effectiveness of FISH, an invasiveness screening tool for non-native freshwater fishes, to perform risk identification assessments in the Iberian Peninsula. Risk Anal 33:1404–1413

Andreu J, Vilá M (2010) Risk analysis of potential invasive plants in Spain. J Nat Conserv 18:34–44

Beinart WJ, Middleton K (2004) Plant transfers in historical perspective: a review article. Environ Hist 10(1):3–29

Bennett BM (2010) A global history of Australian trees. J Hist Biol 44(1):125–145

Braudel F (1972) The Mediterranean and the Mediterranean World in the age of Philip II, vol 1. William Collins Sons, London

Bright C (1998) Life out of bounds. Bioinvasions in a borderless world. W.W. Norton, New York

Capinha C, Essl F, Seebens H, Moser D, Pereira HM (2015) The dispersal of alien species redefines biogeography in the Anthropocene. Science 348:1248–1251

Carruthers J (2011) Trouble in the garden: South African botanical politics ca. 1870–1950. S Afr J Bot 77:258–267

Chew M (2015) Ecologists, environmentalists, experts, and the invasion of the 'second greatest threat'. Int Rev Environ Hist 1:7–40

Clavero M, Nores C, Kubersky-Piredda S, Centeno-Cuadros A (2015) Interdisciplinarity to reconstruct historical introductions: solving the status of cryptogenic crayfish. Biol Rev. https://doi.org/10.1111/brv.12205

Coates P (2007) American perceptions of immigrant and invasive species. University of California Press, Oakland

Callicott JB, Rozzi R, Delgado L, Monticino M, Acevedo M, Harcombe P (2007) Biocomplexity and conservation of biodiversity hotspots: three case studies from the Americas. Philos Trans Biol Sci 362(1478):321–333

Comaroff J, Comaroff JL (2001) Naturing the nation: Aliens, Apocalypse and the Postcolonial State. J S Afr Stud 27(3):627–651

Crowley SL, Hinchcliffe S, Macdonald RA (2016) Invasive species management will benefit from social impact assessment. J Appl Ecol. https://doi.org/10.1111/1365-2664.12817

Cronon W (1995) The trouble with wilderness; or, getting back to the wrong nature. In: Cronon W (ed) Uncommon ground: rethinking the human place in nature. W. W. Norton & Co, New York, pp 69–90

Crosby AW (1972) The columbian exchange: the biological and cultural consequences of 1492. Greenwood Press, Westport, CT
Crosby AW (1986) Ecological imperialism: the biological expansion of Europe, 900–1900. Cambridge University Press, New York
Daehler CC, Gordon DR (1997) To introduce or not to introduce: trade-offs of nonindigenous organisms. Trends Ecol Evol 12(11):424–425
Davis MA, Chew MK, Hobbs RJ, Lugo AE, Ewel JJ, Vermeij GJ, Brown JH et al (2011) Don't judge species on their origins. Nature 474(7350):153–154
Essl F, Dullinger S, Rabitsch W, Hulme PE, Hülber K, Jarošík V, Kleinbauer I, Krausmann F, Kühn I, Nentwig W, Vilà M, Genovesi P, Gherardi F, Desprez-Loustau M-L, Roques A, Pyšek P (2011) Socioeconomic legacy yields an invasion debt. Proc Natl Acad Sci USA 108:203–207
Frawley J, McCalman I (eds) (2014) Rethinking Invasion ecologies from the environmental humanities. Routledge, London
Gippoliti S, Amori G (2004) Mediterranean island mammals: are they a priority for biodiversity conservation? Biogeographia 25:135–144
Goldwyn AJ, Silverman, RM (2016) Introduction: Fernand Braudel and the invention of a modernist's Mediterranean. In: Goldwyn AJ, Silverman, RM (eds) Mediterranean modernism. Intercultural Exchange and Aesthetic Development, 1–26. Palgrave Macmillan US, New York
Grove AT, Rackham O (2001) The nature of Mediterranean Europe: an ecological history. Yale University Press, New Haven
Hall M (2003) Editorial: the native, naturalized and exotic—plants and animals in human history. Landscape Res 28(1):5–9
Hoffmann BD, Courchamp F (2016) Bioinvasions and natural colonisations: are they that different? NeoBiota 29:1–14. https://doi.org/10.3897/neobiota.29.6959
Horden P, Purcell N (2000) The corrupting sea: a study of Mediterranean history. Blackwell, Oxford
Hughes JD (2003) Europe as consumer of exotic biodiversity: Greek and Roman times. Landscape Res 28(1):21–31
Hughes JD (2005) Mediterranean. An environmental history. ABC Clio, Santa Barbara
Hughes JD (2014) Environmental problems of the Greeks and Romans: ecology in the ancient Mediterranean. Johns Hopkins University Press, Baltimore
Hulme PE (2012) Weed risk assessment. A way forward or a waste of time? J Appl Ecol 49:10–19
Keeley J, Bond WJ, Bradstock RA, Pausas JG, Rundel PW (2012) Fire in Mediterranean ecosystems. Cambridge University Press, Cambridge
Keller RP, Cadotte MW, Sandiford G (eds) (2015) Invasive species in a globalized world. The University Chicago Press, Chicago
Kull CA, Rangan H (2008) Acacia exchanges: wattles, thorn trees, and the study of plant movements. Geoforum 39(3):1258–1272
Lachiver M (1988) Vins, vignes et vignerons: histoire du vignoble français. Fayard, Paris
Lidström S, West S, Katzschner T, Pérez-Ramos MI, Twidle H (2015) Invasive narratives and the inverse of slow violence: Alien species in science and society. Environ Humanit 7(1):1–40
Lövbrand E, Becker S, Chilvers J, Forsyth T, Hedren J, Hulme M, Lidskog R, Vasileiadou E (2015) Who speaks for the future of Earth? How critical social science can extend the conversation on the Anthropocene. Glob Environ Change 32:211–218
Lynas M (2011) The God species: saving the planet in the age of humans. National Geographic Books, Washington DC
Macedo M (2011) Port wine landscape: railroads, phylloxera, and agricultural science. Agric Hist 85(2):157–173
Marris E (2011) Rambunctious garden: saving nature in a post-wild world. Bloomsbury, New York
Martins CA (1991) A filoxera na viticultura nacional. Análise Soc 26(112/113):653–688
Masseti M (1998) Holocene endemic and anthropochorous wild mammals of the Mediterranean islands. Anthropozoologica 28:3–20

Masseti M (2006) Domestic Fauna and Anthropocorous Fauna. Hum Evol 21(2). In: Masseti M (ed) Special issue on the Anthropochorous Fauna, 85–94
Masseti M (2007) Ancient historical faunae of continental and insular Asia Minor, and their relations with the western Mediterranean, with particular reference to the Italian peninsula. Int J Anthropol 22(3–4):177–195
McNeill JR (1992) The mountains of the Mediterranean: an environmental history. Cambridge University Press, Cambridge
McNeill JR (2003) Europe's Place in the global history of biological exchange. Landscape Res 28 (1):33–39
McNeely JA (2001) An introduction to human dimensions of invasive alien species. In: McNeeley JA (ed) The great reshuffling: human dimensions of invasive alien species, IUCN, Gland, pp 5–20
Oestreicher A (2005) La filoxera en España: bases para su estudio y consecuencias socio-económicas en la Región de Murcia. Revista murciana de antropología 12:199–208
Pan-Montojo J (1994) La bodega del mundo: La vid y el vino en España, 1800–1936. Alianza Universidade, Madrid
Pereira MG, Hayes JP, Miller C, Orenstein DE (2016) Fire on the hills: an environmental history of fires and fire policy in Mediterranean-type ecosystems. In: Vaz E, de Melo CJ, Pinto LMC (eds) Environmental History in the making, vol 1. Springer, Berlin, pp 145–170
Pfeiffer JM, Vouks RA (2008) Biological invasions and biocultural diversity: linking ecological and cultural systems. Environ Conserv 35(4):281–293
Pooley S (2013) Historians are from Venus, Ecologists are from Mars. Conserv Biol 27(6):1481–1483
Pooley S (2014) Burning table mountain: an environmental history of fire on the Cape Peninsula. Palgrave, Basingstoke
Pyne SJ (1997) Vestal fire: an environmental history, told through fire, of Europe and Europe's encounter with the World. University of Washington Press, Seattle
Ritvo H (2012) President's lecture: going forth and multiplying: animal acclimatization and invasion. Environ Hist 17(2):404–414
Richardson DM (2002) Review: biodiversity versus biomonotony. Divers Distrib 8(2):126–128
Richardson DM (ed) (2011) Fifty years of invasion ecology: the legacy of Charles Elton. Wiley-Blackwell, Oxford
Ritzer G (1993) The McDonaldization of society. Pine Forge Press, Thousand Oaks
Rotherham ID, Lambert RA (eds) (2011) Invasive & introduced plants & animals: human perceptions, attitudes and approaches to management. Earthscan, London
Russell JC, Blackburn TM (2017) The rise of invasive species denialism. Trends Ecol Evol 32 (1):3–6
Seebens H, Blackburn TM, Dyer EE, Genovesi P, Hulme PE, Jeschke JM, Pagad S, Pyšek P, et al (2017) No saturation in the accumulation of alien species worldwide. Nat Commun 8. https://doi.org/10.1038/ncmms14435
Simberloff D (2015) Non-native invasive species and novel ecosystems. F1000Prime Reports 7. http://doi.org/10.12703/P7-47
Smout TC (2003) The Alien species in 20th-century Britain: constructing a new vermin. Landscape Res 28(1):11–20
Valery L, Fritz H, Lefeuvre J-C, Simberloff D (2008) In search of a real definition of the biological invasion phenomenon itself. Biol Invasions 10:1345–1351
Vicente JR, Pinto AT, Araújo MB, Verburg PH, Lomba A, Randin CF, Guisan A, Honrado JP (2013) Using life strategies to explore the vulnerability of ecosystem services to invasion by alien plants. Ecosystems 16:678–693
Woods M, Moriarty PV (2001) Strangers in a strange land: the problem of exotic species. Environ Values 10(2):163–191

Chapter 2
Mare Nostrum, Mare Quod Invaditur—The History of Bioinvasions in the Mediterranean Sea

Bella S. Galil, Agnese Marchini and Anna Occhipinti-Ambrogi

Abstract Intentional and unintentional introductions of marine species into the Mediterranean date back to the 15th century, when maritime shipping expanded with geographical discoveries and new mercantile horizons. Hull fouling on ships was recognized as a vector in the late 19th century, yet early introductions were overlooked because they had occurred centuries before the onset of biogeographical studies. The opening of the Suez Canal in 1869 drew scientific attention to the introduction of marine species and records of shipping and mariculture-introduced species kept appearing in the scientific literature, yet their number and impact were considered inconsequential. Concerns were raised only in the 1980s following conspicuous outbreaks with dire environmental and economic impacts in the Mediterranean and the neighboring Black Sea. The natural and cultural history of marine bioinvasions in the Mediterranean Sea is outlined, as framed by scientists and legislators, whereby perceptions of unintentionally introduced species evolved from ignorance and indifference to 'biological pollution' impairing potentially exploitable marine resources and harming indigenous species and habitats. The anthropogenic vectors at the root of the introductions into the Mediterranean Sea are summarized, and cases of environmental transformation, ecological damage, economic impact and human health issues are highlighted. The development of the legislative response, reflecting cultural attitudes to introductions and economic priorities, is briefly documented.

Keywords Environmental transformation · Ecological damage
Economic impact · Human health · Legislation

B. S. Galil (✉)
The Steinhardt Museum of Natural History, Tel Aviv University, Tel Aviv, Israel
e-mail: galil@post.tau.ac.il

A. Marchini · A. Occhipinti-Ambrogi
Department of Earth and Environmental Sciences, University of Pavia, 27100 Pavia, Italy

A. I. Queiroz and S. Pooley (eds.), *Histories of Bioinvasions in the Mediterranean*,
Environmental History 8, https://doi.org/10.1007/978-3-319-74986-0_2

2.1 The Mediterranean Sea—A Brief Introduction

Located between Europe, Africa and Asia, the Mediterranean Sea covers approximately 2500,000 km^2. It connects with the Atlantic Ocean through the Strait of Gibraltar, with the Sea of Marmara and the Black Sea through the Bosphorus Strait, and with the Red Sea and the Indian Ocean through the Suez Canal. The Strait of Sicily divides the sea into two major basins.

The Mediterranean Sea is a remnant of the Tethys Ocean, which in the Cretaceous formed an uninterrupted equatorial belt with richly diverse warm-water biota. That primeval ocean has shrunk in size since the Oligocene: it became separated from the Indo-Pacific Ocean by the formation of the Suez isthmus ca 10 Ma BP and disconnected from the Atlantic Ocean ca 6 Ma BP with the sealing of the precursor of the Strait of Gibraltar. The progressive closure of its Atlantic connections changed the physical and chemical properties of Mediterranean water and profoundly impacted on its ecosystems. At its nadir, termed the 'Messinian salinity crisis', the isolation of the sea led to desiccation and the creation of evaporitic basins. The once high biodiversity was severely reduced but repopulation by species of Atlantic origin occurred with the re-opening of the Straits of Gibraltar ca 5 Ma BP. The sea's subsequent geological history included drastic changes in climate, sea level, and salinity and oxygen levels which resulted in immigration waves of boreal and subtropical Atlantic biota (Thunell et al. 1977; Rögl 1997).

The sea's present main hydrologic features are: (i) a microtidal regime; (ii) scarce freshwater inputs and evaporation compensated for by inflow of Atlantic surface water; (iii) an eastern basin with high saline (38–39.5) concentration due to evaporation; (iv) oligotrophy, with organic carbon inputs 15–80 times lower in the Eastern than in the Western basin and extremely low concentrations of chlorophyll-a in surface offshore waters (ca 0.05 μg l^{-1}); (v) high homothermy from ca 300–500 m downwards, bottom temperatures ca 12.8–13.5 °C in the Western basin and 13.5–15.5 °C in the Eastern basin. Cooler, lower salinity Atlantic superficial waters enter through the Gibraltar Strait and flow eastwards, getting more saline and thus denser, and therefore sinks and flows westwards as the Levantine Intermediate Current, exiting the sea over the sill of Gibraltar at greater depth than the inflow. The dense, winter-chilled surface waters of the Gulf of Lions, the Adriatic and the Northern Aegean form the highly oxygenated deep-water of the Mediterranean. With an average depth of ca 1500 m, climate-driven thermohaline circulation and short water residency (75–100 years), the Mediterranean is vulnerable to climate change and warming.

The recent marine biota in the Mediterranean Sea is primarily derived from the Atlantic Ocean. The sea is known to host more than 17,000 described marine species and contributes an estimated 7% of the world's marine biodiversity, including a high percentage of endemic species (Coll et al. 2010). However, recent rapid human population growth of coastal resident and transient recreational populations (244 million in 2000, 590 million expected in 2050, *fide* Tosun 2011; 270 million tourists in 2010 and 346 million expected in 2020), coupled with intensification of anthropogenic activities, are driving unprecedented changes in the Mediterranean Sea (Micheli et al. 2013; EEA 2015). Symptoms of complex and fundamental alterations to native species populations, habitats and ecosystems proliferate, including increases in non-indigenous species (NIS). The biota across wide swathes of the Mediterranean Sea, including marine protected areas, has already been altered with dire ecological, economical and human health impacts (Galil et al. 2017c).

2.2 *Mare Quod Invaditur*—The Mediterranean Sea as a Hotspot of Marine Invasions

Records of new marine species introductions in the Mediterranean Sea, and the information about the increase in both NIS presence and abundance in many of the countries bordering its coasts provide a worrying picture of a besieged sea (Galil 2000) (Fig. 2.1). Seven hundred and thirty-six multicellular NIS have been recorded in the Mediterranean Sea to date (the baseline used is the 1800s), with new introductions registered on a monthly basis.

The main vectors of NIS introduction into the Mediterranean are the Suez Canal (63%), shipping (commercial and recreational) and mariculture (Fig. 2.2).

The Mediterranean Sea serves as a major shipping route and is ringed by numerous ports and marinas. Combining global cargo ship movements with port environmental conditions and biogeography to quantify the probability of new primary invasions by ballast water, Seebens et al. (2013) considered that invasion risk in the Mediterranean is high and the Suez Canal to be among the sites of highest invasion probability. In addition to large ports serving transoceanic vessels, many smaller ports and marinas serve as local transport hubs.

Modern industrial marine farming in the Mediterranean began in the 1970s (Plan Bleu 2014). Its development was facilitated by an opportune coastal geography, the temperate-warm climate and the proximity to viable markets. Three of the main farmed shellfish species originate in east and southeast Asia and importation and transfer of their broodstock and seed facilitated the introduction of numerous epibionts [organism that lives on the body surface of another], mainly in the shellfish farms of France, Spain and Italy.

The Suez Canal, a sea-level waterway connecting the Red Sea to the Mediterranean Sea through the Isthmus of Suez, was opened in 1869, and reduced the travel distance between Europe and its Asian colonies. Initially 8 m deep with a cross section of 304 m^2, it was successively enlarged and at present it is 24 m deep with a cross section of 5200 m^2. In 2014 the Egyptian government embarked on a project, termed the 'New Suez Canal', to double the shipping lanes along the canal (Suez Canal Authority 2017).

Other vectors contribute fewer NIS, but a minority of those have had a disproportionate impact. Notorious examples include the green alga *Caulerpa*

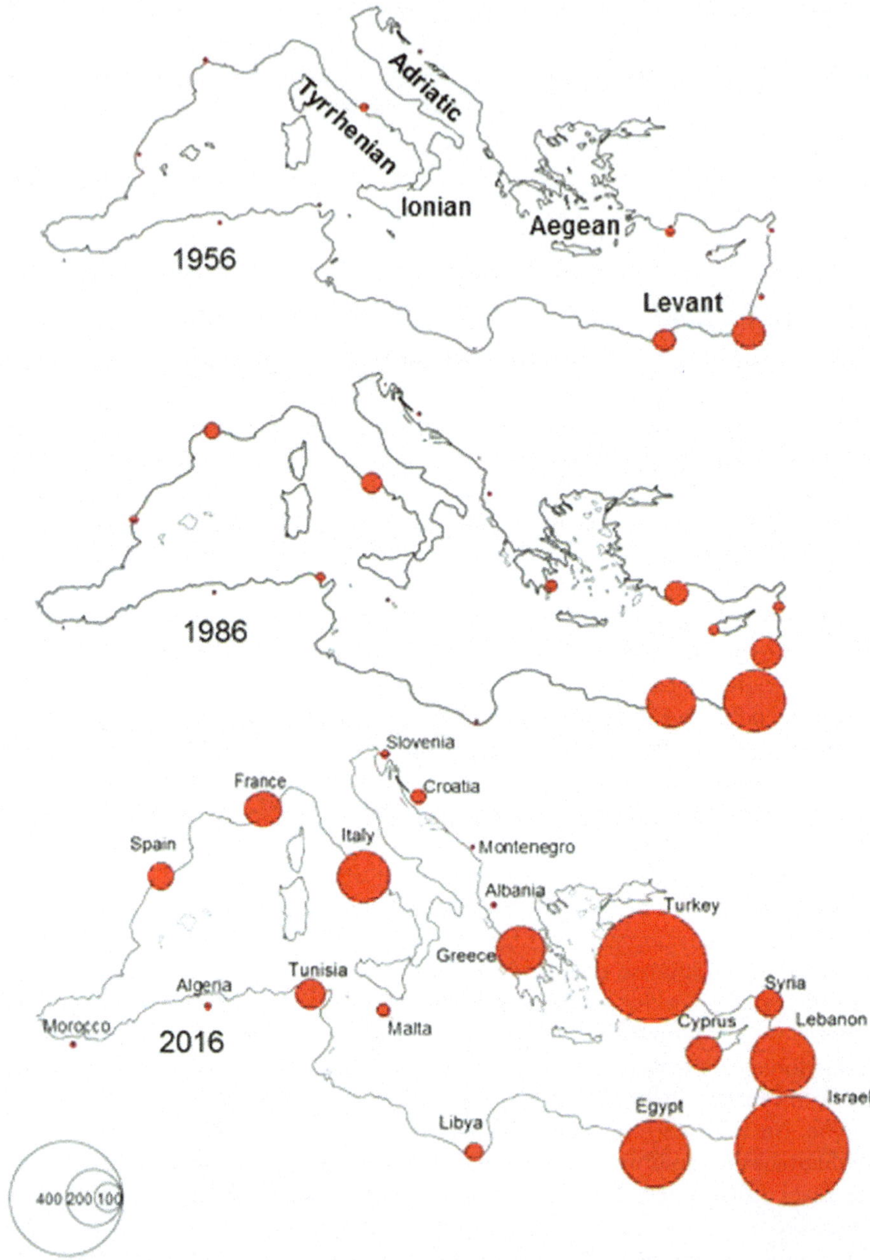

Fig. 2.1 Map of the Mediterranean Sea showing number of non-indigenous species (NIS) per country in 1956, 1986 and 2016. Size of circles is proportional to the number of NIS

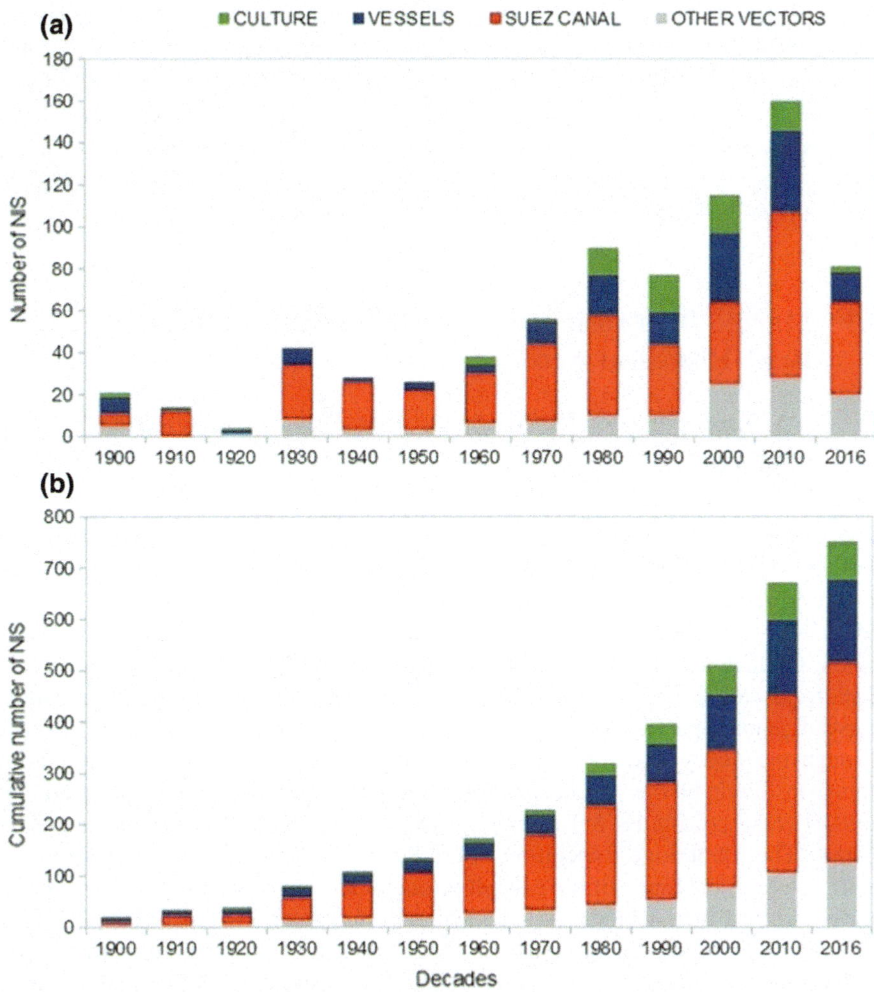

Fig. 2.2 **a** Number of NIS in the Mediterranean Sea introduced in the decades between 1990 and 2016 by different vectors; **b** cumulative number of NIS

taxifolia, originating in Australia and introduced with aquarium spill-over (Meinesz and Hesse 1991), or the brown alga *Fucus spiralis*, introduced from the Atlantic to the Mediterranean in the packaging of fishing baits (Sancholle 1988). The marine aquarium trade is considered to have increased in importance as an introduction vector in the past decade (Zenetos et al. 2016); similarly, the live seafood trade may grow in importance as a vector of introductions (Ferrario et al. 2013; Cecere et al. 2016).

In the early 2000s online inventories of Mediterranean NIS laid bare the magnitude of the problem (CIESM Atlas of Exotic Species in the Mediterranean 2017;

Delivering Alien Invasive Species Inventories for Europe 2017). These pioneering efforts motivated the compilation of country level NIS inventories (e.g. Galil 2007; Sciberras and Schembri 2007; Katsanevakis et al. 2009; Zenetos et al. 2009; Çinar et al. 2011; Occhipinti-Ambrogi et al. 2011; Baziri et al. 2013; Pećarević et al. 2013; Ounifi-Ben Amor et al. 2016). The rise of journals dedicated to marine biodiversity and bioinvasion encouraged research and the number of Mediterranean NIS records multiplied (Galil et al. 2014).

2.3 Marine Bioinvasions in the Mediterranean Sea—A Brief Historical Account

2.3.1 Shipping Precedes Science: The Pseudo-indigenous Species

The seafaring Phoenicians are considered to have initiated maritime voyages between the Mediterranean Sea and the Atlantic coasts of Europe and Africa, but transoceanic voyages commenced only at the end of the 15th century. The naturalists who described the marine biota of the Mediterranean subsequently (Pierre Belon, Guillaume Rondelet, Ippolito Salviani, Conrad Gesner, Ulisse Aldrovandi), assumed it to be native to the sea. Only in the late 19th–early 20th century non-indigenous species (NIS) were recognized as such: Red Sea species introduced through the newly excavated Suez Canal (see Sect. 2.3.4), and fouling [marine organisms living on the surface of submerged man-made objects] transported on the hulls of vessels (Catta 1876; Lindegg 1934). Fifty-five years after the opening of the Canal, 'The Cambridge Expedition to the Suez Canal' embarked on an investigation "…why some forms of life can get through the Canal and others cannot, and … throw light on marine migrations in general" (Gardiner 1924, p. 521). Scientific research on the phenomenon, however, remained scanty and sporadic until the late 20th century when the impacts of an invasive alga were noted off the Mediterranean coast of Europe (Meinesz and Hesse 1991).

The availability of molecular tools in the late 20th century, beginning with DNA-based genetic analyses, enabled research into the geographic origins of some species long believed to be indigenous to the Mediterranean. The term 'pseudo-indigenous' has been coined to account for alien species wrongly assumed to be native to the region when first described (Carlton 2009). The gastropod *Littorina saxatilis*, originally described from the Lagoon of Venice and long believed to be an indigenous species surviving in a glacial refugium, is now considered a human-mediated introduction, likely dating back to the 15th century merchant shipping in the heyday of the Republic of Venice. This was revealed by studies of haplotype diversity of the Atlantic and Venetian populations (Panova et al. 2011). The spaghetti bryozoan *Amathia verticillata*, first described from Naples, Italy, was assumed to be an indigenous species, notwithstanding its cosmopolitan distribution and association with harbors (Marchini et al. 2015a). Recently, a re-appraisal of its ecology and biogeography suggests it is most likely native to the Caribbean (Galil and Gevili 2014). Molecular tools are instrumental in uncovering the phylogeography of suspected (and some unsuspected) pseudo-indigenous species.

2.3.2 Vessels and Non-indigenous Species

Catta (1876) noticed the presence of exotic organisms on the hull of a vessel arriving from India to Marseille, France, and predicted that fouling biota were set to expand with shipping. Globalization, transoceanic canals (Suez, and Panama from 1914) and advances in ship building contributed to a dramatic increase in shipping-introduced NIS (Ruiz et al. 1997). In particular, the replacement of 'dry ballast' by water in the early 20th century favored transport of taxa with pelagic life stages and holoplanktonic taxa, which previously were unlikely to survive in transport. The increase in vessel traffic during and following the second world war left its mark in new cohorts of NIS: turf-forming macroalgae, blue crabs, veined rapa whelks and many more, recorded with increasing frequency near or in harbors (for reviews, see Boudouresque and Verlaque 2002; Galil et al. 2017b). Two remarkable outbreaks of (likely) ballast-introduced predators, the rapa whelk *Rapana venosa* and the comb jelly *Mnemiopsis leidyi* in the 1940s and 1980s respectively, had dire environmental and economic impacts in the neighboring Black Sea (Zolotarev 1996; Shiganova 1998; Saglam and Düzgünes 2013) and concentrated attention on this vector.

Once established, NIS may spread further, transported by the multitudinous vessels plying the sea, both commercial and recreational, other vectors (e.g. mariculture), and with currents. Some (e.g. the swimming crab *Callinectes sapidus*, the reef building tubeworm *Ficopomatus enigmaticus*, the sea squirt *Styela plicata*) have attained a Mediterranean-wide distribution, while others are expanding rapidly (Cvitković et al. 2013; Picciotto et al. 2016). Molecular studies of some vessel-associated NIS revealed unexpectedly high genetic diversity, possibly indicating multiple introductions from various sources, their populations thus continuously re-shuffled by new arrivals (Cabezas et al. 2014; Maltagliati et al. 2016).

Recreational boating plays an important role both in the leisure activities of residents and in nautical tourism in the Mediterranean: 55% of 'superyachts' worldwide register a Mediterranean port as their home port (Superyacht Business Happening 2015). Often they are berthed next to commercial ports, in the vicinity of shellfish farms, or close to marine protected areas, thus serving as an overlooked and unregulated vector for secondary spread of fouling NIS and posing a threat to cherished habitats (Minchin et al. 2006). The swift spread of the aquarium-escapee *C. taxifolia* in the 1980s and 1990s in the western Mediterranean, mainly near major tourist destinations, suggests that recreational boating may have played a significant role—algal fragments may have been transported by boat anchors (Boudouresque et al. 1995).

Mediterranean boaters have largely ignored that threat (Mineur et al. 2008; Ferrario et al. 2016). A survey of Italian boaters disclosed that low frequency of hull cleaning and application of anti-fouling paints, combined with long-distance voyages, is not uncommon, risking transfer of NIS (Ferrario et al. 2016). Unsurprisingly, records of NIS from marinas are increasing (Savini et al. 2006; Ros et al. 2013; Marchini et al. 2015a; Marić et al. 2017; Steen et al. 2017).

2.3.3 Mariculture: Intentional and Unintentional Introductions

Pliny the elder reported that "oyster ponds were first invented by Sergius Orata on the Gulf of Baiae…and he made a great profit" (Pliny 1983, p. 277). Indeed, to this day shellfish culture has remained a mainstay of mariculture in the Mediterranean (FAO 2014), although at present the most widely cultivated oysters are non-indigenous species. The first transoceanic transplantation may date back to the 16th century. The so-called 'Portuguese' oyster, *Crassostrea angulata*, which has flourished in southwestern Portugal and southern Spain for hundreds of years (Fabioux et al. 2002), was likely transported from Asia by the Portuguese (Grade et al. 2016).

Large-scale transplantations ensued: *C. angulata* was imported to restock the oyster beds along the Atlantic coast of France and subsequently introduced to the Mediterranean Sea (Raimbault 1964; Fabioux et al. 2002; Dridi et al. 2014). When production of *C. angulata* was threatened by a viral disease, it was replaced by the Japanese oyster, *Crassostrea gigas*, initially introduced as spat from Japan and brood stock from British Columbia, Canada (Grizel and Héral 1991), and has since been massively introduced across the Mediterranean. Raimbault (1964), who documented the transfer of spat [shellfish spawn] from Portugal and the Atlantic coast of France to the Mediterranean, in 1962–3, does not mention any zoosanitary precautions.

That insouciance has somewhat lessened following the mass mortalities of *C. angulata* in the early 1970s. Grizel and Héral (1991, p. 399), although admitting to "The unofficial nature of some of the first importations" of *C. gigas*, insisted that "Zoosanitary precautions were taken by the use of numerous histological and cytological controls…immersing the spat and the collectors in fresh water reduced substantially the introduction of an associated flora and fauna", congratulated themselves (Grizel and Héral 1991, p. 402) that "… only a few [NIS] species survive in very restricted areas." This could not be further from the truth.

Profound biological changes in sites of oyster farming were described as "abnormal" (Bini 1983), "Japanese biological garden" (Boudouresque et al. 2011, and "melting pot of non-indigenous biota" (Marchini et al. 2015b, p. 2953). Mineur et al. (2014) compiled a list of 48 northwestern Pacific introductions into Europe likely associated with transfer of shellfish, with the main oyster farming regions (e.g., Thau Lagoon, Venice Lagoon and Taranto Seas) as hubs for further dispersal. Two parasitic copepods, *Mytilicola orientalis* and *Mytilicola ostreae*, have been accidentally introduced with *C. gigas* from Japan to the Atlantic coast of France and hence to the shell-farming lagoons of Languedoc in 1979 (Clanzig 1989). Both species are able to infect the indigenous oyster, *Ostrea edulis*, whereas *M. orientalis* also infects the blue mussel, *Mytilus edulis* (Stock 1993).

Introduction of oyster pathogens and parasites has likely been the reason for the continuous demand of new oyster stocks that has in turn enhanced new introductions, in a positive feedback loop described by Mineur et al. (2014). Mineur et al. (2007, 2014, p. 1672) suggest "that the constant discovery rate of non-native marine species is more likely the result of regular arrivals of new introduced species and could be considered as forensic evidence that unreported imports and relaying of oysters from the North Pacific into Europe" have continued to this day, though possibly in smaller numbers during the last decade.

The Manila clam, *Ruditapes philippinarum*, native of the western Pacific, was introduced into the lagoons of Languedoc, France, in the late 1970s (Bodoy et al. 1981), and subsequently to other locations in the Mediterranean. Beginning in 1983, the clam has been intentionally introduced to the Po delta lagoons in the northern Adriatic Sea, to supplement the overexploited native *Ruditapes decussatus* fishery (Cesari and Pellizzato 1985; Occhipinti-Ambrogi 2002; Pranovi et al. 2006), and subsequently across the Italian coastline, including Sardinia and Sicily. Clam seed was imported from a succession of U.K., U.S.A., Spanish and French hatcheries (Breber 2002), with no mention of zoosanitary precautions. Within a few years *R. philippinarum* supplanted the native *R. decussatus* (Breber 2002), and its wild catch surpassed the farmed produce. The intensive, occasionally illegal, fishing effort for clams, blamed on lack of management, resulted in habitat degradation and ecological damage, a reduction of the clam stock and the destruction of nursery areas and feeding grounds for many Lagoon species, including commercial fish (Solidoro et al. 2003; Nunes and Markandya 2008).

Mediterranean countries have been notoriously insouciant about biosecurity hazards relating to movement of stock, feed, and equipment that may result in introduction of marine species (CIESM 2007; Golani et al. 2015; Marchini et al. 2016) as well as illegal introductions. The larger penaeid prawns are widely cultivated. One, the prized tiger prawn, *Penaeus japonicus*, was officially introduced for mariculture in the Mediterranean in the 1960s (Tournier 1972). Yet, the appearance of five non-indigenous prawns, all of commercial interest and newly recorded in the past decade, raises a suspicion of direct human introduction, particularly as these species have been found in the vicinity of fish and shellfish farms. The bilaterally ablated female banana prawn, *Penaeus merguiensis*, collected in the Bay of Iskenderun, Turkey, in 2006, is certainly an escape or an inadvertent release from an aquaculture facility, since unilateral and bilateral eyestalk ablation is commonly used in aquaculture for inducing maturation of gonads (Özcan et al. 2006). The first record *Penaeus aztecus* in the Mediterranean Sea dates to 2009, from Turkey (Deval et al. 2010, as *Farfantepenaeus aztecus*). Soon after, it was found in trawl catches along the Mediterranean coast of Turkey and subsequently from locations in the Aegean, Ionian, Adriatic, Tyrrhenian and Levant seas (Galil et al. 2017a). Unregulated international trade in live prawns continue unabated as revealed by cases of White Spot Disease in Italian shrimp farm stocked with infected shrimps originating from Turkey (Stentiford and Lightner 2011). As in the case of the clandestine importation of *Penaeus pulchricaudatus* from Turkey into Italy, and *Penaeus merguiensis* into Turkey (Özcan et al. 2006), no reporting has been made to the competent authorities of importation of *P. aztecus*. Kevrekidis (2014) raised concern of potential impacts if the population of *P. aztecus* increases to the point where it competes with the commercially important native prawn *Melicertus kerathurus*.

The European Union (EU) established a legal framework to limit the environmental risks related to the introduction and translocation of non-native species in aquaculture (Council Regulation 708/2007 concerning use of alien and locally absent species in aquaculture), but as it pertains only to Member States, and unevenly regulated even in those, illegal introductions and intra-national translocation of shellfish stocks (and their associated biota) continue to contribute to the introduction and spread of marine NIS in the Mediterranean Sea (Bakir and Aydin 2016).

2.3.4 The Suez Canal—A Corridor for Thermophilic Species

The Suez Canal is the main pathway of introduction of non-indigenous species into the Mediterranean Sea. The successive enlargements of the Suez Canal have raised concern over increasing propagule pressure resulting in continuous introductions of new non-indigenous species and associated degradation and loss of native populations, habitats and ecosystem services (Galil et al. 2017b). The opening of the Suez Canal in 1869 focused scientific attention on the introduction of marine species into the Mediterranean Sea. Even before the Canal was fully excavated, a French malacologist, Vaillant (1865) argued that the breaching of the Suez Isthmus would provide a valuable opportunity to observe the mixing of faunas. Vaillant advocated what today would be considered a 'baseline study' and raised provocative and prescient questions. To this end he studied the molluscs in the vicinity of Port Suez, and intended to examine the environs of Port Said, for he believed those localities will provide the first emigrants and receive the first settlers. Fischer (1865, 1870, 1871), who studied molluscs from Port Said, Suez and the Canal's Bitter Lakes, determined that the faunas of the Red Sea and the Mediterranean were entirely distinct, but that the Suez Canal will have the effect of modifying the faunas of both shores of the Isthmus.

The opening of the Suez Canal engendered debates on its impact on the Red Sea and Mediterranean biotas, yet for the next 50 years the documentation of the biota in the Canal itself and the changes in the bordering marine environments were left to learned amateurs and natural scientists. Keller sought evidence for the presence of Red Sea and Mediterranean species in the Canal, which he considered a caravan route for species proceeding from both ends of the canal (1883, Fig. 12). The importance of his work lies in confirming the role of the canal as a 'corridor' for the movement of species and in the transformation of the adjacent marine biotas (Keller 1883). Two avid amateurs, Bavay and Tillier, employees of the 'Compagnie du canal maritime' are to be thanked for assiduously collecting molluscs and fish (Bavay 1897, 1898; Tillier 1902; Tillier and Bavay 1905, 1906).

Travellers provided sporadic glimpses into the movement of species along the Canal: Martens (1887) received molluscs collected by Drs. Gottsche and Pastor; Giesbrecht's (1897) records are based on a single plankton haul taken in the Bitter Lakes by a naval physician, Dr. Krämer; Jordan and Hubbs (1917) record of the silverside *Atherinomorus lacunosus* (as *Atherina forskali*) is based on a day's collection at Port Said. The first planktonic Copepoda described from the Canal were collected in the bathroom of a passing steamer—Professor Herdman's ingenious collecting method consisted of a tow-net fixed to a running sea-water tap (Thompson and Scott 1903). Steinitz (1919), another learned amateur, recognized the scientific significance of the movement of biota through the Canal and noted that no scientific institute had taken on a comprehensive study of biotic transfer. He raised questions as to the exchange of faunas between the two oceans and whether the Red Sea species cause changes in the fauna of the eastern basin of the Mediterranean.

Fifty-five years after the opening of the Canal, 'The Cambridge Expedition to the Suez Canal', consisting of three research students and an amateur field naturalist "of the good old school" (Calman 1950, p. 587), embarked on an investigation of the "intermingling of the Mediterranean and Red Sea organisms in the Suez Canal" (Gardiner 1924, p. 520). The expedition was charged with ascertaining "what forms have passed through the Canal zone from the Mediterranean to the Red Sea and vice versa, when they passed through, whether in the prehistoric period, in the times of the earlier or of the present canal, how they passed through, whether by swimming, by drifting, by attachment to ships, or by other means. All these facts lead up to the question why some forms of life can get through the Canal and others cannot, and it is hoped that the expedition will throw light on marine migrations in general, the area being one which can be periodically investigated at small cost" (Gardiner 1924, p. 521).

The results filled a hefty volume and constitute the only comprehensive study on the biota of the Suez Canal to date. In his concluding remarks Fox (1929, p. 844) stated "… there is a need of an investigation of the coastal faunas of Northern Egypt and Palestine".

Indeed, the advent of the Erythraean bioinvasion[1] in the Mediterranean was traced in numerous faunistic studies of the Levant biota throughout the 20th century. Yet the sole multidisciplinary, multitaxa survey to investigate the spread of the Erythraean biota in the Levant and its impact on the native biota was a joint program by the Smithsonian Institution, the Hebrew University of Jerusalem, and the Sea Fisheries Research Station, Haifa, in the late 1960s.

The resulting list enumerated 140 Erythraean alien species (Steinitz 1970, son of W. Steinitz), forming the base for the compendium prepared by Por (1978), a zoologist at the Hebrew University, Israel.

Since the Levant Sea was recognized early on as vulnerable to bioinvasion (Steinitz 1927), it was *ipso facto* considered to have been "biologically enriched" by it (Tortonese 1973, p. 327), being "biologically underexploited by marine life … a sort of ecological vacuum where many ecological niches are available" (Oliverio and Taviani 2003, p. 314). Por (1971, p. 446) postulated that "The Lessepsian migration is therefore, a phenomenon with a rather clearly set frame which is rapidly approaching its fulfilment". Subsequently, he attempted to put a good face on it by claiming that "The Lessepsian migrants may be considered, in a figurative sense 'welcome guests' in the impoverished, subtropical cul-de-sac" (Por 1978, p. 123). He was proved wrong on both counts when scientists realized that the influx of Erythraean NIS increased with the successive enlargements of the Suez Canal (Fig. 2.2) and profoundly altered the composition of the biota of the eastern Mediterranean Sea, causing major shifts in community structure, function and services (see Sect. 2.4.1). It is now widely believed that "If we do not understand and mitigate the ecological risks associated with the expansion of the Suez Canal, the integrity of a large part of the Mediterranean ecosystem could be in jeopardy." (Samaha et al. 2016, p. 1806).

2.4 Marine NIS and Their Impacts: The Known and the Unknown

2.4.1 Environmental Impacts

Despite the centuries-long history of introductions in the Mediterranean Sea, the scientific community failed to study their environmental impacts until recent decades. Not until negative effects impinged on a prized unique habitat of the endemic seagrass meadows, *Posidonia oceanica* in the 1980s, did the invasion of the "killer alga" *C. taxifolia* raise the concern of the scientific community and the general public (Meinesz et al. 2001). The small population of that Indo-Pacific alga discovered in 1984 next to the Oceanographic Museum of Monaco (see Sect. 2.2) rapidly spread along the shallow shelf of the western Mediterranean, replacing

[1]Referring to species originating in the Red sea, Arabian sea and the Persian gulf, which have been introduced to the Mediterranean Sea through the Suez Canal.

meadows of the native seaweed *P. oceanica* (Meinesz and Hesse 1991). Other *Caulerpa* species (*C. taxifolia* var. *distichophylla* and *C. cylindracea*), which are believed to be introduced by vessels or aquaculture, are now widely spread in the eastern and western Mediterranean (Jongma et al. 2013; Montefalcone et al. 2015).

The impact of *Caulerpa* species on coastal habitats has been widely studied (Piazzi and Ceccherelli 2002; Dumay et al. 2002; Bulleri et al. 2010). In particular, the impact of *Caulerpa* on *Posidonia* has been a subject of controversy, because the seagrass meadows were in decline before their spread, and experimental and observational investigations failed to provide clear conclusions as to the reason for decline (e.g. Bulleri and Piazzi 2015). However, effects on the trophic chain (Terlizzi et al. 2011; Deudero et al. 2011; Felline et al. 2014; Alomar et al. 2016), nutrient cycles (Gennaro et al. 2015), sediments (Balata et al. 2015), sessile and vagile[2] biota, including meiofauna[3] (Žuljević et al. 2011; Musco et al. 2015; Pusceddu et al. 2016) were reported.

In the eastern Mediterranean algae-dominated rocky habitats, including *Cystoseira* meadows, have been decimated by large populations of herbivorous fish introduced through the Suez Canal. The two voracious grazers, the Erythraean rabbitfish *Siganus luridus* and *S. rivulatus*, transformed lush rocky reefs into "barrens" (Sala et al. 2011; Giakoumi 2014; Vergés et al. 2014).

Medium to large-scale modifications of submarine landscapes have been induced by other non-indigenous algae, seagrass, bivalves and serpulid worms (Lindegg 1934; Piazzi and Cinelli 2000; Ballesteros et al. 2007; Sghaier et al. 2011; Cilia and Deidun 2012). In shellfish farming lagoons, some Pacific algae introduced with oysters have developed massive biomass, thus contributing to the oxygen consumption and eutrophication processes of these enclosed environments (Sfriso and Facca 2013).

Commercial exploitation of NIS has also had an impact: the fishing of cultivated Manila clams by vibrating rakes began in the late 1980s and expanded greatly in the mid-1990s (Canu et al. 2011). It has altered the physical structure, nutrient cycling and biological processes of the Northern Adriatic Lagoons

While the striking environmental impacts induced by *Caulerpa* spp. and *Siganus* spp. were subject to well-designed experimental studies, impacts of most Mediterranean NIS are known from observational reports, expert judgment, or plain conjecture (Katsanevakis et al. 2014). For the greatest majority of NIS, mainly medium to small sized invertebrates, no information exists as to their ecological impacts; even basic information on the biology and ecology of some widespread NIS is lacking (Cardeccia et al. 2017). Ignorance of the ecological role NIS play in *Mare Nostrum* does not translate to lack of impact. Rather, it must be acknowledged that the impacts observed thus far are but a fraction of multiple, complex environmental changes upending the evolutionary thread of the Mediterranean biota.

---

[2]Freely motile.

[3]Small interstitial organisms.

2.4.2 The Economic Impact of Erythrean NIS

The economic importance of the Suez Canal was not limited to shipping. The Suez Canal company considered the Erythraean invasion as commercially advantageous for the populations of the southern Levant (Gruvel 1936). Indeed, by the early 1900s, the swimming crab *Portunus segnis*, an early Erythraean NIS, was offered in markets across the Levant (Fox 1924; Calman 1927; Gruvel 1936). The Erythraean penaeids, chiefly *P. pulchricaudatus*, were greatly valued: "…the most magnificent example of shrimp we have ever encountered…" (Gruvel 1936, p. 181, translated from the French). By mid-century the commercially harvested NIS populations serve an excellent indicator of their prevalence.

In the Levantine fishing grounds the bulk of the trawler catch in the 1950s was comprised of three species—the native red bream and hake, and the Erythraean goldband goatfish, *Upeneus moluccensis*. While total catch of goatfish had remained constant, the goldband goatfish "almost completely replaced the Mediterranean species, the red mullet, *Mullus barbatus* in the trawl catch" (Perlmutter 1956, p. 4). In 1955, another Erythraean NIS, the lizardfish, *Saurida lessepsianus* became an important part of the trawl catch across the Levant (Oren 1957; Gorgy 1966; El-Zarka and Koura 1965). By the 1960s *S. rivulatus*, was "… one of the most firmly established erythraean immigrants … one of the commonest fishes of St. George Bay [Lebanon] entering the catches of trammel net, hook and line and seine fishermen in large numbers" (George and Athanassiou 1967, p. 254).

Investigations carried out on the Egyptian shelf in 1970–71 showed that the Erythraean rainbow sardine *Dussumieria elopsoides* accounted for half of the sardine yield on the Egyptian shelf (Al-Kholy and El-Wakeel 1975), having replaced the native clupeids.

In the past half century the number of Erythraean NIS in commercial fisheries in the Levant increased substantially. A recent study of commercial trawl hauls off the Israeli coastline, based on examination of more than 200,000 fishes belonging to 111 species, revealed that Erythraean NIS comprised 54% of individuals and 70% of the biomass at 20 m depth and 67% of individuals and 45% of the biomass at 40 m (Goren et al. 2016).

The share of Erythraean NIS in fishery landings off the Mediterranean coast of Turkey has been around 20% until 1980s, when their percentage decreased. That apparent reduction was caused by changes in the fishing fleet and grounds. Until the 1980s the Mediterranean fleet comprised small coastal trawlers equipped to exploit shelf stocks—where Erythraean NIS were relatively numerous. Following the collapse of the Black Sea fisheries in the 1990s a substantial proportion of the northern fishing fleet, composed of larger and more powerful boats, moved to the Mediterranean and exploited the deeper fishing grounds, where Erythraean NIS were scarce (EastMed 2010). The proportion of Erythraean NIS in the hauls of coastal trawlers has actually increased. In fact, surveys carried out in a 'fishery restricted area' (FRA) on the Mediterranean coast of Turkey, indicate that the number and biomass of Erythraean NIS increased. The authors postulate that "If the percentage of the Lessepsian fishes increases with the same trend, it may be speculated that in the next 50 years the fish fauna of FRAs will be composed solely of Lessepsian species leaving no room for native ones" (EastMed 2010, p. 99). Even in the southern Dodecanese, Greece, where few Erythraean NIS are exploited commercially, fishers using trammel-nets land a large part of the biomass in Erythraean NIS (rabbitfish *S. luridus* 47%, toadfish *Lagocephalus sceleratus* 26%, cornetfish *Fistularia commersonii* 11%, and native fish only 16%) (EastMed 2010).

The increase of Erythraean NIS in fishery landings is not an unalloyed gain. Some of the most commercially important NIS (e.g. the fishes *U. moluccensis*, *S. lessepsianus*, *Siganus* spp., the shrimp *P. pulchricaudatus* and the crab *P. segnis*) have replaced native species, while some, like *P. segnis* and *Lagocephalus* spp., interfere with local artisanal fisheries, damaging gear and spoiling already netted catch (J. Ben Souissi pers. com.). Other Erythraean NIS are simply nuisance species. Coastal trawling and purse-seine fishing are disrupted for the duration of jellyfish *Rhopilema nomadica* swarming due to net clogging and inability to sort the catch: "It is not uncommon that fishermen, especially purse seines, discard entire hauls due to the overwhelming presence of poisonous medusae in their nets" (Golani and Ben Tuvia 1995, p. 287) and sometimes fishermen are obliged to abandon their nets due to the excessive weight of the jellyfish (EastMed 2010). A few Erythraean NIS are venomous and poisonous (see Sect. 2.4.4).

2.4.3 The Economic Impact of Intentionally Introduced NIS—A Silver Lining?

Some Mediterranean countries have benefited economically from the intentional introduction of *C. gigas* and *R. philippinarum* (see Sect. 2.3.3) in terms of increased employment and revenues. In France, the leading producer of *C. gigas* in the Mediterranean Sea, 12,000–13,000 tons are harvested annually in the Étang de Thau, the main farming site (Boudouresque et al. 2011), generating 20–30 million € year^{-1} (EC 2006b) and employing 2000 workers (Rayssac et al. 2012). Italy is the main producer of *R. philippinarum*, harvesting up to 50,000–60,000 tons year^{-1}, generating about 200 million € year^{-1} and employing 4000–5000 workers (Turolla 2008). Major farming sites are concentrated in the lagoons of the Northern Adriatic Sea (Venice, Grado-Marano and Po River Delta). In the Venice Lagoon alone, it accounts for about 85–100 million € year^{-1}, providing employment for 1300 clam harvesters (Canu et al. 2011).

These industries are inherently unstable. Mortalities of *C. gigas* were noted since the mid-2000s, but never exceeding 20% of the stocks until 2008, when herpesvirus disease infected all French oyster farms, causing over 70% juvenile mortality (Segarra et al. 2010). An Asian origin has been postulated for the herpesvirus, as well as for other oyster pathogens that have been jeopardizing European oyster industry (Mineur et al. 2014). Importation of new stocks was considered the only viable solution, as for the 1970s crisis of *C. angulata* (see Sect. 2.3.3). However, new importations entail the risk of introductions of additional oyster-associated pathogens, as well as fouling species potentially threatening oyster production (Davis and Davis 2009). The spatial planning of oyster farms, currently focused entirely on achieving the maximum sustainable yield, should stipulate minimum distance between farms, density regulation and compartmentalization in order to reduce spread of disease (Pernet et al. 2016).

At its peak, in 1999, 64,000 tons of *R. philippinarum* were harvested in Italy, with 40,000 tons in Venice lagoon alone, but the landings declined in the early 2000s (Pellizzato and Da Ros 2005). The initial lack of regulation and enforcement encouraged adoption of a catch-as-catch-can approach which, combined with illegal harvesting, caused an overexploitation of important habitats and nursery areas (Boscolo-Brusà et al. 2013). Wild stocks of *R. philippinarum* decreased due to eutrophication and anoxic events, as well as destructive harvesting (Bartoli et al. 2001; Pranovi et al. 2003; Turolla 2008). As a consequence, a culture which had relied almost entirely on wild spat, now depends in part on importation of clam seeds from foreign hatcheries (Città metropolitana di Venezia 2014).

2.4.4 Human Health Hazards

A handful of the NIS occurring in the Mediterranean Sea are venomous or poisonous and pose a threat to human health (Table 2.1). They comprise recent records (e.g. the striped eel catfish *Plotosus lineatus*, silverstripe toadfish *L. sceleratus*, long-spined sea urchin *Diadema setosum*, stonefish *Synanceia verrucosa*),

Table 2.1 Species with impact on human health introduced to the Mediterranean Sea

Species	Phylum	First Mediterranean record	Country of first record	Number of Mediterranean countries reported from (June 2017)
Rhopilema nomadica	Cnidaria	1976	Israel	10
Macrorhynchia philippina	Cnidaria	1991	Lebanon	3
Diadema setosum	Echinodermata	2006	Turkey	4
Lagocephalus sceleratus	Chordata	2003	Turkey	13
Plotosus lineatus	Chordata	2001	Israel	5
Pterois miles	Chordata	1991	Israel	7
Synanceia verrucosa	Chordata	2010	Israel	3

or species that have greatly increased their spread in the last decade (e.g. the lionfish *Pterois miles* and jellyfish *Rhopilema nomadica*).

The silverstripe toadfish was first recorded in the Mediterranean in 2003, and soon after its populations increased greatly and spread across the Mediterranean Sea from Spain to the Black Sea. Ingestion of the fish, particularly its gonads and liver, causes tetradotoxin (TTX) poisoning. Between 2005 and 2008, 13 victims of TTX poisoning were hospitalized in Israel alone (Eisenman et al. 2008; Bentur et al. 2008). Other severe cases followed (Chamandi et al. 2009; Awada et al. 2010; Kheifets et al. 2012). The full extent of the phenomenon is unknown as many cases go unpublished (Souissi et al. 2014). Despite official prohibitions it is being sought by recreational fishers and occasionally marketed commercially (Nader et al. 2012; Beköz et al. 2013; Souissi et al. 2014; Farrag et al. 2015, 2016). The results of a survey in Turkey, which examined awareness concerning the danger of consuming the fish among fishers, fish-mongers, restaurant staff, medical staff and the public, are shocking (Beköz et al. 2013). The fishermen admitted selling it to local markets and hotels, restaurant managers and fish mongers denied selling the pufferfish though specimens were displayed for sale, and most customers had not heard of the pufferfish, nor were unaware of its lethality. Of the physicians surveyed 76% were unaware that pufferfish are poisonous, and 98% were unfamiliar with the symptoms of pufferfish poisoning. It is clear that both naïve local and tourist consumers are at serious risk.

The population of the striped eel catfish increased greatly along the Levant coastline where it constitutes a sizable portion of the shallow shelf trawl fishery discards (Doğdu et al. 2016). A survey of injuries from marine organisms conducted among Israeli professional fishermen in 2003–2004 noted that 10% of fish-related injuries were caused by the venom bearing spines of *P. lineatus* (Gweta et al. 2008). The recent spate of lionfish records in the Levant and as far west as Tunisia (Dailianis et al. 2016) is of concern since injuries from its venom bearing spines produce intense pain and swelling. The brittle needle-spined sea urchin *D. setosum* inflicts painful injuries on the unwary swimmers, divers and fishers. The spines may inflict deep penetrating wounds and break off easily to become embedded in the tissue. Their venom may cause inflammation, swelling, and acute pain. It has been recorded in Lebanon, Turkey, and Greece (Tsiamis et al. 2015).

The feathery stinging hydroid *Macrorhynchia philippina* has been known in the Levant since the 1990s (Bitar and Bitar-Kouli 1995). A brush with its nematocyst-laden branches may cause a stinging sensation, rash, blisters and raised itchy wheals which may last for days. The nomadic jellyfish has spread throughout the Levant and was recently reported from Sicily and Sardinia, Italy, Malta and Tunisia, where in 2013 densities as high as 10–100 individuals/1000 m^3 were observed (Balistreri et al. 2017). In the southeastern Levant it forms huge swarms, 100 km long, each summer since the early 1980s. In summer 2009 alone 815 hospitalizations due to jellyfish envenomations were recorded along the southeastern coast of Turkey (Öztürk and İşinibilir 2010). The annual swarms off the Israeli coast resulted in envenomation victims suffering adverse effects that have lasted weeks and even months after the event. Though most envenomations result in painful skin eruptions, severe systemic manifestation may occur, including anaphylactic reaction (Friedel et al. 2016). The countries affected by the nomadic jellyfish swarming and the other venomous and poisonous NIS are major seaside recreation destinations in the Mediterranean Sea and thus are exposed to mass envenomation events (Öztürk and İşinibilir 2010).

The toxic dinoflagellate *Ostreopsis* cf. *ovata*, discovered at Villefranche-sur-Mère, France, in 1972, bloomed in western Italy in the 1990s, producing toxic aerosols which manifested in respiratory and dermatologic syndromes in beach goers (Gallitelli et al. 2005). Recent research indicated that it may have originated in the Pacific Ocean (Sato et al. 2011). Other cryptogenic toxic microalgae are hazardous to human health (Penna 2014).

These fragmentary records nonetheless attest to the increasing spread and abundance of NIS of human health concern in the Mediterranean Sea. There is an urgent need to assemble nation-wide, if not region-wide, quantitative data on medically-treated health impacts, as well as acquainting and forewarning the public, health professionals and decision makers.

2.5 The Legislative Response to Introductions into the Mediterranean Sea

Despite more than a century of scientific documentation of marine bioinvasions in the Mediterranean Sea, the development and implementation of management policies has been a mostly reactive process, evolving from restricted and voluntary to widespread and mandatory. The United Nations Convention on the Law of the Sea (UNCLOS) was the first to call for "States … take all measures necessary to prevent, reduce and control … the intentional or accidental introduction of species, alien or new, to a particular part of the marine environment, which may cause significant and harmful changes thereto" (UN 1982, p. 101). In 1991 the Marine Environment Protection Committee, a subsidiary body of the Council of the International Maritime Organisation (IMO) which is empowered to consider any matter concerned with prevention and control of pollution from ships, adopted 'International Guidelines for preventing the introduction of unwanted aquatic organisms and pathogens from ships' ballast water and sediment discharges' (IMO 1991)—voluntary nonbinding guidelines, which proved insufficient to stem the transport of propagules in ballast. The 'Convention for the Control and Management of Ships' Ballast Water and Sediments' was adopted in 2004 and entered into force in 2017 (IMO 2016).

The 'Convention for the Protection of the Mediterranean Sea against Pollution' (Barcelona Convention), adopted in 1976, made no mention of NIS. Amended and renamed 'the Convention for the Protection of the Marine Environment and the Coastal Region of the Mediterranean', it came into force in 2004. The Barcelona Convention and its protocols, together with the Mediterranean Action Plan (MAP), form part of the United Nations Environment Programme (UNEP) Regional Seas Programme. The signatories to the convention adopted an 'Action Plan concerning species introductions and invasive species in the Mediterranean Sea' in 2003 (Regional Activity Center for Specially Protected Areas 2017). The preamble to the Action Plan states that "it is imperative to take immediate steps to prevent the introduction of non-indigenous species, control the spread of those already introduced and endeavor to mitigate the damage they cause to the marine ecosystem." (UNEP-MAP-RAC/SPA 2005, p. 7). The recently adopted Action Plan concerning "Species Introductions and Invasive Species" repeats (nearly verbatim) the 2003 'Actions' which mostly concerned data gathering and dissemination, but no prevention, control and mitigation measures (UNEP/MAP 2016).

A 'Draft Guidelines for controlling the vectors of introduction into the Mediterranean of non-indigenous species and invasive marine species' (UNEP-MAP 2007) states that "the greatest influx of invaders resulted from the opening of the Suez Canal in 1869 that allowed entry of Indo-Pacific and Erythraean biota." Yet, the guidelines are limited to shipping and aquaculture. The 'Draft decision on the Ecosystems Approach including adopting definitions of Good Environmental Status (GES) and targets' (UNEP/MAP 2013) even sought to exclude introduction through the Suez Canal from NIS linked to human activities and main human related pathways and vectors of NIS introduction. A careful reading of the UNEP-MAP-RAC/SPA documents reveals procrastination and unwillingness to undertake the necessary actions to prevent and control NIS introductions in the Mediterranean Sea.

Since 9 of the 23 states bordering the Mediterranean are EU Member States, EU legislation has to be considered as well. Following decades of shellfish importation and translocation that had introduced and spread scores of NIS, trade and transfers of shellfish have been regulated by legislation which authorizes all movements inside Europe, with temporarily implemented restrictions during disease outbreaks (EC 1991, 1995, 2003, 2006a). The regulation concerning use of alien and locally absent species in aquaculture is clear about the legislators' priorities: "Aquaculture has benefited economically from the introduction of alien species and translocation of locally absent species in the past… and the policy objective for the future is to optimise benefits associated with introductions and translocations while at the same time avoiding alterations to ecosystems… and restricting the spread of non-target species and detrimental impacts on natural habitats" (EC 2007, p. 1). The regulation's effectiveness in the Mediterranean is yet to be assessed (Mineur et al. 2014; Cruscanti et al. 2015; Marchini et al. 2016).

The Marine Strategy Framework Directive (MSFD) [DIRECTIVE 2008/56/EC] came into force in 2008. Of the 11 high level descriptors of 'good environmental status' outlined in the Directive, the second states "Non-indigenous species introduced by human activities are at levels that do not adversely alter the ecosystems." (EU 2008, p. 34). MSFD is arguably the single most important policy measure taken by the European Union concerning marine NIS. However, in order for it to achieve its desired aim, the EU needs to undertake coordination of policies and compliance measures with the non-EU signatories to the Barcelona Convention (Galil et al. 2017b).

EU Regulation 1143/2014 on the prevention and management of the introduction and spread of invasive alien species (IAS Regulation)—a binding legal tool for Member States—entered into force in 2015. The Regulation lays down rules to prevent, minimize and mitigate the adverse impacts of the introduction and spread, both intentional and unintentional, of a tiny subset of IAS of 'Union concern' (49 species to date, none marine) on biodiversity and the related ecosystem services. This approach, centered on single species, is hardly likely to succeed in marine ecosystems since the level of quantitative data required for a species to be placed on this list is generally available only for established and widespread species. However outcomes of eradication and control actions of marine NIS reveal they are successful only when the species has limited spatial distribution and is confined to a habitat where damage to non-target species is deemed acceptable (Galil 2009). Vectors management is the most effective strategy for reducing the risk of marine bioinvasions and the best precautionary approach (Williams et al. 2013). Of a suite of 13 marine NIS listed in an 'horizon scanning' exercise to identify the most threatening new and emerging IAS (Roy et al. 2015), 5 species are already well established in the Mediterranean Sea and thus unpromising candidates for management efforts.

A significant milestone towards vector management is the entry into force of the Convention for the Control and Management of Ships' Ballast Water and Sediments in 2017, which is expected to reduce both primary and secondary ballast introductions in the Mediterranean. The Convention is a key international measure for environmental protection and management of marine NIS (IMO 2016). It resulted from the successful, if occasionally disputatious, cooperation of scientists, stakeholders, national and international organizations. Such partnership is urgently needed in the Mediterranean in order to confront a difficult issue and avoid passing the environmental, economic and societal burden to future generations.

Acknowledgements BSG is grateful to Prof. Randall T. Schuh, American Museum of Natural History (AMNH), New York, for gracious hosting, and to the librarians of the AMNH for their unfailing help.

References

Al-Kholy AA, El-Wakeel SK (1975) Fisheries of the south-eastern Mediterranean Sea along the Egyptian coast, Soviet-Egyptian expedition 1970–1971. Bull Inst Ocean Fish 5:1–300

Alomar C, Deudero S, Andaloro F et al (2016) *Caulerpa cylindracea* Sonder invasion modifies trophic niche in infralittoral rocky benthic community. Mar Environ Res 120:86–92

Awada A, Chalhoub V, Awada L et al (2010) Coma profond aréactif reversible après intoxication par des abats d'un poisson méditerranéen [Deep non-reactive reversible coma after a Mediterranean neurotoxic fish poisoning]. Rev Neurol 166:337–340

Bakir K, Aydin I (2016) New localities in the Aegean Sea for alien shrimps *Penaeus aztecus* (Ives, 1891) and *Metapenaeus affinis* (H. Milne Edwards, 1837). Acta Adriat 57(2):273–279

Balata D, Piazzi L, Bulleri F (2015) Sediment deposition dampens positive effects of substratum complexity on the diversity of macroalgal assemblages. J Exp Mar Biol Ecol 467:45–51

Balistreri P, Spiga A, Deidun A (2017) Further spread of the venomous jellyfish *Rhopilema nomadica* Galil, Spanier & Ferguson, 1990 (Rhizostomeae, Rhizostomatidae) in the western Mediterranean. Bioinvasions Rec 6(1):19–24

Ballesteros E, Cebrian E, Alcoverro T (2007) Mortality of shoots of *Posidonia oceanica* following meadow invasion by the red alga *Lophocladia lallemandii*. Bot Mar 50(1):8–13

Bartoli M, Nizzoli D, Viaroli P et al (2001) Impact of *Tapes philippinarum* farming on nutrient dynamics and benthic respiration in the Sacca di Goro. Hydrobiol 455:203–212

Bavay A (1897) Au sujet du passage d'un mollusque de la Mer Rouge dans la Méditerranée. Bull Soc Zool Fr 22:199

Bavay A (1898) Note sur les mollusques du Canal de Suez. Bull Soc Zool Fr 23:161–164

Baziri H, Sghaier YR, Benamer I et al (2013) Alien marine species of Libya: first inventory and new records in El-Kouf National Park (Cyrenaica) and the neighbouring areas. Med Mar Sci 14(2):451–462

Beköz AB, Beköz S, Yilmaz E et al (2013) Consequences of the increasing prevalence of the poisonous *Lagocephalus sceleratus* in southern Turkey. Emerg Med J 30:954–955

Bentur Y, Ashkar J, Lurie Y et al (2008) Lessepsian migration and tetrodotoxin poisoning due to *Lagocephalus sceleratus* in the eastern Mediterranean. Toxicon 52(8):964–968

Bini G (1983) Immissione antropica di molluschi esotici nel Mediterraneo: I—Il lago di Caprolace. Studi per l'Ecologia del Quaternario Firenze 5:113–125

Bitar G, Bitar-Kouli S (1995) Impact de la pollution sur la répartition des peuplements de substrat dur à Beyrouth (Liban-Mediterranée Orientale). Rapports et Procès-Verbaux des Réunions, Vol XIV (Nouvelle série) Comm Int Expl Mer Médit 34:19

Bodoy A, Maître-Allain T, Riva A (1981) Croissance comparée de la palourde européenne *Ruditapes decussatus* et de la palourde japonaise *Ruditapes philippinarum* dans un écosystème artificiel méditerranéen. Vie Marine 2:39–51

Boscolo-Brusà R, Cacciatore F, Ponis E et al (2013) Clam culture in the Venice lagoon: stock assessment of Manila clam (*Venerupis philippinarum*) populations at a nursery site and management proposals to increase clam farming sustainability. Aquat Living Resour 26:1–10

Boudouresque CF, Verlaque M (2002) Assessing scale and impact of ship-transported alien macrophytes in the Mediterranean Sea. CIESM Workshop Monographs 2:53–61

Boudouresque JF, Meinesz A, Ribera MA et al (1995) Spread of the green alga *Caulerpa taxifolia* (Caulerpales, Chlorophyta) in the Mediterranean: possible consequences of a major ecological event. Sci Mar 59:21–29

Boudouresque CF, Klein J, Ruitton S et al (2011) Biological invasion: the Thau Lagoon, a Japanese biological island in the Mediterranean Sea. In: Ceccaldi H-J, Dekeyser I, Girault M, Stora G (eds) Global change: mankind-marine environment interactions. Springer, Netherlands, pp 151–156

Breber P (2002) Introduction and acclimatisation of the Pacific carpet clam, *Tapes philippinarum*, to Italian waters. In: Leppäkoski E, Gollasch S, Olenin S (eds) Invasive aquatic species of Europe: distributions, impacts and management. Kluwer, Dordrecht, pp 120–126

Bulleri F, Piazzi L (2015) Variations in importance and intensity of competition underpin context dependency in the effects of an invasive seaweed on resident assemblages. Mar Biol 162(2): 485–489

Bulleri F, Balata D, Bertocci I, Tamburello L, Benedetti-Cecchi L (2010) The seaweed *Caulerpa racemosa* on Mediterranean rocky reef: from passenger to driver of ecological change. Ecology 91:2205–2212

Cabezas MP, Xavier R, Branco M et al (2014) Invasion history of *Caprella scaura* Templeton, 1836 (Amphipoda: Caprellidae) in the Iberian Peninsula: multiple introductions revealed by mitochondrial sequence data. Biol Invasions 16(10):2221–2245

Calman WT (1927) Report on the Crustacea Decapoda (Brachyura). Zoological Results of the Cambridge Expedition to the Suez Canal, 1924. XIII. Trans Zool Soc London 22:211–217

Calman WT (1950) Obituary, Dr. Robert Gurney. Nature 165:587–588

Canu DM, Campostrini P, Riva SD et al (2011) Addressing sustainability of clam farming in the Venice Lagoon. Ecol Soc 16(3):26

Cardeccia A, Marchini A, Occhipinti-Ambrogi A et al (2017) Assessing biological invasions in European Seas: Biological traits of the most widespread non-indigenous species. Estuar Coast Shelf Sci. https://doi.org/10.1016/j.ecss.2016.02.014

Carlton JT (2009) Deep invasion ecology and the assembly of communities in historical time. In: Rilov G, Crooks JA (eds) Biological invasions in marine ecosystems. Ecological Studies 204, Springer-Verlag Berlin Heidelberg, pp 13–56

Catta JD (1876) Note sur quelques crustacés erratiques. Ann Sci Nat Zool 3:1–33

Cecere E, Belmonte M, Portacci G et al (2016) Live seafood importation as a potential vector for alien introduction in the Taranto Seas (Southern Italy, Mediterranean Sea). Rapp Comm Int Expl Mer Médit 41:428

Cesari P, Pellizzato M (1985) Molluschi pervenuti in Laguna di Venezia per apporti volontari o casuali. Acclimazione di *Saccostrea commercialis* (Iredale and Roughely, 1933) e di *Tapes philippinarum* (Adams and Reeve, 1850). Boll Malacol 21(10–12):237–274

Chamandi SC, Kallab K, Mattar H et al (2009) Human poisoning after ingestion of puffer fish caught From Mediterranean Sea—a case report. Middle East J Anesthesiology 20(2):285–288

CIESM (2007) Impact of mariculture on coastal ecosystems. In: Briand (ed) CIESM Workshop Monograph n°32, CIESM Publisher, Monaco

CIESM Atlas of Exotic Species in the Mediterranean (2017) http://www.ciesm.org/online/atlas/intro.htm. Accessed 3 Aug 2017

Cilia DP, Deidun A (2012) Branching out: mapping the spatial expansion of the Lessepsian invader mytilid Brachidontes pharaonis around the Maltese Islands. Mar Biodiv Rec 5:e28

Çinar ME, Bilecenoglu M, Özturk B et al (2011) An updated review of alien species on the coasts of Turkey. Mediterr Mar Sci 2(2):257–315

Città metropolitana di Venezia (2014) Rendiconto della gestione del Gral all'esame della commissione http://www.provincia.venezia.it/notizie/rendiconto-della-gestione-del-gral-allesame-dellacommissione.html. Accessed 3 Aug 2017

Clanzig S (1989) Introduction récente d'invertébrés dans les lagunes Mediterranéennes du Languedoc-Roussilon (France). Bull Soc Zool Fr 114(3):151–152

Coll M, Piroddi C, Steenbeek J et al (2010) The biodiversity of the Mediterranean Sea: estimates, patterns, and threats. PLoS ONE 5(8):e11842

Cruscanti M, Innocenti G, Bremer A et al (2015) First report of the brown shrimp *Penaeus aztecus* (Crustacea, Decapoda, Penaeidae) in the Tyrrhenian Sea. Mar Biodiv Rec 8:e81

Cvitković I, Despalatović M, Grubelić I et al (2013) Occurrence of *Paraleucilla magna* (Porifera: Calcarea) in the eastern Adriatic Sea. Acta Adriat 54(1):93–100

Dailianis T, Akyol O, Babali N et al (2016) New Mediterranean biodiversity records (July 2016). Mediterr Mar Sci 17(2):608–626

Davis MH, Davis ME (2009) *Styela clava* (Tunicata, Ascidiacea): a new threat to the Mediterranean shellfish industry? Aquat Invasions 4:283–289

Delivering Alien Invasive Species Inventories for Europe (2017) Alien by country. In: European summary. http://www.europe-aliens.org/europeSummary.do. Accessed 3 Aug 2017

Deudero S, Box A, Alós J et al (2011) Functional changes due to invasive species: food web shifts at shallow *Posidonia oceanica* seagrass beds colonized by the alien macroalga *Caulerpa racemosa*. Estuar Coast Shelf Sci 93(2):106–116

Deval MC, Kaya Y, Güven O et al (2010) An unexpected find of the western Atlantic shrimp, *Farfantepenaeus aztecus* (Ives, 1891) (Decapoda, Penaeidae) in Antalya Bay, eastern Mediterranean Sea. Crustaceana 83(12):1531–1537

Doğdu SA, Uyan A, Uygur N et al (2016) First record of the Indo-Pacific striped eel catfish, *Plotosus lineatus* (Thunberg, 1787) from Turkish marine waters. Natural and Engineering Sciences 1(2):25–32

Dridi S, Romdhane MS, Elcafsi M (2014) Gametogenic cycle of *Crassostrea gigas* in contrasting Mediterranean habitats: marine (Gulf of Tunis) and continental (Bizert lagoon) culture sites. J Biol Res (Thessalon) 21:13

Dumay O, Fernandez C, Pergent G (2002) Primary production and vegetative cycle in *Posidonia oceanica* when in competition with the green algae *Caulerpa taxifolia* and *Caulerpa racemosa*. J Mar Biol Assoc UK 82(3):379–387

EastMed (2010) Report of the sub-regional technical meeting on the Lessepsian migration and its impact on Eastern Mediterranean fishery. Scientific and institutional cooperation to support responsible fisheries in the Eastern Mediterranean. GCP/INT/041/EC—GRE—ITA/TD-04, Athens, pp 138

EC [European Commission] (1991) Council directive of 28 January 1991 concerning the animal health conditions governing the placing on the market of aquaculture animals and products (91/67/EEC). Off J Eur Union L46:1–18

EC [European Commission] (1995) Council directive of 22 December 1995 introducing minimum Community measures for the control of certain diseases affecting bivalve mollusks (95/70/EC). Off J Eur Union L332:33–44

EC [European Commission] (2003) Commission decision of 14 November 2003 laying down the animal health conditions and certification requirements for imports of mollusks, their eggs and gametes for further growth, fattening, relaying of human consumption (2003/804/EC). Off J Eur Union L302:22–33

EC [European Commission] (2006a) Council directive (2006/88/EC) of 24 October 2006 on animal health requirements for aquaculture animals and products thereof, and on the prevention and control of certain diseases in aquatic animals. Off J Eur Union L328:14–56

EC [European Commission] (2006b) Assessment of the status, development and diversification of fisheries-dependent communities. Sète-France. Fish/2006/09. In: Regional social and economic impacts of change in fisheries-dependent communities (https://ec.europa.eu/fisheries/documentation/studies/regional_social_economic_impacts_en)

EC [European Community] (2007) Council regulation (708/2007) of 11 June 2007 concerning use of alien and locally absent species in aquaculture. Off J Eur Union L168/1:1–17

EC [European Community] (2008) Directive 2008/56/EC of the European Parliament and of the Council of 17 June 2008 establishing a framework for community action in the field of marine environmental policy (Marine Strategy Framework Directive). Off J Eur Union L 164: 19–40

EEA [European Environment Agency] (2015) State of Europe's Seas. Technical report No. 2/2015, European Environment Agency, Copenhagen, pp 220

Eisenman A, Rusetski V, Sharivker D et al (2008) An odd pilgrim in the holyland. Am J Emerg Med 26(3):383.e3–6

El-Zarka S, Koura R (1965) Seasonal fluctuations in production of the principal edible fish in the Mediterranean Sea off the United Arab Republic. Proc Gen Fish Coun Medit 8:227–259

Fabioux C, Huvet A, Lapègue S et al (2002) Past and present geographical distribution of populations of Portuguese (*Crassostrea angulata*) and Pacific (*C. gigas*) oysters along the European and north African Atlantic coasts. Haliotis 31:33–44

FAO (Food and Agriculture Organization) (2014) Fishery and aquaculture statistics: Global Aquaculture Production 1950–2012. Available from http://www.fao.org/fishery/statistics/global-aquaculture-production/query/en. Accessed 3 Jan 2017

Farrag MMS, El Haweet AAK, Akel A et al (2015) Stock status of pufferfish *Lagocephalus sceleratus* (Gmelin, 1789) along the Egyptian coast, eastern Mediterranean Sea. Am J Life Sci 3:81–91

Farrag MMS, El Haweet AAK, Akel A et al (2016) Occurrence of puffer fishes (Tetraodontidae) in the eastern Mediterranean, Egyptian coast-filling in the gap. BioInvasions Rec 5(1):47–54

Felline S, Mollo E, Ferramosca A et al (2014) Can a marine pest reduce the nutritional value of Mediterranean fish flesh? Mar Biol 161:1275–1283

Ferrario J, Savini D, Lodola A et al (2013) Risk of non-indigenous species introduction via international seafood trade: the case of Chioggia fish market. Biol Mar Mediterr 20(1):236–237

Ferrario J, Marchini A, Borrelli P et al (2016) A fuzzy 'boater'model to detect fouling and spreading risk of non-indigenous species by recreational boats. J Environ Manag 182:198–207

Fischer P (1865) Note sur les faunes conchyliologiques des deux rivages de l'isthme de Suez. J Conch 13:241–248

Fischer P (1870) Sur la faune conchyliologique marine des baies de Suez et de l'Akabah. J Conch 18:161–179
Fischer P (1871) Sur la faune conchyliologique marine de la baie de Suez. J Conch 19:209–226
Fox HM (1924) The migration of a Red Sea crab through the Suez Canal. Nature (London) 113:714–715
Fox HM (1929) Summary of results. XXXIX. Zoological results of the Cambridge expedition to the Suez Canal. 1924. I. Trans Zool Soc Lond 22(6):843–863
Friedel N, Scolnik D, Adir D et al (2016) Severe anaphylactic reaction to Mediterranean jellyfish (*Rhopilema nomadica*) envenomation: case report. Toxicol Rep 3:427–429
Galil BS (2000) A sea under siege–alien species in the Mediterranean. Biol Invasions 2(2):177 EU Regulation 1143/2014 186
Galil BS (2007) Seeing red: alien species along the Mediterranean coast of Israel. Aquat Invasions 2(4):281–312
Galil B (2009) Control and eradication of invasive aquatic invertebrates. In Gherardi F, Corti C, Gualtieri M (eds) Biodiversity conservation and habitat management, vol. II. Encyclopedia of Life Support Systems (EOLSS)
Galil BS, Gevili R (2014) *Zoobotryon verticillatum* (Bryozoa: Ctenostomatida: Vesiculariidae), a new occurrence on the Mediterranean coast of Israel. Mar Biodiv Rec 7:e17
Galil BS, Marchini A, Occhipinti-Ambrogi A et al (2014) International arrivals: widespread bioinvasions in European Seas. Ethol Ecol Evol 26(2–3):152–171
Galil BS, Innocenti G, Douek J et al (2017a) Foul play? On the rapid spread of the brown shrimp *Penaeus aztecus* Ives, 1891 (Crustacea, Decapoda, Penaeidae) in the Mediterranean, with new records from the Gulf of Lion and the southern Levant. Mar Biodiv. https://doi.org/10.1007/s12526-016-0518-x
Galil BS, Marchini A, Occhipinti-Ambrogi A et al (2017b) The enlargement of the Suez Canal—Erythraean introductions and management challenges. Manag Biol Invasions (in press)
Galil BS, Marchini A, Occhipinti-Ambrogi A (2017c) East is East and West is West? Management of marine bioinvasions in the Mediterranean Sea. Estuar Coast Shelf Sci. https://doi.org/10.1016/j.ecss.2015.12.021
Gallitelli M, Ungaro N, Addante LM et al (2005) Respiratory illness as a reaction to tropical algal blooms occurring in a temperate climate. JAMA 293(21):2595–2600
Gardiner JS (1924) The biology of the Suez Canal. Nature 114:520–521
Gennaro P, Piazzi L, Persia E et al (2015) Nutrient exploitation and competition strategies of the invasive seaweed *Caulerpa cylindracea*. Eur J Phycol 50(4):384–394
George CJ, Athanassiou V (1967) A two year study of the fishes appearing in the seine fishery of St. George Bay, Lebanon. Ann Mus Civ St Nat Genova 76:237–294
Giakoumi S (2014) Distribution patterns of the invasive herbivore *Siganus luridus* (Rüppell, 1829) and its relation to native benthic communities in the central Aegean Sea, Northeastern Mediterranean. Mar Ecol 35(1):96–105
Giesbrecht W (1897) Ueber pelagische Copepoden des Rothen Meeres. Zool Jahrb, Abt Syst. Geogr Biol Tiere 9:315–328
Golani D, Ben Tuvia A (1995) Lessepsian migration and the Mediterranean Fisheries of Israel. In: Armantrout NB (ed) Condition of the world's aquatic habitats. Proceedings of the World Fisheries Congress. Theme 1. Science Publishers, Lebanon, New Hampshire, pp 279–289
Golani D, Sonin O, Rubinstein G (2015) Records of *Paralichthys lethostigma* and *Sciaenops ocellatus* in the Mediterranean and *Channa micropeltes* in Lake Kinneret (Sea of Galilee), Israel. Mar Biodiv Rec 8:e39
Goren M, Galil BS, Diamant A et al (2016) Invading up the food web? Invasive fish in the southeastern Mediterranean Sea. Mar Biol 163:180. https://doi.org/10.1007/s00227-016-2950-7
Gorgy S (1966) Les pêcheries et le milieu marin dans le secteur Méditerranéen de la République Arabe Unie. Rev Trav Inst Pêches Marit 30(1):25–92
Grade A, Chairi H, Lallias D et al (2016) New insights about the introduction of the Portuguese oyster, *Crassostrea angulata*, into the North East Atlantic from Asia based on a highly polymorphic mitochondrial region. Aquat Living Resour 29(4):404

Grizel H, Héral M (1991) Introduction into France of the Japanese oyster (*Crassostrea gigas*). J Cons Int Explor Mer 47:399–403

Gruvel A (1936) Contribution à l'étude de la bionomie générale et de l'exploitation de la faune du Canal de Suez. Mém Inst Égypte 29:1–229

Gweta S, Spanier E, Bentur Y (2008) Venomous fish injuries along the Israeli Mediterranean coast: scope and characterization. Isr Med Assoc J 10(11):783–788

IMO (International Maritime Organization) (1991) International guidelines for preventing the introduction of unwanted aquatic organisms and pathogens from ships' ballast water and sediment discharges, London. Resolution MEPC 50(31):16

IMO (International Maritime Organization) (2016) Global treaty to halt invasive aquatic species to enter into force in 2017. http://www.imo.org/en/mediacentre/pressbriefings/pages/22-bwm-.aspx. Accessed 3 Aug 2017

Jongma DN, Campo D, Dattolo E et al (2013) Identity and origin of a slender *Caulerpa taxifolia* strain introduced into the Mediterranean Sea. Bot Mar 56(1):27–39

Jordan DS, Hubbs CL (1917) XVIII. Notes on a collection of fishes from Port Said, Egypt. Ann Carnegie Mus 11:461–468

Katsanevakis S, Tsiamis K, Ioannou G et al (2009) Inventory of alien marine species of Cyprus (2009). Mediterr Mar Sci 10(2):109–134

Katsanevakis S, Wallentinus I, Zenetos A et al (2014) Impacts of invasive alien marine species on ecosystem services and biodiversity: a pan-European review. Aquat Invasions 9(4):391–423

Keller C (1883) Die Fauna im Suez-Kanal und die Diffusion der Mediterranean und Erythräischen Tierwelt. Neue Denkschriften der allgemeinen Schweizerischen Gesellschaft für die gesamten naturwissenschaften Ser. 3, 28:1–39

Kevrekidis K (2014) The occurrence of the Atlantic penaeid prawn *Farfantepenaeus aztecus* (Ives, 1891) in the Thermaikos Gulf (Aegean Sea, Eastern Mediterranean): considerations on the potential establishment and impact on the autochthonous *Melicertus kerathurus* (Forskål, 1775). Crustaceana 87:1606–1619

Kheifets J, Rozhavsky B, Girsh Solomonovich Z et al (2012) Severe Tetrodotoxin poisoning after consumption of *Lagocephalus sceleratus* (pufferfish, fugu) fished in Mediterranean Sea, treated with Cholinesterase inhibitor. Case Rep Crit Care 2012:782507

Lindegg G (1934) La "*Mercierella enigmatica*" Fauvel nello stagno di Cabras in Sardegna. Natura, Milano 25:135–145

Maltagliati F, Lupi L, Castelli A, Pannacciulli FG (2016) The genetic structure of the exotic ascidian *Styela plicata* (Tunicata) from Italian ports, with a reappraisal of its worldwide genetic pattern. Mar Ecol 37(3):492–502

Marchini A, Ferrario J, Minchin D (2015a) Marinas may act as hubs for the spread of the pseudoindigenous bryozoan *Amathia verticillata* (Delle Chiaje, 1822) and its associates. Sci Mar 79(3):355–365

Marchini A, Ferrario J, Sfriso A et al (2015b) Current status and trends of biological invasions in the Lagoon of Venice, a hotspot of marine NIS introductions in the Mediterranean Sea. Biol Invasions 17(10):2943–2962

Marchini A, Ferrario J, Occhipinti-Ambrogi A (2016) The relative importance of aquaculture and shipping as vectors of introduction of marine alien species: the case of Olbia (Sardinia). Rapp Comm Int Expl Mer Médit 41:430

Marić M, Ferrario J, Marchini A, Occhipinti-Ambrogi A, Minchin D (2017) Rapid assessment of marine non-indigenous species on mooring lines of leisure craft: new records in Croatia (eastern Adriatic Sea). Mar Biodiv. https://doi.org/10.1007/s12526-016-0541-y

Marten E (1887) Conchylien aus dem Suezkanal. Sitzungs-Bericht der Gesellschaft naturforschender Freunde zu Berlin 6:89–96

Meinesz A, Hesse B (1991) Introduction et invasion de l'algue tropicale *Caulerpa taxifolia* en Méditerranée nord-occidentale. Oceanol Acta 14(4):415–426

Meinesz A, Belsher T, Thibaut T et al (2001) The introduced green alga *Caulerpa taxifolia* continues to spread in the Mediterranean. Biol Invasions 3(2):201–210

Micheli F, Halpern BS, Walbridge S et al (2013) Cumulative human impacts on Mediterranean and Black Sea marine ecosystems: assessing current pressures and opportunities. PLoS ONE 8(12):e79889

Minchin D, Floerl O, Savini D, Occhipinti-Ambrogi A (2006) Small craft and the spread of exotic species. In: Davenport J, Davenport JL (eds) The ecology of transportation: managing mobility for the environment. Springer, Netherlands, pp 99–118

Mineur F, Belsher T, Johnson MP et al (2007) Experimental assessment of oyster transfers as a vector for macroalgal introductions. Biol Conserv 137:237–247

Mineur F, Johnson MP, Maggs CA (2008) Macroalgal introductions by hull fouling on recreational vessels: Seaweeds and sailors. Environ Manage 42:667–676

Mineur F, Le Roux A, Maggs CA et al (2014) Positive feedback loop between introductions of non-native marine species and cultivation of oysters in Europe. Conserv Biol 28(6):1667–1676

Montefalcone M, Morri C, Parravicini V et al (2015) A tale of two invaders: divergent spreading kinetics of the alien green algae *Caulerpa taxifolia* and *Caulerpa cylindracea*. Biol Invasions 17(9):2717–2728

Musco L, Andaloro F, D'Anna G et al (2015) Impatto di *Caulerpa taxifolia* var. *distichophylla* su macro- e meiofauna associate a *Posidonia oceanica*. Biol Mar Mediterr 22(1):136–137

Nader M, Indary S, Boustany L (2012) The pufferfish *Lagocephalus sceleratus* (Gmelin, 1789) in the eastern Mediterranean. GCP/INT/041/EC—GRE—ITATD—10, FAO EastMed, pp 41

Nunes PALD, Markandya A (2008) Economic value of damage caused by marine bio-invasions: lessons from two European case studies. ICES J Mar Sci 65:775–780

Occhipinti Ambrogi A (2002) Current status of aquatic introductions in Italy. In: Leppäkoski E, Gollasch S, Olenin S (eds) Invasive aquatic species of Europe: distributions, impacts and management. Kluwer, Dordrecht, pp 311–324

Occhipinti-Ambrogi A, Marchini A, Cantone G et al (2011) Alien species along the Italian coasts: an overview. Biol Invasions 13(1):215–237

Oliverio M, Taviani M (2003) The Eastern Mediterranean Sea: tropical invasions and niche opportunities in a "Godot Basin". Biogeographia 24:313–327

Oren OH (1957) Changes in the temperature of the eastern Mediterranean Sea in relation to the catch of the Israel trawl fishery during the years 1954–55 and 1955–56. Bull Inst Océanogr 1102:1–13

Ounifi-Ben Amor K, Rifi M, Ghanem R (2016) Update of alien fauna and new records from Tunisian marine waters. Medit Mar Sci 17(1):124–143

Özcan T, Galil BS, Bakir K et al (2006) The first record of the banana prawn *Fenneropenaeus merguiensis* (De Man, 1888) (Crustacea: Decapoda: Penaeidae) from the Mediterranean Sea. Aquat Invasions 1:286–288

Öztürk B, Işinibilir M (2010) An alien jellyfish *Rhopilema nomadica* and its impacts to the Eastern Mediterranean part of Turkey. J Black Sea/Medit Environ 16(2):149–156

Panova M, Blakeslee AMH, Miller AW et al (2011) Glacial history of the North Atlantic marine snail, *Littorina saxatilis*, inferred from distribution of mitochondrial DNA lineages. PLoS ONE 6(3):1–14

Pećarević M, Mikus J, Cetinic AB et al (2013) Introduced marine species in croatian Waters (eastern Adriatic Sea). Mediterr Mar Sci 14(1):224–237

Pellizzato M, Da Ros L (2005) Clam farming quality as a management tool: a proposal based on recent studies in Northern Adriatic lagoons. Aquacult Int 13:57–66

Penna A (2014) Potential allochthonous microalgae in the Mediterranean Sea: a contribution addressing to this statement. Biol Mar Mediterr 21(1):52–56

Perlmutter A (1956) Sea fisheries. Report to the government of Israel. Technical assistance program USOM agricultural report No 13, pp 58

Pernet F, Lupo C, Bacher C et al (2016) Infectious diseases in oyster aquaculture require a new integrated approach. Phil Trans R Soc B 371:20150213

Piazzi L, Ceccherelli G (2002) Effects of competition between two introduced *Caulerpa*. Mar Ecol Progr Ser 225:189–195

Piazzi L, Cinelli F (2000) Effects of the spread of the introduced Rhodophyceae *Acrothamnion preissii* and *Womersleyella setacea* on the macroalgal community of *Posidonia oceanica* rhizomes in the western Mediterranean Sea. Cryptog Algol 21(3):291–300

Picciotto M, Bertuccio C, Giacobbe S et al (2016) *Caulerpa taxifolia* var. *distichophylla*: a further stepping stone in the western Mediterranean. Mar Biodiv Rec 9(1):73

Plan Bleu (2014) Economic and social analysis of the uses of the coastal and marine waters in the Mediterranean, characterization and impacts of the Fisheries, Aquaculture, Tourism and recreational activities, Maritime transport and Offshore extraction of oil and gas sectors. Technical Report UNEP(DEPI)/MED WG.401/6 4th Meeting of the EcAp Coordination Group, Athens, Greece, 9–10 Oct 2014. Plan Bleu, Valbonne

Pliny (1983) Natural history, iii:8–11. Loeb classical library no. 353. Harvard University Press, Cambridge, MA, USA, p 616

Por FD (1971) One hundred years of Suez Canal—A century of lessepsian migration: retrospect and viewpoints. Syst Zool 20(2):138–159

Por FD (1978) Lessepsian Migration— the influx of Red Sea Biota into the Mediterranean by way of the Suez Canal. Ecological Studies vol 23, Springer-Verlag, Berlin-Heidelberg-New York, pp 228

Pranovi F, Franceschini G, Casale V et al (2006) An ecological imbalance induced by a non-native species: the Manila clam in the Venice Lagoon. Biol Invasions 8:595–609

Pranovi F, Libralato S, Raicevich S et al (2003) Mechanical clam dredging in Venice Lagoon: effects on ecosystem stability evaluated with a trophic mass-balance model. Mar Biol 143:393–403

Pusceddu A, Fraschetti S, Scopa M et al (2016) Meiofauna communities, nematode diversity and C degradation rates in seagrass (*Posidonia oceanica* L.) and unvegetated sediments invaded by the algae *Caulerpa cylindracea* (Sonder). Mar Environ Res 119:88–99

Raimbault R (1964) Croissance des huitres atlantiques élevées dans les eaux méditerranéennes françaises. Science et Pêche 126:1–10

Rayssac N, Pérignon A, Gervasoni E et al (2012) Projet PRONAMED 2010–2011. Évaluation du potentiel d'approvisionnement naturel en naissains d'huîtres creuses en Méditerranée—Rapport final. http://archimer.ifremer.fr/doc/00088/19959/17626.pdf. Accessed 11 Jan 2017

Regional Activity Center for Specially Protected Areas (2017) Non-indigenous species. http://www.rac-spa.org/aliens. Accessed 3 Aug 2017

Rögl F (1997) Palaeogeographic Considerations for Mediterranean and Paratethys Seaways (Oligocene to Miocene). Ann Naturhist Mus Wien Ser A Mineral Petrogr Geol Palaontol Anthropol Prahist 99:279–310

Ros M, Vázquez-Luis M, Guerra-García JM (2013) The role of marinas and recreational boating in the occurrence and distribution of exotic caprellids (Crustacea: Amphipoda) in the Western Mediterranean: Mallorca Island as a case study. J Sea Res 83:94–103

Roy HE, Adriaens T, Aldridge DC et al (2015) Invasive Alien species—prioritising prevention efforts through horizon scanning. ENV.B.2/ETU/2014/0016. European Commission, pp 231

Ruiz GM, Carlton JT, Grosholz ED et al (1997) Global invasions of marine and estuarine habitats by non-indigenous species: mechanisms, extent, and consequences. Am Zool 37(6):621–632

Saglam H, Düzgünes E (2013) Feeding behaviour of invasive whelk, *Rapana venosa* Valenciennes, 1846 from the south-eastern Black Sea, Turkey. Rapp Comm Int Expl Mer Médit 40:587

Sala E, Kizilkaya Z, Yildirim D et al (2011) Alien marine fishes deplete algal biomass in the eastern Mediterranean. PLoS ONE 6(2):e17356

Samaha C, zu Dohna H, Bariche M (2016) Analysis of Red Sea fish species' introductions into the Mediterranean reveals shifts in introduction patterns. J Biogeog 43(9):1797–1807

Sancholle M (1988) Présence de *Fucus spiralis* (Phaeophyceae) en Méditerranee occidentale. Cryptog Algol 9:157–161

Sato S, Nishimura T, Uehara K et al (2011) Phylogeography of *Ostreopsis* along West Pacific Coast, with special reference to a novel clade from Japan. PLoS ONE 6(12):e27983

Savini D, Marchini A, Forni G et al (2006) Touristic harbours and secondary spread of alien species. Biol Mar Mediterr 13:760–763

Sciberras M, Schembri PJ (2007) A critical review of records of alien marine species from the Maltese Islands and surrounding waters (Central Mediterranean). Mediterr Mar Sci 8(1):41–66

Seebens H, Gastner MT, Blasius B (2013) The risk of marine bioinvasion caused by global shipping. Ecol Lett 16(6):782–790

Segarra A, Pépin JF, Arzul I et al (2010) Detection and description of a particular Ostreid herpesvirus 1 genotype associated with massive mortality outbreaks of Pacific oysters, *Crassostrea gigas*, in France in 2008. Virus Res 153(1):92–99

Sfriso A, Facca C (2013) Annual growth and environmental relationships of the invasive species *Sargassum muticum* and *Undaria pinnatifida* in the lagoon of Venice. Estuar Coast Shelf Sci 129:162–172

Sghaier YR, Zakhama-Sraieb R, Benamer I, Charfi-Cheikhrouha F (2011) Occurrence of the seagrass *Halophila stipulacea* (Hydrocharitaceae) in the southern Mediterranean Sea. Bot Mar 54(6):575–582

Shiganova TA (1998) Invasion of the Black Sea by the ctenophore *Mnemiopsis leidyi* and recent changes in pelagic community structure. Fish Oceanogr 7(3/4):305–310

Solidoro C, Melaku Canu D, Rossi R (2003) Ecological and economic considerations on fishing and rearing of *Tapes philippinarum* in the lagoon of Venice. Ecol Model 170:303–318

Souissi JB, Rifi M, Ghanem R et al (2014) *Lagocephalus sceleratus* (Gmelin, 1789) expands through the African coasts towards the Western Mediterranean Sea: a call for awareness. Manag Biol Invasions 5(4):357–362

Steen F, Aragay J, Zuljevic A et al (2017) Tracing the introduction history of the brown seaweed *Dictyota cyanoloma* (Phaeophyceae, Dictyotales) in Europe. Eur J Phycol. https://doi.org/10.1080/09670262.2016.1212998

Steinitz W (1919) Memorial on the founding of a sea-station on the coast of Palestine for zoological investigations. Manuscript, pp 16

Steinitz W (1927) Beiträge zur Kenntnis der Küstenfauna Palästinas. I. Pubbl Staz Zool Napoli 8 (3–4):311–353

Steinitz H (1970) A critical List of Immigrants via the Suez Canal. Biota of the Red Sea and Eastern Mediterranean, pp 59–63 [mimeo]

Stentiford GD, Lightner DV (2011) Cases of white spot disease (WSD) in European shrimp farms. Aquaculture 319:302–306

Stock JH (1993) Copepoda (Crustacea) associated with commercial and non-commercial Bivalvia in the East Scheldt, The Netherlands. Bijdr Dierk 63:61–64

Suez Canal Authority (2017) Stages in developing Suez Canal. In: Canal characteristics. http://www.suezcanal.gov.eg/English/About/SuezCanal/Pages/CanalCharacteristics.aspx. Accessed 3 Aug 2017

Superyacht Business Happening (2015) Mediterranean as a strategic area. In: Industry outlook. http://www.yarenetworking.com/wp-content/uploads/2015/01/YR15_Catalogue_Low_EXE.pdf. Accessed 3 Aug 2017

Terlizzi A, Felline S, Lionetto MG et al (2011) Detrimental physiological effects of the invasive alga *Caulerpa racemosa* on the Mediterranean white sea bream *Diplodus sargus*. Aquat Biol 12:109–117

Thompson IC, Scott A (1903) Report on the Copepoda collected by Professor Herdman, at Ceylon, in 1902. Ceylon Pearl Oyster Fisheries—1903—Supplementary reports 7: 227–307

Thunell RC, Williams DF, Kennet JP (1977) Late Quaternary paleoclimatology, stratigraphy and sapropel history in eastern Mediterranean deep-sea sediments. Mar icropaleontol 2:371–388

Tillier JB (1902) Le Canal de Suez et sa faune ichthyologique. Mem Soc Zool France 15(3–4): 279–318

Tillier JB, Bavay A (1905) Les mollusques testacés du Canal de Suez. Bull Soc Zool Fr 30:170–181

Tillier JB, Bavay A (1906) Au sujet des mollusques testacés du Canal de Suez. Bull Soc Zool Fr 31:129–131

Tortonese E (1973) Facts and perspectives related to the spreading of Red Sea organisms into the eastern Mediterranean. Ann Mus Civ St Nat Giacomo Doria 79:322–329
Tosun MS (2011) Demographic divide and labor migration in the Euro-Mediterranean Region. Institute for the Study of Labor, Discussion Paper No 6188. pp 23
Tournier H (1972) Conditions d'acclirriatation des crevettes *Penaeus kerathurus* et *P. japonicus* dans les eaux du littoral languedocien. Science et Pêche 213:1–13
Tsiamis K, Aydogan Ö, Bailly N et al (2015) New Mediterranean biodiversity records (July 2015). Mediterr Mar Sci 16(2):472–488
Turolla E (2008) La venericoltura in Italia. In: Lovatelli A, Farías A, Uriarte I (eds) Estado actual del cultivo y manejo de moluscos bivalvos y su proyección futura: factores que afectan su sustentabilidad en América Latina. Taller Técnico Regional de la FAO, 20–24 Aug 2007, Puerto Montt, Chile. FAO Actas de Pesca y Acuicultura, no 12, FAO, Rome, pp 177–188
UN (United Nations) (1982) United Nations Convention on the Law of the Sea. pp 202
UNEP/MAP (2007) Annex XIII. Draft Guidelines for Controlling the Vectors of Introduction into the Mediterranean of Non-Indigenous Species and Invasive Marine Species. In: Report of the eighth meeting of national Focal Points for SPAs, UNEP(DEPI)/MED WG.308/14
UNEP/MAP (2013) Draft decision on the Ecosystems Approach including adopting definitions of Good Environmental Status (GES) and targets. Meeting of the MAP Focal Points. UNEP (DEPI)/MED WG.387/7, pp 46
UNEP/MAP (2016) Annex III. Updated action plan concerning species introductions and invasive species in the Mediterranean Sea. In: Report of the 19th ordinary meeting of the Contracting Parties to the Convention for the protection of the marine environment and the coastal region of the Mediterranean and its protocols, UNEP(DEPI)/MED IG.22/28 (Decision IG.22/15), pp 602–611
Vaillant L (1865) Recherches sur la faune malacologique de la baie de Suez. J Conch 13:97–127
Vergés A, Tomas F, Cebrian E et al (2014) Tropical rabbitfish and the deforestation of a warming temperate sea. J Ecol 102(6):1518–1527
Williams SL, Davidson IC, Pasari JR et al (2013) Managing multiple vectors for marine invasions in an increasingly connected world. Bioscience 63(12):952–966
Zenetos A, Pancucci-Papadopoulou M, Zogaris S et al (2009) Aquatic alien species in Greece (2009): tracking sources, patterns and effects on the ecosystem. J Biol Res (Thessalon) 12:135–172
Zenetos A, Apostolopoulos G, Crocetta F (2016) Aquaria kept marine fish species possibly released in the Mediterranean Sea: first confirmation of intentional release in the wild. Acta Icht Pisc 46(3):255–262
Zolotarev V (1996) The Black Sea ecosystem changes related to the introduction of new mollusc species. Mar Ecol 17(1–3):227–236
Žuljević A, Thibaut T, Despalatović M et al (2011) Invasive alga *Caulerpa racemosa* var. *cylindracea* makes a strong impact on the Mediterranean sponge *Sarcotragus spinosulus*. Biol Invasions 13(10):2303–2308

Chapter 3
The Apparently Relentless Spread of the Major Decapod Alien Species in the Mediterranean Basin and European Inland Waters

Chiara Manfrin, Catherine Souty-Grosset, Pedro Anastácio, Julian Reynolds and Piero G. Giulianini

Abstract Among the many alien crustacean species so far recorded in the Mediterranean basin are numerous decapods—crayfish, crabs and prawns—that have in common the reputation of being both aggressive and adaptable to various environments. Their entrance to the Mediterranean basin has often been fostered by human actions, such as the transfer of new species for commercial purposes or accidentally, as bycatch or because they were present in discarded ships' ballast water. Marine alien species enter the Mediterranean basin from two main directions: from the Atlantic and beyond, and from the Indian Ocean via the Suez Canal. For freshwaters, the extensive European canal network is an important aid to migration, but human discards are also significant. Protection of native fauna and ecosystems is at best partial. Freshwater ecosystems particularly at risk include lagoons, river estuaries, ponds and marshes, while comparable marine systems are shallow muddy coastal regions, rocky reefs, estuaries/bays and salt marshes. Local communities, principally anglers and people working with aquatic resources, usually are curious at the appearance of a new species and the possibility of establishing a new business is often the first thought. However, freshwater decapod alien species cause major

C. Manfrin (✉) · P. G. Giulianini
Department of Life Sciences, University of Trieste, via L. Giorgieri, 5, 34127 Trieste, Italy
e-mail: cmanfrin@units.it

C. Souty-Grosset
Laboratoire Ecologie & Biologie des Interactions-UMR CNRS 7267, Equipe Ecologie Evolution Symbiose, Université de Poitiers, 5, Rue Albert Turpin, TSA 51106, 86073 Poitiers Cedex 9, France

P. Anastácio
MARE—Marine and Environmental Sciences Centre, Departamento de Paisagem, Ambiente e Ordenamento, Escola de Ciências e Tecnologia, Universidade de Évora, Rua Romão Ramalho, 59, 7000-671 Évora, Portugal

J. Reynolds
Emeritus, Department of Zoology, Trinity College, University of Dublin, Dublin 2, Ireland

A. I. Queiroz and S. Pooley (eds.), *Histories of Bioinvasions in the Mediterranean*, Environmental History 8, https://doi.org/10.1007/978-3-319-74986-0_3

damage to agriculture production, infrastructures, riverbanks, irrigation systems and fish production, leading to consistent economic losses. In particular, they are known to be injurious to plant production, causing increasing losses to agriculture, consuming young seedlings and seeds and causing substantial water losses from field areas. Less information is available for seawater species to date, but their increasing number and growing size of populations requires the attention of the scientific community to evaluate and predict their influence on the marine environment and/or on local fish and bivalve production. Both freshwater and marine alien decapods thus represent a serious threat to native European species and have a negative impact on species richness and habitats. The economic losses caused by these species require reliable and cost-effective options for managing their abundance and if possible for their eradication.

Keywords Alien decapods · Mediterranean basin · Socio-economic impacts

3.1 Biodiversity, Heritage Status and Socio-economic Activities of Mediterranean Decapod Crustaceans

The highly biodiverse Mediterranean Basin has been impacted for millennia by human agriculture and settlement, leading to species losses and even extinctions. This chapter focuses on larger crustaceans (Order Decapoda) in fresh or salt water, including lobsters, crayfish, crabs and shrimps, a group of invertebrates with many forms of economic importance valued at millions of euros, as human food, bait, or as a nuisance. Some native species are in decline, and a large number of species have been introduced, accidentally or deliberately, from elsewhere.

3.1.1 Ecology of Lobsters, Crayfish, Crabs and Shrimps (Decapoda)

Key environmental factors affecting native crustaceans include temperature, salinity and habitat. Temperature determines crustacean activity levels, moulting and breeding, regulated by hormones. Satellite data show a steady increase in surface temperatures of the Mediterranean Sea. This is leading to 'Meridionalisation', the retreat of some northern species and a northward advance of southern species, and provides opportunities for invading alien forms.

Freshwater and estuarine crustaceans are all benthic (crabs, crayfish) or nektobenthic[1] (shrimps), in lakes, streams, estuaries and wetlands. These habitats and their biota are under extreme stress from human overuse, leading to local extinction.

[1]Organisms that swim near the bottom of a sea or lake.

Estuarine forms including shrimps, prawns, crabs and crayfish must adapt to muddy conditions with salinity variation and sometimes reduced oxygen. Freshwater invaders in the Mediterranean Basin include narrow-clawed crayfish *Astacus leptodactylus* and North American crayfish species, particularly red swamp crayfish *Procambarus clarkii*, both tolerating low levels of salinity. The Chinese mitten crab *Eriocheir sinensis* is mostly freshwater but requires access to the ocean for breeding.

3.1.2 Protection of Mediterranean Habitats and Fauna

The rate of endemism, i.e. the number of species living exclusively in the Mediterranean, is relatively high, averaging more than one-quarter of the whole Mediterranean biota (Tortonese 1965; Lejeusne et al. 2010; Giaccone 1999; Coll et al. 2010). The IUCN (IUCN-Med 2007) noted that species richness of native marine fishes is greatest in the western Mediterranean, as far east as Italy. Endemic fish species are found particularly in the northern part of the basin (Spain, France, Italy, Adriatic). To what extent this is paralleled by crustaceans is not known.

The Mediterranean basin comes under two policy zones: EU, including parts of Portugal, Spain, France, Italy, former Yugoslavia and Greece, and Non-EU (Levant and Maghreb), with fewer treaties. The Levant includes Turkey (just extending into the mouth of the Black Sea), Syria, Lebanon, Israel/Palestine and Egypt (with the Suez Canal, an important source of alien species), while the Maghreb takes in Libya, Tunisia, Algeria and Morocco.

Mediterranean marine species were poorly protected until the 1970s. After 1991, France protected 6 marine invertebrates, including the Mediterranean slipper lobster *Scyllarides latus*. In 1992, the EU Habitats Directive listed 622 animals and plants needing Special Areas of Conservation (SACs). It included no crustaceans or marine plants. The Barcelona Convention in 1996 listed 55 marine species including two crustaceans, the tufted ghost crab *Oxypode cursor* and the rare barnacle *Pachylasma giganteum* from the Mediterranean Sea area. The UNEP/MAP-Barcelona Convention 2012 stimulated a Mediterranean-wide assessment of biodiversity, subdividing the Mediterranean region into four geographic areas or sub-regions for analysis purposes: the Western Mediterranean, Ionian Sea and Central Mediterranean, Adriatic Sea, and Eastern Mediterranean.

The European Union (European Comission 2016) has recently listed species of concern, of which 6 are alien freshwater decapod crustaceans—one crab (*Eriocheir sinensis*) and 5 crayfish: *Orconectes limosus*, *Orconectes virilis*, *Pacifastacus leniusculus*, *Procambarus clarkii* and *Procambarus fallax f. virginalis*.

3.1.2.1 Protection of Mediterranean Fresh Waters

There are major impacts on Mediterranean fresh waters through land use disturbance, principally sediment and nutrient loads from agriculture; from altered hydrology, including impoundment and abstraction for consumptive (irrigation) and non-consumptive use; and from increasing numbers of alien invasive forms, introduced either deliberately (e.g. crayfish) or accidentally. The Mediterranean region risks losing 56% of its endemic freshwater fish, 36% of freshwater crabs and crayfish, 29% of amphibians, and 19% of dragonflies during the next decades. Reductions in wetland biodiversity are particularly bad in the Eastern Mediterranean (MWO 2012).

There are different strategies for freshwater protection, including whole catchment management, natural-flow maintenance, exclusion of non-native species, vegetated buffer strips and partial water discharges (e.g. at high water periods), also multiple-use models and use of the river continuum concept. Inclusion in parks is not always protective, and many impacts occur outside the park perimeter (Saunders et al. 2002).

3.1.3 Native Decapods and Their Exploitation

Numerous native species of crustaceans are exploited by local or regional fisheries, for example, the Mediterranean slipper lobster *Scyllarides latus*, chiefly in crab pots off Catalonia and around eastern Italy, and various shrimp and prawn fisheries, e.g. for the common prawn *Palaemon serratus* off Andalucia and Catalonia, red shrimp *Aristeus antennatus* in suitable waters off Catalonia and the Adriatic, and for other species in the Levant. The striped prawn *Melicertus kerathurus* is commercially important to fisheries in the Levant and for pond aquaculture around the Aegean Sea and in the central and western Mediterranean.

In Greek Aegean Sea waters *M. kerathurus* and the rose shrimp *Parapenaeus longirostris* are the most important in terms of landing and economic value (Mytilineou et al. 2001). The main commercial species in Greek waters are the penaeid shrimps *P. longirostris* and *M. kerathurus*, the Norway lobster *Nephrops norvegicus*, the spiny lobster *Palinurus elephas* and the European lobster *Homarus gammarus* (the last two protected in the Mediterranean). Some other species, such as the portunid crab *Liocarcinus depurator*, majid crabs *Maja* spp., the calappoid crab *Calappa granulata*, and the squat lobster *Munida* sp., could be characterized as by-catch. Other shrimp species are locally important, such as the palaemonid shrimp *Palaemon adspersus*, or are potentially marketable, such as the deep-water shrimps (Pandalidae, Aristeidae) (Kapiris et al. 2007). In addition, invasive shrimps and prawns that have spread from Suez to the Levant, are now sustaining fisheries. There are ongoing experiments under laboratory conditions on embryonic development, spawning and larval rearing of the Mediterranean spider crab *Maja squinado*, as a candidate species for a restocking program (Rotllant et al. 2014). No EU Protected Designations of Origin (PDO) have been found for native crustaceans that could be affected by alien species. However, several species are under consideration, e.g. (from official PDO website): possible PDO or PGI (Protected Geographical Indication) for Spain include prawns *Palaemon serratus* from

Andalucia near Cadiz and Cartagena, and red shrimp *Aristeus antennatus* from deep water off Vilanova (Catalonia).

3.2 Alien Decapod Species in the Mediterranean

The main pathways for marine Non-Indigenous Species (NIS) introduction in Europe's seas are shipping (51%) and the Suez Canal (37%). These are followed by aquaculture related activities (17%) and, to a much lesser extent, aquarium trade (3%) and inland canals (2%). This is a pattern observed in all regional seas, except for the Eastern Mediterranean where introductions via the Suez Canal exceed those by shipping. Trends in pathways show an increasing rate of introductions by shipping and corridors (in particular the Suez Canal) since the 1990s, while aquaculture mediated introductions have been decreasing since the 2000s. This can be attributed to the adoption of effective EU regulation (European Environment Agency 2015). Aquarium trade emerges as a lower but increasing pathway since the 2000s (European Environment Agency 2015).

Crustacea are the second largest group among alien species in the Mediterranean Sea, with 159 species (Zenetos et al. 2012b). Appendix 3.1 provides a list of Mediterranean and European marine, oligohaline[2] and freshwater alien decapod species. The list of marine decapods is based on Zenetos et al. (2010) and updates, mainly of the "New Mediterranean biodiversity records" series up to November 2016 (see the Appendix 3.1 for references). For the freshwater and oligohaline species the list is based on Holdich et al. (2009) and on the European Alien Species Information Network—EASIN (https://easin.jrc.ec.europa.eu/). The 'worst' species (i.e. having most serious negative impacts), based on documented impacts on biodiversity and socioeconomics of the basin, are listed in "Alien Marine Species in the Mediterranean—the 100 'Worst Invasives' and their Impact" (Streftaris and Zenetos 2006) or in Delivering Alien Invasive Species Inventories for Europe—DAISIE "100 of the Worst", in terms of their impacts on biodiversity, economy and health (http://www.europe-aliens.org/speciesTheWorst.do).

In the Mediterranean Sea 309 autochthonous species of decapods have been reported (Coll et al. 2010), of which 40 are endemic. Here we list 86 Mediterranean decapod alien species, representing 21.8% of all Mediterranean decapods. Moreover, 49 species (12.4% of the taxon) are reported as "established", with "self-maintaining and self-perpetuating populations unsupported by and independent of humans" (Zenetos et al. 2010).

As far as freshwater/oligohaline species are concerned, 18 alien species are listed with 6 invasive and one 'worst'. The invasive species belong to the North America fauna or, in one case (*A. leptodactylus*), to Eastern Europe. *P. clarkii* is listed among DAISIE's "100 of the Worst" invasives (http://www.europe-aliens.org) and,

---

[2]Characterized by very low salinity, specifically (in later use) in the range 0–0.5%; brackish.

recently, among the 9 European invasive alien species impacting on more than four threatened species (IUCN Red List categories CR, EN, VU), as it affects 5 European native species (Genovesi et al. 2015). In the latter paper other crayfish *Orconectes limosus*, *O. virilis* and *Pacifastacus leniusculus* are listed in Annex 1, each affecting one or two threatened European native species (Genovesi et al. 2015).

Crustacean aquaculture is limited for the Mediterranean, mostly occurring in brackish waters along the coast of Egypt. The dominant farmed species is the kuruma prawn (*Marsupenaeus japonicus*) from the Red Sea and Indo-West Pacific, providing about 90% of total crustacean production, with semi-intensive cultivation in Greece, Spain, Italy and other Mediterranean countries. Some other west Indo-Pacific shrimps are farmed in the Aegean Sea (Türkmen 2007). In addition, extensive aquaculture in Turkish lagoons produces blue crab *Callinectes sapidus* for export, though domestic demand is insignificant.

3.3 Introduction, Spread and Impacts on Biodiversity

More than 70 decapod species have been recorded in the Eastern Mediterranean Sea, most of them Lessepsian (migrating through the Suez Canal, developed by Ferdinand de Lesseps). At least 33 Lessepsian decapods are known along the coasts of Israel (Galil and Shlagman 2011), with new records constantly added. Out of the known Lessepsian decapods, only 9 species are shrimps which belong to the infraorder Caridea (Galil and Shlagman 2011; Galil and Mendelson 2013; Karhan et al. 2013; Rothman et al. 2013).

Occhipinti-Ambrogi and Galil (2010) list some 19 alien decapod species around the Italian coasts. There has been some analysis of economic impact of aliens displacing native forms. Galil and Zenetos (2002), reviewing impacts of exotics on the eastern Mediterranean, reported rapid declines in populations of the striped prawn *Melicertus kerathurus* and the snapping shrimp *Alpheus dentipes*, attributed to invasive displacement by the kuruma prawn *Marsupenaeus japonicus*.

We will focus on 14 invasive decapod species classified by the global invasive species database (Invasive species specialist group, http://www.issg.org/database) in the list of 100 of the "worst" invasive species.

3.3.1 Freshwater Species

The Chinese mitten crab *Eriocheir sinensis* is a catadromous species (Lowe et al. 2000) (Fig. 3.1). Its adults mate and spawn in estuaries, the juveniles move up the rivers and adults migrate back to estuaries to complete the life-cycle and die (Veilleux and De Lafontaine 2007; Dittel and Epifanio 2009). This species is native to Eastern Asia and has a wide invasive distribution across the globe. Transport of

Fig. 3.1 Specimen of Chinese mitten crab. *Credit* Pedro Anastácio

larvae in ballast water is considered the main process of invasion (Herborg et al. 2003) but live trade in the species has also been considered a possible mechanism (Gollasch and Nehring 2006). In Europe it is present since 1912 (Aller river, Germany) and during the early 20th century, population explosions resulted in major efforts to mitigate their impacts in rivers, canals, and associated municipal facilities (Grosholz 2011). This species has extensive burrowing activity, sometimes damaging river bank integrity, and can damage fishing nets. In the Mediterranean area, this species is present in the Tagus River in Portugal (Cabral and Costa 1999) since the late 1980's and in the Guadalquivir in Spain (Garcia-de-Lomas et al. 2010) since 1987, and was first observed in Italy (Venice Lagoon) in 2005 (Mizzan 2005). In France the first Mediterranean occurrence was in 1959, although a population was not successfully established at that date (Petit 1960).

The red swamp crayfish *Procambarus clarkii* (Fig. 3.2) is native to North America (USA and North of Mexico). According to Souty-Grosset et al. (2016), it is considered one of the most invasive species for aquatic ecosystems. Since its introduction into Spain in 1973 for aquaculture, it has spread rapidly through rivers, lakes, ponds and coastal marshes. The species has strong dispersal abilities overland (Gherardi and Barbaresi 2000; Ramalho and Anastácio 2015), but passive dispersal by human activities and animals has also been described (Águas et al. 2014; Banha and Anastácio 2015). The red swamp crayfish is currently recorded from 15 European countries and North Africa (Egypt), but populations are most prevalent in Mediterranean areas of Italy, Portugal, Spain and France.

In addition to being a vector of crayfish plague, lethal to indigenous European crayfish species, it can cause severe impacts on aquatic ecosystems such as

Fig. 3.2 Adults of Red swamp crayfish. *Credit* Catherine Southy-Grosset

estuaries, marshes, canals, artificial lakes (dams), gravel pit lakes, ponds and fishponds due to its short life-history, rapid growth, burrowing activities and high population density. This invasive crayfish is a polytrophic keystone species that can exert multiple pressures on lake littoral ecosystems (Gherardi and Acquistapace 2007). It is considered an agricultural pest in rice-growing Mediterranean territories, mainly due to its digging behaviour but also consuming young rice plants (Anastácio et al. 2000; Anastácio and Marques 1997; Anastácio et al. 2005).

In 2006 an unusual find of *P. clarkii* was made within a protected brackish Mediterranean biotope "Palude di Torre Flavia" (Rome, Italy) (Scalici et al. 2010). The investigation showed that this species can damage brackish habitats and human livelihoods. It also showed its high ecological plasticity and its adaptation to the atypical thermal habitat of about 12 °C below its optimum water temperature (Peruzza et al. 2015), indicating the need to intervene before the species spreads further in brackish water.

3.3.2 Marine Species

3.3.2.1 Shrimps

Indo-Pacific prawn species, in particular *Marsupenaeus japonicus*, *Metapenaeus monoceros*, and *Penaeus semisulcatus*, are highly valued and have evidently benefited Levantine fisheries, but can have high discard rates (Duruer et al. 2008). According to FAO statistics, in the Mediterranean area *Marsupenaeus japonicus* is produced in aquaculture since 1985 and there are wild captures since 1997. *Metapenaeus monoceros* has been captured since 2001 and there is no FAO aquaculture production statistics. FAO statistics on *Penaeus semisulcatus* are also scarce, with only one record of aquaculture in 1996 and no records of wild captures.

The kuruma shrimp *M. japonicus*, is native to the Indo-Pacific region from the Red Sea and East Africa to Fiji. The kuruma shrimp has entered the Levantine Basin through the Suez Canal, but some introductions have been attributed to escapes from aquaculture (Gravili et al. 2010). Farmed prawns can also create major problems for native shrimp species and wild fishes by transferring parasites and disease with escapes from aquaculture facilities. Galil and Zenetos (2002), reviewing impacts of exotics on the eastern Mediterranean, reported the rapid decrease of populations of the native prawn *Melicertus kerathurus* and the shrimp *Alpheus dentipes*, attributed to invasive displacement by the kuruma shrimp. The native prawn was commonly caught by trawlers along the Israeli coastal shelf on sandy or sandy-mud bottoms, and supported a commercial fishery throughout the 1950s (Holthuis and Gottlieb 1958). It has since nearly disappeared, its habitat overrun by Erythrean penaeid prawns. Geldiay and Kocatas (1972) reported that it has also been replaced by *M. japonicus* in fisheries catches off the southern coast of Turkey. D'Udekem d'Acoz (1999) reported that *M. japonicus* has almost evicted the native *P. kerathurus* from the easternmost part of the Mediterranean. The rapid advent of *M. monoceros* into the Gulf of Gabes, Tunisia, has raised concerns over the fate of *M. kerathurus* fisheries there (Chaouachi et al. 1998).

The Peregrine shrimp *Metapenaeus stebbingi* is a native species in the Indian Ocean which invaded the Mediterranean Sea in the second half of the 20th Century and was first spotted in 1924 in Egypt. Currently the impacts of this invasive shrimp on the native fauna in areas where it has been introduced are uncertain, but the peregrine shrimp may have an advantage over the native shrimp *M. kerathurus* in competing for food resources.

The penaeid shrimp *Melicertus hathor* is also native to the Indian Ocean. *M. hathor* was first reported in the Mediterranean Sea in 1999 and in 2002 was first recorded on the eastern Mediterranean coasts of Turkey. *M. hathor*, an immigrant species from the Red Sea, was seen in the Gulf of Antalya in March–April 2005 (Gokoglu and Kaya 2006). It became established in the eastern Mediterranean (Gokoglu and Kaya 2006; Çinar et al. 2011) and recently expanded to Greece (Siokou et al. 2013). It lives on sandy bottoms down to 80 m depth. The shrimp is reported as locally common and of some commercial importance in Iskenderun Bay and it is considered a threat to biodiversity (Streftaris and Zenetos 2006; Zenetos et al. 2005).

The green tiger prawn *Penaeus semisulcatus* is economically important in its Indo-West Pacific native distribution. FAO statistics about it are scarce, with only one record of aquaculture in 1996 and no records of wild captures. It is caught in the Mediterranean by Turkish, Israeli and Egyptian trawlers. The species has reached the eastern Mediterranean through the Suez Canal; it is now found all along the coasts of Egypt, Israel, Lebanon, Syria and southern Turkey (FAO and Holthuis 1980). It is a carrier for numerous pathogens such as *Thelohania*, baculoviral mid-gut necrosis virus, *Cephaloidophora setnai*, hepatopancreatic parvo-like virus of shrimps, infectious hypodermal and hematopoietic necrosis virus, *Penaeus monodon* HPV (Hepatopancreatic Parvo Virus), and White spot syndrome virus (Forbes 2007).

3.3.2.2 Crabs

Introduced crabs constitute among the best examples of introduced marine and estuarine species that have had significant impacts on coastal habitats and economies. The Mediterranean Sea has the highest number of alien brachyuran[3] species as many have invaded through the Suez Canal, making the Mediterranean the meeting place of Atlantic and Indo-West Pacific faunas (Brockerhoff and McLay 2011). Some 39 alien Brachyura species of Red Sea/Indo-Pacific origin, belonging to 19 families, have been recorded in the Mediterranean Sea, mainly in the eastern part (Galil 2011; Zenetos et al. 2012a; Karhan et al. 2013; Galil and Mendelson 2013; Zaouali et al. 2013a). In this area, their pathway of introduction is the Suez Canal, but penetration has been either unintentional (Lessepsian migration) or ship-mediated.

The flower crab *Portunus segnis*, also known as the blue swimmer crab, is native to the Indian and West Pacific Oceans. Galil (2011) considers it the first crustacean invader in the Mediterranean Sea, originally recorded at Port Said, Egypt, in 1898 (Galil and Zenetos 2002) and later in eastern Sicily, Italy (Crocetta 2006). Recently it has been collected in the Tyrrhenian Sea and in the Aegean Sea, off Turkey (Yokes et al. 2007; Corsini-Foka et al. 2004). Global warming is expected to favour this tropical species (Galil 2007), found primarily within estuaries and inshore coastal waters on sandy or muddy substrates and from the intertidal down to 55 meters depth, where it preys on slow-moving invertebrates (Lai et al. 2010). *P. segnis* has been commercially valuable since the early 1920s in the Levant (CIESM 2014, http://www.ciesm.org/online/atlas), commercially exploited in Egypt (Abdel-Razek 1987) and in smaller numbers in Turkey (Duruer et al. 2008; Zenetos et al. 2010).

---

[3]Designating or of certain crabs with a short abdomen folded beneath the main body. It represents a suborder of crustaceans.

The Sally Lightfoot crab *Percnon gibbesi* is the most invasive decapod species in the Mediterranean. Native to west and east coasts of America and the eastern Atlantic from Madeira to the Gulf of Guinea, it was first collected in the Mediterranean Sea in 1999 from the Balearic Islands, and Sicily and its adjacent islands (Relini et al. 2000; Garcia and Reviriego 2000; Mori and Vacchi 2002). Later reports documented its dispersal northwards along the Tyrrhenian coast of Italy, along the Ionian coast of Calabria, and recent records from Greece confirm its eastwards spread. The species was reported for the first time from the southern coast of Turkey in 2005–2006 (Yokes and Galil 2006b). Its feeding habits (it consumes primarily algae, but also other crabs, polychaetes, gastropods, crustaceans and jellyfish) may affect the structure of benthic communities, particularly algal assemblages, and it may compete with native species for food and shelter (Katsanevakis et al. 2011). Exclusion of native crabs may occur in some areas: its habitat overlaps with the omnivorous native crab *Pachygrapsus marmoratus* and the carnivorous pebble crab *Eriphia verrucosa*.

Two mechanisms of introduction of *P. gibbesi* have been suggested, either through shipping (Galil and Zenetos 2002) or by larval drift with the Atlantic surface current that enters the Mediterranean Sea through the Strait of Gibraltar (Abelló et al. 2003). Other possible vectors are aquarium trade, as for other Mediterranean aliens (Padilla and Williams 2004) and fishing nets. Eradication may be impossible in practice as this species is now too widespread in the Mediterranean Sea.

The blue crab *Callinectes sapidus* (Fig. 3.3) is native to the western Atlantic Ocean and Gulf of Mexico and was transported into the Mediterranean in ballast tanks from the northeast coast of the USA around 1949. Together with high fecundity and dispersal capacity, and fast growth, its broad environmental tolerances predispose *C. sapidus* as a likely successful invader (Hill et al. 1989). As is the case with many swimming crab species, it has the tendency to significantly expand its native range. This is partially caused by long planktonic development periods, during which developing larvae may be carried hundreds of miles by ocean currents.

The first Mediterranean report was from the north Adriatic Sea in 1949; subsequently, the species has been widely recorded in the Mediterranean, especially in its eastern basin (Castejón and Guerao 2013; Nehring 2011; Galil 2008; Galil et al. 2002). Reports from the Ebro delta, Spain (Tuncer and Bilgin 2008) and Guadalquivir, southern Iberia (Cabal et al. 2006) confirm the active southwards expansion of this species along western estuaries in the Iberian Peninsula (Ribeiro and Veríssimo 2014). The presence of *C. sapidus* has also been ascertained along Maltese coasts (Sciberras and Schembri 2007), in the Ligurian Sea (Bisconti and Silvi 2005), in the Aegean-Levantine Sea basin (Zaitsev and Öztürk 2001; Bisconti and Silvi 2005; Enzenrob et al. 1997; Serbetis 1959), and in the Ionian Sea (Thessalou-Legaki et al. 2012; Molnar et al. 2008). In the Adriatic Sea, the first record of *C. sapidus* dates back to 1949 near the Grado Lagoon (Giordani Soika 1951), one of the northernmost Adriatic Sea locations. Recently it has reappeared at the same site after almost 70 years of absence (Manfrin et al. 2016). In Greece, Perdikaris et al. (2016) described the invasion of *C. sapidus* in a significant number of estuaries, lagoons, and transitional waters along the entire coastal zone of the

Fig. 3.3 Blue crab. *Credit* Chiara Manfrin

Ionian Sea. This range expansion will likely continue due to active (swimming) spread and passive (ballast water) transportation.

Blue crabs perform a variety of ecosystem functions and can play a major role in energy transfer within estuaries and lagoons. At various stages in the life cycle, blue crabs serve as both prey and as consumers of plankton, small invertebrates, fish, and other crabs (Nehring 2012). *Callinectes sapidus* readily feeds on clams, mussels and oysters and has been reported to feed on fish caught in traps and to damage nets. Potential impacts may therefore include some loss of aquaculture, commercial, or recreational harvest. Despite the nomination of blue crab as one of the 100 'Worst Invasive Alien species in the Mediterranean' (Streftaris and Zenetos 2006), up to now no definite long term impacts of *C. sapidus* on non-native environments are known. Negative effects are mainly related to damage of fishing nets and fish harvesting difficulties in lagoons.

Say's mud crab *Dyspanopeus sayi*, native to the North American Atlantic coast, became established in the Mediterranean Sea since the 1970s. *D. sayi* was reported as very abundant in the lagoon of Venice, Italy (Occhipinti-Ambrogi 2000) where it was first recorded in the Mediterranean in 1992 (Froglia and Speranza 1993), but it had been in the Lagoon since the 1980s, according to local fishermen (Occhipinti-Ambrogi 2000). By 2001 it had almost disappeared in the canal of Venice for unknown reasons (Occhipinti-Ambrogi 2001) and now its presence is limited to the marine sectors of the lagoon near the mouths (Mizzan et al. 2005).

D. sayi in the Venice lagoon feeds on bivalves and may affect local clam farming, although its presence is now regarded as limited (Mizzan et al. 2005; Occhipinti-Ambrogi 2001). It has also been detected in the Ebro delta on the Iberian Peninsula (Schubart et al. 2012).

The swimming crab *Charybdis helleri* is a Lessepsian sublittoral portunid crab species (Falciai and Minervini 1995) native to the Indo-Pacific, currently found in the eastern Mediterranean Sea. It was first recorded off the coast of Israel in 1924–1925 (Steinitz 1929), and was also reported off Egypt, Lebanon, Syria, Turkey and Cyprus (Galil and Zenetos 2002).

Another swimming crab *Charybdis longicollis* is a small crab native to the western Indian Ocean and established in the Mediterranean Sea since 1954 (Lewinsohn and Holthuis 1964). This is very abundant on sandy-muddy bottoms from 25 to 80 m in the southeastern Mediterranean and it is considered a pest by fishermen using bottom trawls. In 2009 near Yumurtalık in Turkey *C. longicollis* was the most abundant species (63% of the total catch) with a bottom trawl at 30 m depth (Çinar et al. 2011). Off the coast of Israel, *C. longicollis* is common at 80 m and is found down to 250 m; no other Erythraean alien species has yet established a population at that depth (Innocenti et al. 2017).

The one-horned spider crab *Menaethius monoceros*, native to the Indo-Pacific, was first reported in the Mediterranean in 2002, off the island of Punta Sparviero near Punta Ala (central Tyrrhenian, Italy), and was presumably introduced by ship bilge water (Falciai 2003).

3.3.3 Impacts on Economic Activities, Health and Heritage

3.3.3.1 Consequences to Economic Activities

Biological invasions represent one of the most important direct drivers of biodiversity loss and a major pressure on several types of ecosystems (MEA 2005) with also economic impacts such as agriculture, fisheries, and international trade (Lovell et al. 2006). It is known that alien invasives have damaged economies by diminishing fisheries, fouling ships' hulls, and clogging intake pipes (Molnar et al. 2008).

Blue crabs *C. sapidus* support an important fishery in their native range along the Atlantic coast of North America as well as in their introduced range in the eastern Mediterranean Sea. However, due to climate change and its supposed positive effects on the occurrence of blue crabs, *C. sapidus* might well become a candidate target species in commercial fisheries elsewhere, for example in the Adriatic Sea, the European Atlantic coast and in the North Sea (Nehring 2011). In this context it will be interesting to see if *C. sapidus* will significantly reduce stocks of the introduced Pacific oyster (*Crassostrea gigas*), commercially used in several European countries, because adult blue crabs prefer molluscs such as oysters as their primary food sources (Hill et al. 1989). Blue crabs are also reported to mutilate fish caught in traps and trammel nets, and tear the nets (Banoub 1963; Beqiraj and Kashta 2010).

Two freshwater decapods (*P. clarkii* and *E. sinensis*) also provide well known examples of potential economic issues. Both species can cause the collapse of river or levee banks due to burrowing activity. Since its introduction in Europe in 1912, *E. sinensis* has also been reported to cause considerable damage to fisheries by consuming netted fish and by cutting nets. Consumption of rice plants by *P. clarkii* may lead to heavy losses in rice production, as occurred in the Mondego river area, Portugal, during the early stages of invasion in 1990 and 1991 (Marques et al. 1992). The species is now widespread in European rice production areas of Portugal, Spain, France and Italy, including three Italian areas of Protected Designation of Origin for rice: Riso del delta del Po, Riso di Baraggia Biellese e Vercellese and Riso Vialone Nano Veronese. In spite of its potential for damage there is some crayfish industrial processing activity in the South of Spain, based on crayfish captures from Spain and Portugal.

3.3.3.2 Human Health Effects

The occurrence of blue crabs could have harmful impacts on human health as well as on tourism because they have been implicated as carriers of strains of the bacterium *Vibrio cholera*, responsible for outbreaks of human cholera (Hill et al. 1989). Prey preference for filter-feeding organisms, such as clams and mussels, make the blue crab a candidate for algal toxin (microcystins) contamination, so that this commercially important edible crab species is a potential vector of these toxins to humans (Garcia et al. 2010).

3.3.3.3 Cultural Heritage Effects

In some countries, there have been shifts in gastronomy. Either a native crayfish species was replaced by another (e.g. white-clawed crayfish *Austropotamobius pallipes italicus* progressively replaced by the invasive red swamp crayfish *P. clarkii* in Italy which started in 1990 (Barbaresi and Gherardi 2000), or an invasive species (e.g. *P. clarkii*) became the source of new dishes of local relevance. The latter was the case in Portugal with the creation of invasive crayfish festivals in at least two different locations (Ferreira do Zêzere first, and S. Pedro do Corval) (Anastácio et al. in press). Additionally, it is relevant to notice that *Austropotamobius pallipes italicus* has been traditionally considered a native species in the Iberian Peninsula but intense scientific debate on its status contradicts this. The species was introduced from Italy to Spain in 1588 as a diplomatic gift from Francesco I de' Medici to King Philip II of Spain (Clavero et al. 2016). It has been important for inland fisheries and gastronomy mostly until the late 1970's when its populations were heavily reduced by the crayfish plague (Alonso et al. 2016).

3.4 Management of Invasive Alien Decapods

3.4.1 Current State of Invasive Decapod Eradication and Management

Freshwater, estuarine and coastal environments are among the most invaded ecosystems, as well as the most altered (Galil 2009). Prevention and control of the introduction and spread of harmful Invasive Alien Species (IAS) is now a priority for the European community, since the costs associated with them in Europe amount to some € 12.5 billion annually (Kettunen et al. 2009). IAS represent a major threat to wetlands and their native plants and animals, causing damage worth billions of Euros every year (4th Strategic Plan of the Ramsar Convention). A recent EU LIFE European project (RARITY, http://www.life-rarity.eu/) on conservation and re-stocking of the white-clawed crayfish and on the control of the red swamp crayfish (*P. clarkii*), cost 2,675,000 € over 3 years in one of the 20 Italian regions. The RARITY project contributed to removal of about 20,000 red swamp crayfish from the environment, with a consequent decrease in the overall number of populations and a reduction of 87% of the *P. clarkii* population at one site. Another ongoing LIFE project (SOS Tuscan Wetlands, http://www.life-sostuscanwetlands.eu/), aimed at significant reduction of density or local eradication of red swamp crayfish populations in the Lake of Sibolla and in the wetlands of Poggioni (Italy), has requested resources over four years of 1,375,000 €.

It is thus clear that eradication campaigns are extremely costly, so prevention by imposing restrictions on species considered of European Union concern, as well as supporting an early warning and rapid eradication approach for those species, are priority actions to avoid damage caused by the spread of IAS (European Union 2014).

3.4.1.1 Coordinated Actions

Coordination of activities, and a variety of control "tools", are needed to contain, remove, and assess IAS populations and guide management decisions. Currently, control of IAS includes physical restraints, removal, pesticides, biological controls or interference with reproduction.

Identification of source populations and reconstruction of possible pathways of invasion are key issues in our understanding of the invasion process and especially in the design of effective measures to minimise introduction and spreading of alien species (Galil 2011). Molecular markers provide effective tools to investigate invasion histories, as shown for *E. sinensis* in European and North American waters (Herborg et al. 2007; Hänfling and Weetman 2003). Genetic analyses based on older vouchers as well as on living specimens from native and non-native occurrences would be important for understanding the invasion history of *C. sapidus* in Europe. This method is successfully applied to many different species (Richman et al. 2015; Froufe et al. 2015; Czerniejewski et al. 2012).

Information concerning the distribution, abundance, rates of spread, and impacts of species is critical to regional planning and management efforts, particularly in the

Mediterranean. Also, understanding the diverse ecological, economic, and social impacts of IAS is important to prioritizing control and management options. Katsanevakis et al. (2016) have developed a standardised, quantified method (CIMPAL) to assess and map cumulative impacts of invasive alien species on marine ecosystems of the Mediterranean Sea using data retrieved from the European Alien Species Information Network; the CIMPAL index showed strong spatial heterogeneity, the patterns varying depending on the initial pathway of introduction. Considering the importance of understanding change over time, re-introductions and spread, this index takes in consideration the abundance and the relative abundances of the population of IAS. Moreover, it considers their impacts on native species, on ecosystem processing and functioning, and on keystone species too. The most important pathways of initial introduction in the Mediterranean Sea have been identified in: (1) the Canal Suez especially for species of Indo-Pacific origins, (2) shipping, principally for the transportation of predominantly sedentary species, and (3) aquaculture, which refers either to commercial species that were introduced to be cultured or to species accidentally introduced together with imported target species. CIMPAL has provided a baseline that can be built on with future improved information. Such analysis and mapping of regional impacts is urgently requested by policy makers and managers, e.g. for the implementation of the 2008 EU Marine Strategy Framework Directive.

3.5 Conclusions

The warming of the western and eastern Mediterranean Sea will continue to dominate throughout the 21st century, with a sea surface temperature increase (2.5–3 °C) expected by the end of this century. Marine ecosystems will become more 'tropical', favouring invasive decapods. As circulation changes and temperatures rise there will also be modification of pathways from outside the Mediterranean Sea ecosystems, such as invasive marine crustaceans travelling through the Suez Canal from the Red Sea. Increasing water temperatures will support the continued invasion of warm water species from the eastern to western basins. Climate change is likely to affect the structure of marine communities and provide further opportunities for alien species to spread and out-compete native species.

Appendix 3.1

List of alien decapod species in the Mediterranean Sea and in European inland waters. In the column "FIRST RECORD IN MED. BASIN", references older than the 1960 refer to "CIESM Atlas of Exotic Crustaceans in the Mediterranean Sea". est = established, cas = casual, cry = cryptogenic, ques = questionable. FW = freshwater, MAR = marine; OLI = oligohaline. 'Worst' refers to species identified as having particularly negative impacts, as listed in Streftaris and Zenetos (2006).

Species	Environment	Authority	Origin	First record in med. Basin	Status	Worst impacts	Refs.
Actaea savignii	MAR	H. Milne Edwards, 1834	Indo-Pacific	2010 (Ünsal Karhan et al. 2013)	est		Crocetta et al. (2015)
Actumnus globulus	MAR	Heller, 1861	Red Sea	1978 (Galil et al. 2006)	cas		Zenetos et al. (2010)
Alpheus audouini	MAR	Coutiere, 1905	Indo West Pacific	1924 (http://www.ciesm.org/atlas/Alpheusaudouini.php)	est		Zenetos et al. (2010)
Alpheus inopinatus	MAR	Holthuis and Gottlieb, 1958	Indian/Red Sea	1936 (http://www.ciesm.org/atlas/Alpheusinopinatus.php)	est		Zenetos et al. (2010)
Alpheus migrans	MAR	Lewinsohn and Holthuis, 1978	Red Sea	1975 (Lewinsohn and Holthuis 1978)	est		Zenetos et al. (2010)
Alpheus rapacida	MAR	de Man, 1908	Indo West Pacific	1960 (Lewinsohn and Holthuis 1964)	est		Zenetos et al. (2010)
Ashtoret lunaris	MAR	Forskål, 1775	Indo West Pacific	1987 (Galil and Golani 1990)	cas		Zenetos et al. (2010)
Astacus astacus	FW	Linnaeus, 1758	E Europe	–	est		Souty-Grosset et al. (2006), EASIN (16/1/2017)
Astacus leptodactylus	FW/OLI	Eschscholtz, 1823	E Europe	–	inv		Holdich and Pöckl (2007), Souty-Grosset et al. (2006), EASIN (16/1/2017)
Atergatis roseus	MAR	Rüppell, 1830	Indo-Pacific	1961 (Lewinsohn and Holthuis 1964)	est		Crocetta et al. (2015), Mytilineou et al. (2016), Zenetos et al. (2010), Zenetos et al. (2015)
Atyaephyra desmarestii	FW/OLI	Millet, 1831	S Europe	–	est		EASIN (16/1/2017)
Austropotamobius pallipes	FW	Lereboullet, 1858	C Europe	–	est		Souty-Grosset et al. (2006), EASIN (16/1/2017)

(continued)

(continued)

Species	Environment	Authority	Origin	First record in med. Basin	Status	Worst impacts	Refs.
Austropotamobius torrentium	FW	Schrank, 1803	Danube Basin	–	cas		EASIN (16/1/2017)
Calappa hepatica	MAR	Linnaeus, 1758	Indo-Pacific	1992 (Balkis and Çeviker 2003)	cas		Zenetos et al. (2010)
Calappa pelii	MAR	Herklots, 1851	E Atlantic	1991, 1993 (Pastore 1996)	cas		Zenetos et al. (2010)
Callinectes danae	MAR	Smith, 1869	W Atlantic	1981 (Mizzan 1993)	cas		Zenetos et al. (2010)
Callinectes sapidus	MAR	Rathbun, 1896	W Atlantic	1949 (http://www.ciesm.org/atlas/Callinectessapidus.php)	est	+	Streftaris and Zenetos (2006), Eleftheriou et al. (2011), Katsanevakis et al. (2014), Dailianis et al. (2016), Karachle et al. (2016), Thessalou-Legaki et al. (2012), Zenetos et al. (2010), Manfrin et al. (2016)
Cancer bellianus	MAR	J.Y. Johnson, 1861	E Atlantic	2012 (Mytilineou et al. 2016)	cas		Mytilineou et al. (2016)
Carupa tenuipes	MAR	Dana, 1851	Indo-Pacific	2002 (Galil 2004)	est		Zenetos et al. (2010)
Charybdis feriata	MAR	Linnaeus, 1758	Indo-Pacific	2004 (Abello and Hispano 2006)	est		Karachle et al. (2016), Zenetos et al. (2010)
Charybdis helleri	MAR	A. Milne-Edwards, 1861	Indo West Pacific	1924 (http://www.ciesm.org/atlas/Charybdishelleri.php)	est	+	Streftaris and Zenetos (2006), Zenetos et al. (2010)
Charybdis japonica	MAR	A. Milne-Edwards, 1867	Indo-Pacific	2006 (Froglia 2012)	cas		Zenetos et al. (2010)

(continued)

(continued)

Species	Environment	Authority	Origin	First record in med. Basin	Status	Worst impacts	Refs.
Charybdis longicollis	MAR	Leene, 1938	Indian/Red Sea	1961 (Holthuis 1961)	est	+	Streftaris and Zenetos, (2006), Zenetos et al. (2010)
Charybdis lucifera	MAR	Fabricius, 1798	Indo-Pacific	2006 (Mizzan and Vianello 2009)	cas		Zenetos et al. (2010)
Cherax destructor	FW	Clark, 1936	E Australia	–	est		Souty-Grosset et al. (2006), EASIN (16/1/2017)
Cherax quadricarinatus	FW	von Martens, 1868	Australia	–	cas		Souty-Grosset et al. (2006), Jaklič and Vrezec (2011)
Cherax tenuimanus	FW	Smith, 1912	W Australia	–	cas		Souty-Grosset et al. (2006), EASIN (16/1/2017)
Coleusia signata	MAR	Paulson, 1875	Indo West Pacific	1953 (http://www.ciesm.org/atlas/Coleusiasignata.php)	est		Zenetos et al. (2010)
Cryptosoma cristatum	MAR	Brulle, 1837	Tropical E Atlantic	1987 (Garcia-Raso 1993)	cas		Zenetos et al. (2010)
Daira perlata	MAR	Herbst, 1790	Indo West Pacific	1988 (Enzenross and Enzenross 1995)	cas		Zenetos et al. (2010)
Dorippe quadridens	MAR	Fabricius, 1793	Indo West Pacific	2004 (Galil 2005)	est		Zenetos et al. (2010), Mytilineou et al. (2016)
Dyspanopeus sayi	MAR	Smith, 1869	N E Atlantic	1992 (Froglia and Speranza 1993)	est	+	Streftaris and Zenetos (2006), Katsanevakis et al. (2014), Thessalou-Legaki et al. (2012), Zenetos et al. (2010)
Elamena mathoei	MAR	Desmarest, 1823	Indo West Pacific	2011 (Zaouali et al. 2013b)	cas		Zenetos et al. (2012a, b), Zaouali et al. (2013b)

(continued)

(continued)

Species	Environment	Authority	Origin	First record in med. Basin	Status	Worst impacts	Refs.
Eriocheir sinensis	MAR/OLI	H. Milne Edwards, 1853	Pacific	1959 (Petit 1960)	inv	+	Holdich and Pöckl (2007), Streftaris and Zenetos (2006), Zenetos et al. (2010)
Eucrate crenata	MAR	de Haan, 1835	Indo-Pacific	1924 (http://www.ciesm.org/atlas/Eucratecrenata.php)	est		Zenetos et al. (2010)
Eurycarcinus integrifrons	MAR	de Man, 1879	Indian/Red Sea	2009 (Özcan et al. 2010)	cas		Zenetos et al. (2010)
Farfantepenaeus aztecus	MAR	Ives, 1891	W Atlantic	2009 (Deval et al. 2010)	est		Zenetos et al. (2010)
Fenneropenaeus merguiensis	MAR	de Man, 1888	Indo West Pacific	2006 (Özcan et al. 2006)	cas		Zenetos et al. (2010)
Glabropilumnus laevis	MAR	Dana, 1852	Indian	1924 (http://www.ciesm.org/atlas/Heteropanopelaevis.php)	cas		Zenetos et al. (2010)
Gonioinfradens paucidentata	MAR	A. Milne-Edwards, 1861	Indo-Pacific	2010 (Corsini-Foka et al. 2010)	est		Zenetos et al. (2010)
Grapsus granulosus	MAR	H. Milne Edwards, 1853	Red Sea	2006 (Zaouali et al. 2007)	cas		Zenetos et al. (2010)
Halimede tyche	MAR	Herbst, 1801	Indo West Pacific	1998 (Galil 2000)	cas		Zenetos et al. (2010)
Hemigrapsus sanguineus	MAR	de Haan, 1835	Pacific	2001 (Breton et al. 2002)	cas		Zenetos et al. (2010)
Herbstia nitida	MAR	Manning and Holthuis, 1981	Tropical E Atlantic	2002 (Pastore and Denitto 2002)	cas		Zenetos et al. (2010)
Hyastenus hilgendorfi	MAR	de Man, 1887	Indo West Pacific	1960 (Lewinsohn and Holthuis 1964)	est		Zenetos et al. (2010)

(continued)

(continued)

Species	Environment	Authority	Origin	First record in med. Basin	Status	Worst impacts	Refs.
Ixa monodi	MAR	Holthuis and Gottlieb, 1956	Red Sea	1955 (http://www.ciesm.org/atlas/Ixamonodi.php)	est		Zenetos et al. (2010), Zenetos et al. (2015)
Leptochela aculeocaudata	MAR	Paulson, 1875	Indo West Pacific	1933 (http://www.ciesm.org/atlas/Leptochelaaculeocaudata.php)	est		Zenetos et al. (2010)
Leptochela pugnax	MAR	de Man, 1916	Indo West Pacific	1947 (Holthuis and Gottlieb 1958)	est		Zenetos et al. (2010)
Libinia dubia	MAR	H. Milne Edwards, 1834	W Atlantic	1996 (Enzenross et al. 1997)	est		Zenetos et al. (2010)
Lucifer hanseni	MAR	Nobili, 1905	Indo West Pacific	1924 (http://www.ciesm.org/atlas/Luciferhanseni.php)	ques		Zenetos et al. (2010)
Lysmata kempi	MAR	Chace, 1997	Indo-Pacific	2010 (Froglia and Deval 2013)	cas		Zenetos et al. (2012a, b), Froglia and Deval (2013)
Macrobrachium nipponense	FW	de Haan, 1849	Asia	–	cas		EASIN (16/1/2017)
Macrophthalmus graeffei	MAR	A. Milne Ewdards, 1873	Indo West Pacific	1994 (Enzenross and Enzenross 1995)	est		Zenetos et al. (2010)
Marsupenaeus japonicus	MAR	Bate, 1888	Indo-Pacific	1924 (http://www.ciesm.org/atlas/Marsupenaeusjaponicus.php)	est	+	Streftaris and Zenetos (2006), Zenetos et al. (2010)
Matuta victor	MAR	Fabricius, 1781	Indo-Pacific	2012 (Galil and Mendelson 2013)	est		Crocetta et al. (2015), Dailianis et al. (2016)
Menaethius monoceros	MAR	Latreille, 1825	Indo-Pacific	1978 (Falciai 2003)	cas	+	Streftaris and Zenetos (2006), Zenetos et al. (2010)

(continued)

(continued)

Species	Environment	Authority	Origin	First record in med. Basin	Status	Worst impacts	Refs.
Metapenaeopsis aegyptia	MAR	Galil and Golani, 1990	Indo-Pacific	1987 (Galil and Golani 1990)	est		Zenetos et al. (2010)
Metapenaeopsis mogiensis consobrina	MAR	Nobili, 1904	Indo West Pacific	1996 (Galil 1997)	est		Zenetos et al. (2010)
Metapenaeus affinis	MAR	H. Milne Edwards, 1837	Indo West Pacific	2008 (Aydin et al. 2009)	est		Zenetos et al. (2010)
Metapenaeus monoceros	MAR	Fabricius, 1798	Indo West Pacific	1924 (http://www.ciesm.org/atlas/Metapenaeusmonoceros.php)	est		Zenetos et al. (2010)
Metapenaeus stebbingi	MAR	Nobili, 1904	Indian	1927 (http://www.ciesm.org/atlas/Metapenaeusstebbingi.php)	est	+	Streftaris and Zenetos (2006), Zenetos et al. (2010)
Micippa thalia	MAR	Herbst, 1803	Indo West Pacific	1994 (Enzenross and Enzenross 1995)	est		Zenetos et al. (2010)
Myra subgranulata	MAR	Kossmann, 1877	Indian/Red Sea	1929 (http://www.ciesm.org/atlas/Myrasubgranulata.php)	est		Zenetos et al. (2010)
Necora puber	MAR	Linnaeus, 1767	NW Atlantic	1968 (Zibrowius 1992)	est		Zenetos et al. (2010)
Notopus dorsipes	MAR	Linnaeus, 1758	Indo West Pacific	1962 (Lewinsohn and Holthuis 1964)	cas		Zenetos et al. (2010)
Ocypode cursor	MAR	Linnaeus, 1758	E Atlantic	–	est		Mytilineou et al. (2016)
Ogyrides mjoebergi	MAR	Balss, 1921	Indo West Pacific	1947 (http://www.ciesm.org/atlas/Ogyridesmjoebergi.php)	est		Zenetos et al. (2010)

(continued)

(continued)

Species	Environment	Authority	Origin	First record in med. Basin	Status	Worst impacts	Refs.
Orconectes immunis	FW	Hagen, 1870	N America	–	cas		Souty-Grosset et al. (2006), EASIN (16/1/2017)
Orconectes limosus	FW	Rafinisque, 1817	N America	–	inv		Holdich and Pöckl (2007), Souty-Grosset et al. (2006), EASIN (16/1/2017)
Orconectes rusticus	FW	Hagen, 1870	N America	–	inv		Holdich and Pöckl (2007), Souty-Grosset et al. (2006), EASIN (16/1/2017)
Orconectes virilis	FW	Hagen, 1870	N America	–	inv		Holdich and Pöckl (2007), Souty-Grosset et al. (2006), EASIN (16/1/2017)
Pacifastacus leniusculus	FW	Dana, 1852	N America	–	inv		Holdich and Pöckl (2007), Souty-Grosset et al. (2006), EASIN (16/1/2017)
Palaemon macrodactylus	MAR	Rathbun, 1902	NW Pacific	2005 (Torres et al. 2012)	cas		Zenetos et al. (2012a)
Palaemonella rotumana	MAR	Borradaile, 1898	Indo West Pacific	1948 (http://www.ciesm.org/atlas/Palaemonellarotumana.php)	est		Zenetos et al. (2010)
Palinurus ornatus	MAR	Fabricius, 1798	Indo West Pacific	1988 (Galil et al. 1989)	cas		Zenetos et al. (2010)
Paralithodes camtschaticus	MAR	Tilesius, 1815	NE Pacific	2008 (Faccia et al. 2009)	cas		Zenetos et al. (2010)
Penaeus aztecus	MAR	Ives, 1891	W Atlantic	2009 (Deval et al. 2010)	est		Crocetta et al. (2015), Mytilineou et al. (2016), Zenetos et al. (2015)
Penaeus hathor	MAR	Burkenroad, 1959	Indian	1997 (Galil 1999)	est		Zenetos et al. (2015)

(continued)

(continued)

Species	Environment	Authority	Origin	First record in med. Basin	Status	Worst impacts	Refs.
Penaeus semisulcatus	MAR	de Haan, 1844	Indo West Pacific	1928 (http://www.ciesm.org/atlas/Penaeussemisulcatus.php)	est	+	Streftaris and Zenetos (2006), Zenetos et al. (2010)
Percnon gibbesi	MAR	H. Milne Edwards, 1853	W Atlantic	1999 (Relini et al. 2000)	est	+	Streftaris and Zenetos (2006), Thessalou-Legaki et al. (2012), Nicolaidou et al. (2012) Zenetos et al. (2010)
Periclimenes calmani	MAR	Tattersall, 1921	Indo-Pacific	1924 (http://www.ciesm.org/atlas/Periclimenescalmani.php)	cas		Zenetos et al. (2010)
Pilumnopeus vauquelini	MAR	Audouin, 1826	Indian/Red Sea	1924 (http://www.ciesm.org/atlas/Pilumnopeusvauquelini.php)	est		Zenetos et al. (2010)
Pilumnus minutus	MAR	de Haan, 1835	Indo-Pacific	1936 (http://www.ciesm.org/atlas/Pilumnushirsutus.php)	est		Zenetos et al. (2010)
Plagusia squamosa	MAR	Herbst, 1790	Indo West Pacific	1968 (Shiber 1981)	est		Zenetos et al. (2010)
Portunus segnis	MAR	Forskål, 1775	Indian	1924 (http://www.ciesm.org/atlas/Portunuspelagicus.php)	est	+	Streftaris and Zenetos (2006), Zenetos et al. (2010), Crocetta et al. (2015)
Potamon ibericum	FW	Bieberstein, 1808	SE Europe – CW Asia	–	cas		EASIN (16/1/2017)

(continued)

(continued)

Species	Environment	Authority	Origin	First record in med. Basin	Status	Worst impacts	Refs.
Procambarus acutus	FW	Girard, 1852	N America	–	cas		EASIN (16/1/2017)
Procambarus clarkii	FW/OLI	Girard, 1852	N America	–	inv	+	Holdich and Pöckl (2007), Souty-Grosset et al. (2006), EASIN (16/1/2017)
Procambarus fallax f. virginalis	FW	Hagen, 1870		–	est		Holdich and Pöckl (2007), Souty-Grosset et al. (2006), Chucholl et al. (2012), EASIN (16/1/2017)
Processa macrodactyla	MAR	Holthuis, 1952	Tropical E Atlantic	1985 (Garcia-Raso and Salas Casanova 1985)	est		Zenetos et al. (2010)
Rhithropanopeus harrisii	MAR	Gould, 1841	N Atlantic	1994 (Mizzan and Zanella 1996)	est		Zenetos et al. (2010)
Rimapenaeus similis	MAR	Smith, 1885	W Atlantic	2006 (Abdallah et al. 2010)	est		Zenetos et al. (2010)
Scyllarus caparti	MAR	Holthuis, 1952	Tropical E Atlantic	1977 (Froglia 1979)	cas		Zenetos et al. (2010)
Scyllarus posteli	MAR	Forest, 1963	Temperate-tropical E Atlantic	1982 (García Raso 1982)	cas		Zenetos et al. (2010)
Sirpus monodi	MAR	Gordon, 1953	Tropical E Atlantic	2002 (Pancucci-Papadopoulou and Naletaki 2007)	cas		Zenetos et al. (2010)
Solenocera crassicornis	MAR	H. Milne Edwards, 1837	Indo West Pacific	1981 (Abdel-Razek et al. 1981)	cas		Zenetos et al. (2010)
Sphaerozius nitidus	MAR	Stimpson, 1858	Indo West Pacific	1976 (Ramadan and Dowidar 1976)	cas		Zenetos et al. (2010)
Sternodromia spinirostris	MAR	Miers, 1881	Tropical E Atlantic	1969 (Pastore 1976)	cas		Zenetos et al. (2010)

(continued)

(continued)

Species	Environment	Authority	Origin	First record in med. Basin	Status	Worst impacts	Refs.
Synalpheus tumidomanus africanus	MAR	Crosnier and Forest, 1966	Tropical E Atlantic	–	cry		Zenetos et al. (2010)
Thalamita gloriensis	MAR	Crosnier, 1962	Indo West Pacific	1977 (Relini Orsi and Mori 1979)	cas		Zenetos et al. (2010)
Thalamita indistincta	MAR	Apel and Spiridonov, 1998	Indian	2002 (Hasan and Noël 2008)	cas		Zenetos et al. (2010)
Thalamita poissonii	MAR	Audouin, 1826	Indo West Pacific	1952 (http://www.ciesm.org/atlas/Thalamitapoissonii.php)	cry		Zenetos et al. (2010)
Trachysalambria palaestinensis	MAR	Steinitz, 1932	Red Sea	1924 (http://www.ciesm.org/atlas/Trachysalambria.php)	est		Zenetos et al. (2010)
Urocaridella pulchella	MAR	Yokes and Galil, 2006	Indo-Pacific	2003 (Yokes and Galil 2006a)	est		Zenetos et al. (2010)

References

Abdel-Razek FA (1987) Crab fishery of the Egyptian waters with notes on the bionomics of *Portunus pelagicus* (L.). Acta Adriat 28(1–2):143–154

Abdel-Razek FA, El-Hawary MM, Halim Y, Dobisheva SS (1981) A new indo-pacific penaeid in the Mediterranean, *Solenocera indica* Nataraj. Bull Inst Oceanogr Fish Cairo 7(3):575–577

Abello P, Hispano C (2006) The capture of the Indo-Pacific crab *Charybdis feriata* (Linnaeus, 1758) (Brachyura: Portunidae) in the Mediterranean Sea. Aquat Invasions 1:13–16 (e-journal)

Abelló P, Visauta E, Buci A, Demestre M (2003) Noves dades sobre l'expansió del cranc *Percnon gibbesi* (Brachyura: Grapsidae: Plagusiinae) a la Mediterrània occidental. Bolleti Societat d'Hist Nat Balears 46:73–77

Águas M, Banha F, Marques M, Anastácio PM (2014) Can recently-hatched crayfish cling to moving ducks and be transported during flight? Limnologica—Ecology and Management of Inland Waters 48:65–70. https://doi.org/10.1016/j.limno.2014.07.001

Alonso F, Temiño C, Diéguez-Uribeondo J (2016) Distribución y situación actual del cangrejo de río autóctono, *Austrapotamobius pallipes*, en España. Rev Aquat 11

Anastácio PM, Frias AF, Marques JC (2000) Impact of crayfish densities on wet seeded rice and the inneficiency of a non-ionic surfactant as an ecotechnological solution. Ecol Eng 15(1–2):17–25

Anastácio PM, Marques JC (1997) Crayfish, *Procambarus clarkii*, effects on initial stages of rice growth in the lower Mondego River valley (Portugal). Freshw Crayfish 11(1):608–617

Anastácio PM, Parente V, Correia AM (2005) Crayfish effects on seeds and seedlings: identification and quantification of damage. Freshw Biol 50(4):697–704. https://doi.org/10.1111/j.1365-2427.2005.01343.x

Anastácio PM, Ribeiro F, Sousa R, Capinha C, Banha F, Gama AM, Filipe AF, Rebelo RipI (in press) Animais invasores em ecossistemas dulçaquícolas. In: Invasões Biológicas em Portugal: História, Diversidade e Gestão

Aydin I, Galil BS, Bakir K (2009) The first record of the Jinga Shrimp, Metapenaeus affinis (H. Milne Edwards, 1837) (Decapoda, Penaeidae) from the Mediterranean Sea. Crustaceana 82 (8):1091–1095. https://doi.org/10.1163/156854009X448907

Balkis H, Çeviker D (2003) A new exotic crab species [*Calappa hepatica* (Linnaeus, 1758)] for the Mediterranean fauna. Isr J Zool 49(4):320–321

Banha F, Anastácio PM (2015) Live bait capture and crayfish trapping as potential vectors for freshwater invasive fauna. Limnol—Ecol Manage Inland Waters 51:63–69. https://doi.org/10.1016/j.limno.2014.12.006

Banoub MW (1963) Survey of the blue-crab *Callinectes sapidus* (Rath.), in Lake Edku in 1960. Notes Mem Alex Inst Hydrobiol Egypt 69:1–17

Barbaresi S, Gherardi F (2000) The invasion of the alien crayfish *Procambarus clarkii* in Europe, with particular reference to Italy. Biol Invasions 2:259–264

Abdallah OBHHB, Hamida NBH, Jarboui O, Froglia C (2010) First occurrence of the yellow roughneck shrimp, *Rimapenaeus similis* (Smith, 1885) (Crustacea: Decapoda: Penaeidae) in the Mediterranean Sea (Tunisian waters). Biol Invasions 12(5):999–1001

Beqiraj S, Kashta L (2010) The establishment of blue crab *Callinectes sapidus* Rathbun, 1896 in the Lagoon of Patok, Albania (south-east Adriatic Sea). Aquat Invasions 5:219–221. https://doi.org/10.3391/ai.2010.5.2.16

Bisconti M, Silvi E (2005) Prima segnalazione di *Callinectes sapidus* Rathbun, 1896 (Crustacea, Decapoda, Brachyura) nella provincia di Livorno. Quad Mus Stor Nat Livorno 18:1–6

Breton G, Faasse M, Noël P, Vincent T (2002) A new alien crab in Europe: *Hemigrapsus sanguineus* (Decapoda: Brachyura: Grapsidae). J Crustac Biol 22(1):184–189

Brockerhoff AM, McLay CL (2011) Human-mediated spread of Alien crabs. In: Galil BS, Clark PF, Carlton JT (eds) In the wrong place—Alien Marine crustaceans: distribution, biology and impacts, vol 6. Springer Science, Frankfurt, pp 27–106

Cabal J, Millán JAP, Arronte JC (2006) A new record of *Callinectes sapidus* Rathbun, 1896 (Crustacea: Decapoda: Brachyura) from the Cantabrian Sea, Bay of Biscay, Spain. Aqu Inv 1:186–187

Cabral HN, Costa MJ (1999) On the occurrence of the chinese mitten crab, *Eriocheir sinensis*, in Portugal (Decapoda, Brachyura). Crustaceana 72(1):55–58

Castejón D, Guerao G (2013) A new record of the American blue crab, *Callinectes sapidus* Rathbun, 1896 (Decapoda: Brachyura: Portunidae), from the Mediterranean coast of the Iberian Peninsula. BioInvasion Rec 2:141–143

Chaouachi B, Dos Santos A, Ben Hassine OK (1998) Occurrence of the shrimp *Metapenaeus monoceros* (Fabricius, 1798) in the Gulf of Gabes: a first record for Tunisian coasts. In: IV International Crustacean Congress, Amsterdam (Abstract)

Chucholl C, Morawetz K, Groß H (2012) The clones are coming—strong increase in Marmorkrebs (*Procambarus fallax* (Hagen, 1870) *f. virginalis*) records from Europe. Aquat Invasions 7 (4):511–519

Çinar ME, Bilecenoglu M, Ozturk B, Katagan T, Yokes MB, Aysel V, Dagli E, Açik S, Ozcan T, Erdogan H (2011) An updated review of Alien species on the coasts of Turkey. Mediterr Mar Sci 12(2):257–315

Clavero M, Nores C, Kubersky-Piredda S, Centeno-Cuadros A (2016) Interdisciplinarity to reconstruct historical introductions: solving the status of cryptogenic crayfish. Biol Rev 91:1036–1049

Coll M, Piroddi C, Steenbeek J, Kaschner K, Ben Rais Lasram F, Aguzzi J, Ballesteros E, Bianchi CN, Corbera J, Dailianis T, Danovaro R, Estrada M, Froglia C, Galil BS, Gasol JM, Gertwagen R, Gil J, Guilhaumon F, Kesner-Reyes K, Miltiadis-Spyridon K, Koukouras A, Lampadariou N, Laxamana E, López-Fé de la Cuadra CM, Lotze HK, Martin D, Mouillot D, Oro D, Raicevich S, Rius-Barile J, Saiz-Salinas JI, San Vicente C, Somot S, Templado J, Turon X, Vafidis D, Villanueva R, Voultsiadou E (2010) The Biodiversity of the Mediterranean Sea: Estimates, Patterns, and Threats. PLoS ONE 5(8):e11842. https://doi.org/10.1371/journal.pone.0011842

Corsini-Foka M, Kondylakos G, Economidis PS (2004) Occurrence of the Lessepsian species *Portunus pelagicus* (Crustacea) and Apogon pharaonis (Pisces) in the marine area of Rhodes Island. Mediterr Mar Sci 5(1):83–89. https://doi.org/10.12681/mms.213

Corsini-Foka M, Pancuccipapadopoulou MA, Kondilatos G, Kalogirou S (2010) *Gonioinfradens paucidentatus* (A. Milne Edwards, 1861) (Crustacea, Decapoda, Portunidae): a new alien crab in the Mediterranean Sea. Mediterr Mar Sci 11(2):311–340

Crocetta F (2006) First record of *Portunus pelagicus* (Linnaeus, 1758) (Decapoda: Brachyura: Portunidae) in the northern Tyrrhenian Sea. Crustaceana 79(9):1145–1148. https://doi.org/10.1163/156854006778859632

Crocetta F, Agius D, Balistreri P, Bariche M, Bayhan YK, Çakir M, Ciriaco S, Corsini-Foka M, Deidun A, El Zrelli R, Ergüden D, Evans J, Ghelia M, Giavasi M, Kleitou P, Kondylato G, Lipej L, Mifsud C, Özvarol Y, Pagano A, Portelli P, Poursanidis D, Rabaoui L, Schembri PJ, Taşkin E, Tiralongo F, Zenetos A (2015) New mediterranean biodiversity records (October 2015). Medit Mar Sci 16(3):682–702. https://doi.org/10.12681/mms.1477

Czerniejewski P, Skuza L, Drotz MK, Berggren M (2012) Molecular connectedness between self and none self-sustainable populations of Chinese mitten crab (*Eriocheir sinensis*, H. Milne Edwards, 1853) with focus to the Swedish lake Vänern and the Oder and Vistula river in Poland. Hereditas 149(2):55–61. https://doi.org/10.1111/j.1601-5223.2012.02246.x

d'Udekem d'Acoz C (1999) Inventaire et distribution des crustacés décapodes de l'Atlantique nord-oriental, de la Méditerranée et des eaux continentales adjacentes au nord de 25°N. Collection Patrimoines Naturels 40 (Muséum national d'Histoire naturelle: Paris):383

Dailianis T, Akyol O, Babali N, Bariche M, Crocetta F, Gerovasileiou V, Chanem R, Gökoğlu M, Hasiotis T, Izquierdo-Muñoz A, Julian D, Katsanevakis S, Lipez L, Mancini E, Mytilineou C, Ounifi Ben Amor K, özgül A, Ragkousis M, Rubio-Portillo E, Servello G, Sini K, Stamouli C, Sterioti A, Teker S, Tiralongo F, Trkov D (2016) New mediterranean biodiversity records (July 2016). Mediterr Mar Sci:1791–6763. https://doi.org/10.12681/mms.1734

Deval MC, Kaya Y, Güven O, Gōkoğlu M, Froglia C (2010) An unexpected find of the western atlantic shrimp, *Farfantepenaeus aztecus* (Ives, 1891) (Decapoda, Penaeidae) in Antalya bay, Eastern Mediterranean Sea. Crustaceana 83(12):1531–1537

Dittel AI, Epifanio CE (2009) Invasion biology of the Chinese mitten crab *Eriochier sinensis*: a brief review. J Exp Mar Biol Ecol 374:79–92

Duruer EC, Kinacigil T, Soykan O, Tosunoğlu Z (2008) Contribution to some biological and fishery aspects of commercial penaeid prawns in Mersin Bay (Northeastern Mediterranean, Turkey). Crustaceana 81:577–585

EASIN The European Alien Species Information Network (EASIN), http://easin.jrc.ec.europa.eu

Eleftheriou A, Anagnostopoulou-Visilia E, Anastasopoulou E, Ates AS, Bachari NEI, Cavas L, Cevik C, Ulha M, Cevik F, Delos AL, Derici OB, Erguden D, Fragopoulu N, Giangrande A, Goksan T, Gravili C, Gurlek M, Hattour A, Kapiris K, Kouraklis P, Lamouti S, Prato E, Papa L, Papantoniou G, Parlapiano I, Poursanidis D, Turan C, Yaglioglu D (2011) New mediterranean biodiversity records (December 2011). Mediterr Mar Sci 12(2):491–508. https://doi.org/10.12681/mms.48

Enzenrob R, Enzenrob L, Bingel F (1997) Occurrence of blue crab, *Callinectes sapidus* (Rathbone, 1896) (Crustacea, Brachyura) on the Turkish Mediterranean and the Adjacent Aegean Coast and its size Distribution in the Bay of Iskenderun. J Turk Zool 21:113–122

Enzenross R, Enzenross L (1995) Erstnachweis indopazifischer Brachyura (Crustacea: Decapoda) von der türkischen Mittelmeerküste. Stuttgarter Beiträge zur Naturkunde 521 (Serie A [Biologie]):1–4

Enzenross R, Enzenross L, El Abed A (1997) Nouvelles mentions de deux Crustacés d'origine non-Méditerranéenne dans les eaux du Golfe de Gabès (Tunisie). Rapports Doc Inst Nat Sci Technol Mer, Salammbo 1:1–13

European Comission E (2016) Commission implementing regulation (EU) 2016/1141 of 13 July 2016 adopting a list of invasive alien species of Union concern pursuant to Regulation (EU) No 1143/2014 of the European Parliament and of the Council. Official Journal of the European Union L 189/4

European Environment Agency E (2015) Pathways of introduction of marine non-indigenous species

European Union E (2014) LIFE and invasive Alien species. Publ Off Eur Union Luxemb. https://doi.org/10.2779/14722

Faccia I, Alyakrinsky A, Bianchi CN (2009) The crab that came in from the cold: first record of *Paralithodes camtschaticus* (Tilesius, 1815) in the Mediterranean Sea. Aquat Invasions 4 (4):715–718. https://doi.org/10.3391/ai.2009.4.4.22

Falciai L, Minervini R (1995) Crustàceos Decàpodos de Europa. Ediciones Omega, Barcelona

Falciai L (2003) First record of *Menaethius monoceros* (Latreille, 1825) (Decapoda, Majidae) in the central Thyrrenian Sea. Crustaceana, Leiden 75(10):1279–1283

FAO, Holthuis LB (1980) Shrimps and prawns of the world. An annotated catalogue of species of interest to fisheries. FAO fisheries synopsis, vol 1. Food and agriculture organization of the United Nations (FAO), Rome (Italy)

Forbes A (2007) *Penaeus monodon* (giant tiger prawn). http://www.cabi.org/isc/datasheet/71093. Accessed 28 Jan 2017

Froglia C (1979) Segnalazione di alcuni Crostacei Decapodi nuovi per la fauna adriatica. Quad Lab Tecnol Pesca 2(4):191–196

Froglia C (2012) First record of *Charybdis japonica* (Crustacea: Decapoda: Portunidae) in the Mediterranean Sea. Mar Biodivers Rec 5:e33

Froglia C, Deval MC (2013) First record of an exotic hippolytid shrimp in the eastern Mediterranean Sea. Mediterr Mar Sci 15(1):168–171

Froglia C, Speranza S (1993) First record of *Dyspanopeus sayi* (Smith, 1869) in the Mediterranean Sea (Crustacea: Decapoda: Xanthidae). Quad Ist Ric Pesca Marittima 5:163–166

Froufe E, Varandas S, Teixeira A, Sousa R, Filipová L, Petrusek A, Edsman L, Lopes-Lima M (2015) First results on the genetic diversity of the invasive signal crayfish *Pacifastacus*

leniusculus (Dana, 1852) in Europe using novel microsatellite loci. J Appl Genet 56(3):375–380. https://doi.org/10.1007/s13353-015-0272-y
Galil B (2008) Alien species in the Mediterranean Sea—which, when, where, why? Hydrobiologia 606(1):105–116
Galil BS (1997) Two lessepsian migrant decapods new to the coast of Israel. Crustaceana 70 (1):111–114
Galil BS (1999) *Melicertus hathor* Burkenroad, 1959—a Red Sea penaeid prawn new to the Mediterranean. Crustaceana 72(9):1226–1228
Galil BS (2000) *Halimede tyche* (Herbst, 1801) (Decapoda; Xanthoidea; Halimedinae) from the Mediterranean coast of Israel, with a review of the genus. J Crustac Biol 20(2):325–331
Galil BS (2004) *Carupa tenuipes* Dana, 1851: an Indo-Pacific swimming crab new to the Mediterranean (Decapoda, Brachyura, Portunidae). Crustaceana 77(2):249–251
Galil BS (2005) *Dorippe quadridens* (Decapoda, Dorippidae)—First record of an alien dorippid in the Mediterranean Sea. Crustaceana 78:497–498
Galil BS (2007) Seeing red: Alien species along the Mediterranean coast of Israel. Aquat Invasions 2:281–312
Galil BS (2011) The alien Crustaceans in the Mediterranean Sea: an historical review. In: Galil BS, Clark PF, Carlton JT (eds) In the wrong place-alien marine Crustaceans: distribution, biology and impacts. Springer, Berlin, pp 377–401
Galil BS, Clark PF, Ng PKL, Falciai L (2006) Actumnus globulus (Crustacea: Decapoda: Pilumnidae)—another Erythrean pilumnid in the Mediterranean. JMBA2—Biodiv Rec 5307:4 (e-journal)
Galil BS, Froglia C, Noël P (2002) CIESM Atlas of exotic species in the Mediterranean, vol 2. Crustaceans: Decapods and Stomatopods. CIESM
Galil BS, Golani D (1990) Two new migrant decapods from the eastern Mediterranean. Crustaceana 58(3):229–236
Galil BS, Mendelson M (2013) A record of the moon crab *Matuta victor* (Fabricius, 1781) (Crustacea: Decapoda: Matutidae) from the Mediterranean coast of Israel. BioInvasions Rec 2:69–71
Galil BS, Pisanty S, Spanier E, Tom M (1989) The Indo-Pacific lobster *Panulirus ornatus* (Fabricius, 1798) (Crustacea: Decapoda): a new Lessepsian migrant to the eastern Mediterranean. Isr J Zool 35(4):241–243
Galil BS, Shlagman A (2011) An annotated list of the Decapod Crustacea of the Mediterranean coast of Israel—half a century later. In: Fransen C, De Grave S, Ng P (eds) Studies on Malacostraca: Lipke Bijdeley Holthuis memorial volume, vol 14. Crustaceana monographs. Brill, Leiden, pp 269–282. https://doi.org/10.1163/9789047427759_018
Galil BS, Zenetos A (2002) A sea change—exotics in the eastern Mediterranean Sea. In: Leppäkoski E, Gollasch S, Olenin S (eds) Invasive aquatic species of Europe. Distribution, impacts and management. Springer, Dordrecht, pp 325–336
Galil SB (2009) Control and eradication of invasive aquatic invertebrates. In: Gherardi F, Corti C, Gualtieri M (eds) Biodiversity conservation and habitat management, vol II. EOLSS Publisher/UNESCO, Oxford, pp 331–344
Garcia-de-Lomas J, Dana E, López-Santiago J, González R, Ceballos G, Ortega F (2010) Management of the Chinese mitten crab, *Eriocheir sinensis* (H. Milne Edwards, 1853) in the Guadalquivir Estuary (Southern Spain). Aquat invasions 5(3):323–330. https://doi.org/10.3391/ai.2010.5.3.11
Garcia-Raso JE (1993) New record of other African species of Crustacea Decapoda, *Cycloes cristata* (Brullé) from European and Mediterranean waters. Bios 1(1):215–221
Garcia-Raso JE, Salas Casanova MC (1985) New record of the species *Processa macrodactyla* Holthuis, 1952 (Decapoda, Caridea) from Spanish waters. Crustaceana 49(1):88–91
Garcia AC, Bargu S, Dash P, Rabalais NN, Sutor M, Morrison W, Walker ND (2010) Evaluating the potential risk of microcystins to blue crab (*Callinectes sapidus*) fisheries and human health in a eutrophic estuary. Harmful Algae 9:134–143. https://doi.org/10.1016/j.hal.2009.08.011

Garcia L, Reviriego B (2000) Presència del cranc subtropical *Percnon gibbesi* (H. Milne Edwards, 1853) (Crustacea, Decapoda, Grapsidae) a les Illes Balears. Primera cita a la Mediterrània occidental. Bol Soc Hist Nat Bal 43:81–89

García Raso JE (1982) Familia Scyllaridae Latreille 1825 (Crustacea, Decapoda) en la Región Sur-mediterránea española. Bol Asoc Esp Entomol 6(1):73–78

Geldiay A, Kocatas A (1972) A report on the occurrence of Penaeidae (Decapoda, Crustacea) along the coast of Turkey from eastern Mediterranean to the vicinity of Izmir, as a resultof migration and its factors. Paper presented at the 17e Congrés international de zoologie, Monte Carlo, Sept 1972

Genovesi P, Carboneras C, Vilá M, Walton P (2015) EU adopts innovative legislation on invasive species: a step towards a global response to biological invasions? Biol Invasions 17(5):1307–1311

Gherardi F, Acquistapace P (2007) Invasive crayfish in Europe: the impact of *Procambarus clarkii* on the littoral community of a Mediterranean lake. Freshw Biol 52:1249–1259

Gherardi F, Barbaresi S (2000) Invasive crayfish: activity patterns of *Procambarus clarkii* in the rice fields of the Lower Guadalquivir (Spain). Arch Hydrobiol 150(1):153–168

Giaccone G (1999) L'origine della biodiversità vegetale del Mediterraneo. Notiziario SIBM 35:35–51

Giordani Soika A (1951) Il *Neptunus pelagicus* (L.) nell'Alto Adriatico. Natura 42:18–20

Gokoglu M, Kaya Y (2006) First record of *Melicertus hathor* (Penaeidae) from the Gulf of Antalya (Mediterranean Sea). Mar Biodivers Rec 1:e17

Gollasch S, Nehring S (2006) National checklist for aquatic alien species in Germany. Aquat Invasions 1(4):245–269. https://doi.org/10.3391/ai.2006.1.4.8

Gravili C, Belmonte G, Cecere E, Denitto F, Giangrande A, Guidetti P, Longo C, Mastrototaro F, Moscatello S, Petrocelli A, Piraino S, Terlizzi A, Boero F (2010) Nonindigenous species along the Apulian coast, Italy. Chem Ecol 26:121–142

Grosholz ED (2011) Crabs. In: Simberloff D, Rejmànek M (eds) Encyclopedia of biological invasions. University of California, p 756

Hänfling B, Weetman D (2003) Characterization of microsatellite loci for the Chinese mitten crab, *Eriocheir sinensis*. Mol Ecol Notes 3(1):15–17

Hasan H, Noël P (2008) First record of *Thalamita indistincta* Apel & Spiridonov, 1998 (Decapoda, Brachyura, Portunidae) in the Mediterranean. Crustaceana 81(2):247–252

Herborg LM, Rushton SP, Clare AS, Bentley MG (2003) Spread of the Chinese mitten crab (*Eriocheir sinensis* H. Milne Edwards) in Continental Europe: analysis of a historical data set. Hydrobiologia 503(1):21–28

Herborg LM, Weetman D, van Oosterhout C, Hänfling B (2007) Genetic population structure and contemporary dispersal patterns of a recent European invader, the Chinese mitten crab, *Eriocheir sinensis*. Mol Ecol 16(2):231–242. https://doi.org/10.1111/j.1365-294X.2006.03133.x

Hill RJ, Watt F, Wilson CM, Fifis T, Underwood PA, Tribbick G, Geysen HM, Thomas JO (1989) Bands, interbands and puffs in native Drosophila polytene chromosomes are recognized by a monoclonal antibody to an epitope in the carboxy-terminal tail of histone H1. Chromosoma 98:411–421

Holdich DM, Pöckl M (2007) Invasive crustaceans in European inland waters. In: Gherardi F (ed) Biological invaders in inland waters: profiles, distribution, and threats. Springer, Dordrecht, pp 29–75

Holdich DM, Reynolds JD, Souty-Grosset C, Sibley PJ (2009) A review of the ever increasing threat to European crayfish from non-indigenous crayfish species. Knowl Manage Aquat Ecosyst 11:394–395

Holthuis LB (1961) Report on a collection of crustacea decapoda and stomatopoda from Turkey and the Balkans. Zooogische Verh 47:1–67

Holthuis LB, Gottlieb E (1958) An annotated list of the decapod crustacea of the Mediterranean coast of Israel, with an appendix listing the decapoda of the eastern Mediterranean. Bullettin of the research council of Israel, vol 7B. State of Israel, Ministry of Agricultures, Division of Fisheries, The Sea Fisheries Research Station, Haifa, Israel

Innocenti G, Stasolla G, Goren M, Stern N, Levitt-Barmats Y, Diamant A, Galil BS (2017) Going down together: invasive host, *Charybdis Longicollis* (Decapoda: Brachyura: Portunidae) and invasive parasite, *Heterosaccus Dollfusi* (Cirripedia: Rhizocephala: Sacculinidae) on the upper slope off the Mediterranean coast of Israel. Mar Biol Res:1–8. https://doi.org/10.1080/17451000.2016.1240873

IUCN-Med (2007) Regional situation analysis—draft 6

Jaklič M, Vrezec A (2011) The first tropical alien crayfish species in European waters: the redclaw *Cherax quadricarinatus* (Von Martens, 1868) (Decapoda, Parastacidae). Crustaceana 84 (5):651–665

Kapiris K, Mytilineou CH, Politou CY, Kavadas S, Conides A (2007) Research on shrimps' resources and fishery in Hellenic waters. In: Papaconstantinou C, Zenetos A, Vassilopoulou V, Tserpes G (eds) State of Hellenic waters. HCMR, Athens, pp 421–432

Karachle PK, Angelidis A, Apostolopoulos G, Ayas D, Ballesteros M, Bonnici C, Brodersen MM, Castriota L, Chalari N, Cottalorda JM, Crocetta F, Deidun A, Đođo Ž, Dogrammatzi A, Dulčić J, Fiorentino F, Gönülal O, Harmelin JG, Insacco G, Izquierdo-Gómez D, Izquierdo-Muñoz A, Joksimović A, Kavadas S, Malaquias MAE, Madrenas E, Massi D, Micarelli P, Minchin D, Önal U, Ovalis P, Poursanidis D, Siapatis A, Sperone E, Spinelli A, Stamouli C, Tiralongo F, Tunçer S, Yaglioglu D, Zava B, Zenetos A (2016) New Mediterranean biodiversity records (March 2016). Mediterr Mar Sci 17:230–252

Karhan SÜ, Yokeş MB, Clark PF, Galil BS (2013) First Mediterranean record of *Actaea savignii* (H. Milne Edwards, 1834) (Crustacea: Decapoda: Brachyura: Xanthidae), an additional Erythraean alien crab. BioInvasions Rec 2:145–148

Katsanevakis S, Acar Ű, Ammar I, Balci BA, Bekas P, Belmonte M, Chintiroglou CC, Consoli P, Dimiza M, Fryganiotis K, Gerovasileiou V, Gnisci V, Gülşahin N, Hoffman R, Issaris Y, Izquierdo-Gomez D, Izquierdo-Munoz A, Kavadas S, Koehler L, Konstantinidis E, Mazza G, Nowell G, Őnal U, Őzen MR, Pafilis P, Pastore M, Perdikaris C, Poursanidis D, Prato E, Russo F, Sicuro B, Tarkan AN, Thessalou-Legaki M, Tiralongo F, Triantaphyllou M, Tsiamis K, Tunçer S, Turan C, Türker A, Yapici S (2014) New Mediterranean biodiversity records (October, 2014). Mediterr Mar Sci 15(3):675–695. https://doi.org/10.12681/mms.1123

Katsanevakis S, Stelzenmüller V, South A, Sorensen TK, Jones PJS, Kerr S, Badalamenti F, Anagnostou C, Breen P, Chust G, D'Anna G, Duijn M, Filatova T, Fiorentino F, Hulsman H, Johnson K, Karageorgis AP, Kröncke I, Mirto S, Pipitone C, Portelli S, Qiu W, Reiss H, Sakellariou D, Salomidi M, van Hoofm L, Vassilopoulou V, Vega Fernández T, Vöge S, Weber A, Zenetos A, ter Hofstede R (2011) Ecosystem-based marine spatial management: review of concepts, policies, tools, and critical issues. Ocean Coast Manage 54:807–820. https://doi.org/10.1016/j.ocecoaman.2011.09.002

Katsanevakis S, Tempera F, Teixeira H (2016) Mapping the impact of alien species on marine ecosystems: the Mediterranean Sea case study. Divers Distrib 22:694–707

Kettunen M, Genovesi P, Gollasch S, Pagad S, Starfinger U (2009) Technical support to EU strategy on invasive alien species (IAS). Assessment of the impacts of IAS in Europe and the EU

Lai JCV, Ng PKL, Davie PJF (2010) A revision of the *Portunus pelagicus* (Linnaeus, 1758) species complex (Crustacea: Brachyura: Portunidae), with the recognition of four species. Raffles Bull Zool 58(2):199–237

Lejeusne C, Chevaldonne P, Pergent-Martini C, Boudouresque CF, Perez T (2010) Climate change effects on a miniature ocean: the highly diverse, highly impacted Mediterranean Sea. Trends Ecol Evol 25:250–260

Lewinsohn C, Holthuis LB (1964) New records of decapod Crustacea from the Mediterranean coast of Israel and the eastern Mediterranean. Zool Meded 40:45–63

Lewinsohn C, Holthuis LB (1978) On a new species of Alpheus (Crustacea, Decapoda, Natantia) from the Eastern Mediterranean. Zoologische Meded 53(7):75–82

Lovell SJ, Stone SF, Fernandez I (2006) The economic impacts of aquatic invasive species: a review of the literature. Agric Res Econ Rev 35:195–208

Lowe S, Browne M, Boudjelas S, De Poorter M (2000) 100 of the world's most invasive alien species. A selection from the global invasive species database

Manfrin C, Comisso G, dall'Asta A, Bettoso N, Chung JS (2016) The return of the blue crab (*Callinectes sapidus*, Rathbun, 1896) after 70 years from its first appearance in the Gulf of Trieste (Northern Adriatic Sea, Italy). Check List J 12(6):7

Marques JC, Adão MH, Anastácio PM Lagostim Vermelho da Louisiana, *Procambarus clarkii* (Girard), espécie exótica em Portugal: bioecologia, impacto e controlo. In: 3ª Conferência Nacional sobre a Qualidade do Ambiente, Aveiro, Portugal, 5–7 Feb 1992

Mea MEA (2005) Ecosystems and human wellbeing: biodiversity synthesis. World Resources Institute, Washington, D.C

Mizzan L (1993) Presence of swimming crabs of the genus Callinectes (Stimpson) (Decapoda, Portunidae) in the Venice Lagoon (North Adriatic Sea—Italy). Boll Mus Civ St Nat Venezia 42:31–43

Mizzan L (2005) *Rhitropanopeus harrisii* (Gould, 1841)(Crustacea, Decapoda Panopeidae) e *Eriocheir sinensis* H. Milne-Edwards, 1854 (Crustacea, Decapoda Grapsidae): due nuovi granchi esotici in laguna di Venezia. Boll Mus Civ St Nat Venezia 56:89–96

Mizzan L, Trabucco R, Tagliapietra G (2005) Nuovi dati sulla presenza e distribuzione di specie alloctone del macrozoobenthos della Laguna di Venezia. Boll Soc Veneziana St Nat Mus Civ St Nat 56:69–88

Mizzan L, Vianello C (2009) First record of *Charybdis (Charybdis) lucifera* (Fabricius, 1798) (Cruscacea, Decapoda, Portunidae) in the Mediterranean Sea. Boll Mus Civ St Nat Venezia 59:27–30

Mizzan L, Zanella L (1996) First record of *Rhithropanopeus harrisii* (Gould, 1841) (Crustacea, Decapoda, Xanthidae) in the Italian waters. Boll Mus Civ St Nat Venezia 46:109–120

Molnar JL, Gamboa RL, Revenga C, Spalding MD (2008) Assessing the global threat of invasive species to marine biodiversity. Front Ecol Environ 6:485–492

Mori M, Vacchi M (2002) On a new occurrence of the alien flat crab, *Percnon gibbesi* (H. Milne Edwards), in the southern Sicily (Central Mediterranean Sea) (Crustacea, Brachyura, Grapsidae). Anal Mus Civ St Nat "Giacomo Doria" 114:295–302

Mwo MWO (2012) First Mediterranean Wetlands observatory report MWO (Mediterranean Wetlands Outlook). Tour du Valat, France

Mytilineou C, Akel ESHK, Babali N, Balistreri M, Bariche M, Boyaci YÖ, Cilenti L, Constantinou C, Crocetta F, Çelik M, Dereli H, Dounas F, Durucan A, Garrido A, Gerovasileiou V, Kapiris K, Kebapcioglu T, Kleitou P, Krystalas A, Lipej L, Maina I, Marakis P, Mavrič B, Moussa R, Peña-Rivas L, Poursanidis D, Renada W, Rizkalla SI, Rosso A, Scirocco T, Sciuto F, Servello G, Tiralongo F, Yapici S, Zenetos A (2016) New Mediterranean biodiversity records (November 2016). Mediterr Mar Sci 17(3):794–821. https://doi.org/10.12681/mms.1976

Mytilineou CH, Maiorano P, Kavadas S, D'Onghia G, Kapiris K, Capezzuto F (2001) Size structure comparison in some demersal species between two areas of different fishing impact in the Eastern-Central Mediterranean NAFO SCR. Paper presented at the deep-sea fisheries symposium, Cuba, 12–14 Sept 2001

Nehring S (2011) Invasion history and success of the American blue crab *Callinectes sapidus* in European and adjacent waters. In: Galil BS, Clark PF, Carlton JT (eds) In the wrong place—Alien marine Crustaceans: distribution, biology and impacts. Springer, Berlin, pp 607–624. https://doi.org/10.1007/978-94-007-0591-3_21

Nehring S (2012) Online database of the European network on invasive Alien species (NOBANIS)—invasive Alien species fact sheet—*Callinectes sapidus*. NOBANIS

Nicolaidou A, Along iG, Aydogan O, Catra M, Cavas L, Cevik C, Dosi A, Circosta V, Giakoumi S, Giménez Casalduer F, Filiz H, Izquierdo-Muñoz A, Kalogirou S, Konstantinidis E, Kousteni V, Kout J, Legaki A, Megalofonou P, Ovalis P, Paolillo G, Paschos I, Perdikaris C, Poursanidis D, Ramos-Esplá AA, Reizopoulou S, Sperone E,

Taskin E, Tripepi S, Vázquez-Luis M (2012) New mediterranean biodiversity records (June 2012). Mediterr Mar Sci 13:162–174
Occhipinti-Ambrogi A (2000) Biotic invasions in a Mediterranean lagoon. Biol Invasions 2:165–176. https://doi.org/10.1023/A:1010004926405
Occhipinti-Ambrogi A (2001) Report on the current status of introductions in Italy (Marine environment). Report of the working group on introductions and transfers of Marine organisms. In: Advisory committee on the Marine environment ICES CM 2002/ACME:08 Ref. EF, Barcelona, March 21–23, 2001. International council for the exploration of the sea, Copenhagen, Denmark, pp 59–63
Occhipinti-Ambrogi A, Galil BS (2010) Marine alien species as an aspect of global change. Adv Oceanogr Limnol 1:199–218
Özcan T, Galil BS, Bakır K, Katağan T (2006) The first record of the banana prawn *Fenneropenaeus merguiensis* (De Man, 1888) (Crustacea: Decapoda: Penaeidae) from the Mediterranean Sea. Aquat Invasions 1(4):286–288 (e-journal)
Özcan T, Katagan T, Ng PKL (2010) First record of *Eurycarcinus integrifrons* de man, 1879 (Decapoda, Pilumnidae) from the Mediterranean Sea. Crustaceana 83(4):507–510
Padilla DK, Williams SL (2004) Beyond ballast water: aquarium and ornamental trades as sources of invasive species in aquatic ecosystems. Front Ecol Environ 2(3):131–138
Pancucci-Papadopoulou MA, Naletaki M (2007) A new alien species in the Mediterranean? On the presence of *Sirpus monodi* Gordon, 1953 (Brachyura, Pirimelidae) in Greece. Mediterr Mar Sci 8(2):91–96
Pastore M (1976) Decapoda Crustacea in the Gulf of Taranto and the Gulf of Catania with a discussion of a new species of Dromidae (Decapoda Brachyura) in the Mediterranean Sea. Thalassia Jugoslav 8(1):105–117
Pastore M (1996) The genus Calappa in the Ionian sea. Oebalia 21:187–196
Pastore M, Denitto F (2002) Occurrence of *Herbstia nitida* manning and Holthuis, 1981 in the Otranto channel. In: 8th Colloquium Crustacea Decapoda Mediterranea, p 80
Perdikaris C, Konstantinidis E, Gouva E, Ergolavou A, Klaoudatos D, Nathanailides C, Paschos I (2016) Occurrence of the invasive crab species *Callinectes sapidus* Rathbun, 1896, in NW Greece. Walailak J Sci Technol (WJST) 13(7):503–510
Peruzza L, Piazza F, Manfrin C, Bonzi LC, Battistella S, Giulianini PG (2015) Reproductive plasticity of a *Procambarus clarkii* population living 10 °C below its thermal optimum. Aquat Invasions 10(2):199–208. https://doi.org/10.3391/ai.2015.10.2.08
Petit G (1960) Le crabe chinois est parvenu en Méditerranée. Vie et Milieu 11(1):133–136
Ramadan SE, Dowidar NM (1976) Brachyura (Decapoda Crustacea) from the Mediterranean waters of Egypt. Thalassia jugoslav 8(1):127–139
Ramalho RO, Anastácio PM (2015) Factors inducing overland movement of invasive crayfish (*Procambarus clarkii*) in a ricefield habitat. Hydrobiologia 746(1):135–146. https://doi.org/10.1007/s10750-014-2052-9
Relini M, Relini Orsi L, Puccio V, Azzurro E (2000) The exotic crab *Percnon gibbesi* (H. Milne Edwards, 1853) (Decapoda, Grapsidae) in the Central Mediterranean. Sci Mar 64:337–340
Relini Orsi L, Mori M (1979) Due reperti mediterranei di *Thalamita gloriensis* Crosnier, 1962 (Crustacea: Decapoda: Portunidae). Oebalia 1:7–13
Ribeiro F, Veríssimo A (2014) A new record of *Callinectes sapidus* in a western European estuary (Portuguese coast). Mar Biodivers Rec 7:e36
Richman NI, Böhm M, Adams SB, Alvarez F, Bergey EA, Bunn JJ, Burnham Q, Cordeiro J, Coughran J, Crandall KA, Dawkins KL, DiStefano RJ, Doran NE, Edsman L, Eversole AG, Füreder L, Furse JM, Gherardi F, Hamr P, Holdich DM, Horwitz P, Johnston K, Jones CM, Jones JP, Jones RL, Jones TG, Kawai T, Lawler S, López-Mejía M, Miller RM, Pedraza-Lara C, Reynolds JD, Richardson AM, Schultz MB, Schuster GA, Sibley PJ, Souty-Grosset C, Taylor CA, Thoma RF, Walls J, Walsh TS, Collen B (2015) Multiple drivers of decline in the global status of freshwater crayfish (Decapoda: Astacidea). Philos Trans R Soc Lond B Biol Sci 370(1662). https://doi.org/10.1098/rstb.2014.0060

Rothman SBS, Shlagman A, Galil BS (2013) *Saron marmoratus*, an Indo-Pacific marble shrimp (Hippolytidae: Decapoda: Crustacea) in the Mediterranean Sea. Mar Biodivers Rec 6:1–3

Rotllant G, Guerao G, Gras N, Estévez A (2014) Larval growth and biochemical composition of the protected Mediterranean spider crab *Maja squinado* (Brachyura, Majidae). Aquat Biol 20:13–21

Saunders DL, Meeuwig JJ, Vincent ACJ (2002) Freshwater protected areas: strategies for conservation. Conserv Biol 16(1):30–41. https://doi.org/10.1046/j.1523-1739.2002.99562.x

Scalici M, Chiesa S, Scuderi S, Celauro D, Gibertini G (2010) Population structure and dynamics of *Procambarus clarkii* (Girard, 1852) in a Mediterranean brackish wetland (Central Italy). Biol Invasions 12:1415–1425

Schubart CD, Guerao G, Abelló P (2012) First record and evidence of an stablished population of the North America mud crab Dyspanopeus sayi (Brachyira: Heterotremata: Panopeidae) in the western Mediterranean. Sci Mar 76(1):79–85

Sciberras M, Schembri P (2007) A critical review of records of Alien Marine species from the Maltese Islands and surrounding waters (central Mediterranean). Mediterr Mar Sci 8:41–66

Serbetis C (1959) Un nouveau crustacé commestible en mer Egeé *Callinectes sapidus* Rath. (Decapod. Brach.). Proc Gen Fish Counc Medit 5:505–507

Shiber JG (1981) Brachyurans from Lebanese waters. Bull Mar Sci 31(4):864–875

Siokou I, Ates AS, Ayas D, Souissi JB, Chatterjee T, Dimiza M, Durgham H, Dogrammatzi K, Erguden D, Gerakaris V, Grego M, Issaris Y, Kadis K, Katağan T, Kapiris K, Katsanevakis S, Kerkhof F, Papastergiadou E, Pešić V, Polychronidis L, Rifi M, Salomidi M, Sezgin M, Triantaphyllou M, Tsiamis K, Turan C, Tziortzis I, D'Udekem D'Acoz C, Yaglioglu D, Zaouali J, Zenetos A (2013) New Mediterranean Marine biodiversity records (June 2013). Mediterr Mar Sci 14(1):238–249. https://doi.org/10.12681/mms.450

Souty-Grosset C, Anastácio PM, Aquiloni L, Banhab F, Choquera J, Chucholl C, Tricarico E (2016) The red swamp crayfish *Procambarus clarkii* in Europe: impacts on aquatic ecosystems and human wellbeing. Limnol 58:78–93

Souty-Grosset C, Holdich DM, Noël PY, Reynolds JD (eds) Haffner, P (2006) Atlas of crayfish in Europe, vol 64. Museum national d'Histoire naturelle, Patrimoines Naturels. Patrimoines naturels, Paris

Steinitz W (1929) Die Wanderung indopazifischer Arten ins Mittelmeer seit Beginn der Quartärperiode. Int Rev Gesamten Hydrobiol Hydrographie 22:1–90

Streftaris N, Zenetos A (2006) Alien Marine species in the Mediterranean-the 100 'worst invasives' and their impact. Mediterr Mar Sci 7(1):87–118

Thessalou-Legaki M, Aydogan O, Bekas P, Bilge G, Boyaci Y, Brunelli E, Circosta V, Crocetta F, Durucan F, Erdem M, Ergolavou A, Filiz H, Fois F, Gouva E, Kapiris K, Katsanevakis S, Kljajic Z, Konstantinidis E, Konstantinou G, Koutsogiannopoulos D, Lamon S, Macic V, Mazzette R, Meloni D, Mureddu A, Paschos I, Perdikaris C, Piras F, Poursanidis D, Ramos-Espla A, Rosso A, Sordino P, Sperone E, Sterioti A, Taskin E, Toscano F, Tripepi S, Tsiakkiros L, Zenetos A (2012) New Mediterranean biodiversity records. Mediterr Mar Sci 13:312–327

Torres AP, Dos Santos A, Cuesta JA, Carbonell A, Massuti E, Alemany F, Reglero P (2012) First record of *Palaemon macrodactylus* Rathbun, 1902 (Decapoda, Palaemonidae) in the Western Mediterranean. Mediterr Mar Sci 13(2):278–282

Tortonese E (1965) La comparsa di *Callinectes sapidus* Rath. (Decapoda, Brachyura) nel Mar Ligure. Ann Mus Civ St Nat Genova-Doriana 4:1–3

Tuncer S, Bilgin S (2008) First record of *Callinectes sapidus* Rathbun, 1896 (Crustacea: Decapoda: Brachyura) in the Dardanelles, Canakkale, Turkey. Aquat Invasions 3:469

Ünsal Karhan S, Yokeş MB, Clark PF, Galil BS (2013) First Mediterranean record of *Actaea savignii* (H. Milne Edwards, 1834) (Crustacea: Decapoda: Brachyura: Xanthidae), an additional Erythraean alien crab. BioInvasions Rec 2(2):145–148

Veilleux É, De Lafontaine Y (2007) Biological synopsis of the Chinese mitten crab (*Eriocheir sinensis*), Fisheries and Oceans Canada, Science Branch, Pacific Region, Pacific Biological Station, p 45

Yokes B, Galil BS (2006a) New records of alien decapods Crustacea from the Mediterranean coast of Turkey, with a description of a new palaemonid species. Zoosystema 28(3):747–755

Yokes B, Galil BS (2006b) Touchdown—first record of *Percnon gibbesi* (H. Milne Edwards, 1853) (Crustacea: Decapoda: Grapsidae) from the Levantine coast. Aquat Invasions 1:130–132

Yokes MB, Karhan SÜ, Okus E, Yüksek A, Aslan-Yilmaz A, Yilmaz IN, Demirel N, Demir V, Galil BS (2007) Alien crustacean decapods from the Aegean coast of Turkey. Aquat Invasions 2:162–168. https://doi.org/10.3391/ai.2007.2.3.2

Zaitsev YP, Öztürk B (2001) Exotic species in the Aegean, Marmara, Black, Azov and Caspian Seas. Turkish Marine Research Foundation, Istanbul

Zaouali J, ben Souissi J, Galil BS, d'Udekem d'Acoz C, ben Abdallah A (2007) Grapsoid crabs (Crustacea: Decapoda: Brachyura) new to the Sirte Basin, southern Mediterranean Sea—the roles of vessel traffic and climate change. JMBA2—Biodiv Rec 5770:4

Zaouali J, Ben Souissi J, Rifi M, d'Udekem d'Acoz C (2013a) First occurrence of a Hymenosomatid crab *Elamena mathoei* (Desmarest, 1823) (Crustacea: Decapoda: Barchyura) in the Mediterranean Sea. Mediterr Mar Sci 14(2):278–281

Zaouali J, Ben Souissi J, Rifi M, D'Udekem d'Acoz C (2013b) First occurrence of a Hymenosomatid crab *Elamena mathoei* (Desmarest, 1823) (Crustacea: Decapoda: Brachyura) in the Mediterranean Sea. Mediterr Mar Sci 14(2):278–281. https://doi.org/10.12681/mms.411

Zenetos A, Akel EHK, Apostolidis C, Bılecenoglu M, Bitar G, Buchet V, Chalari N, Corsini-Foka M, Crocetta F, Dogrammatzi A, Drakulić M, Fanelli G, Giglio G, Imsiridou A, Kapiris K, Karachle PK, Kavadas S, Kondylatos G, Lefkaditou E, Lipej L, Mavrič B, Minos G, Moussa R, Prato E, Pancucci-Papadopoulou MA, Renda W, Rios N, Rizkalla SI, Russo F, Servonnat M, Siapatis A, Sperone E, Theodorou JA, Tiralongo F, Tzovenis I (2015) New Mediterranean biodiversity records (April 2015). Mediterr Mar Sci 16(1):266–284

Zenetos A, Ballesteros E, Verlaque M (2012a) Alien species in the Mediterranean Sea by 2012. A contribution to the application of European Union's Marine strategy framework directive (MSFD). Part 2. introduction trends and pathways. Mediterr Mar Sci 13(2):328–352

Zenetos A, Gofas S, Morri C, Rosso A, Violanti D, Garcia Raso JE, Çinar ME, Almogi-Labin A, Ates AS, Azzurro E, Ballesteros E, Bianchi CN, Bilecenoglu M, Gambi MC, Giangrande A, Gravili C, Hyams-Kaphzan O, Karachle PK, Katsanevakis S, Lipej L, Mastrototaro F, Mineur F, Pancucci-Papadopoulou MA, Ramos-Esplà A, Salas C, San Martin G, Sfriso A, Streftaris N, Verlaque M (2012b) Alien species in the Mediterranean Sea by 2012. A contribution to the application of European Union's Marine strategy framework directive (MSFD). Part 2. introduction trends and pathways. Mediterr Mar Sci 13:328–352

Zenetos A, Çinar ME, Pancucci-Papadopoulou MA, Harmelin JG, Furnari G, Andaloro F, Bellou N, Streftaris N, Zibrowius H (2005) Annotated list of Marine Alien species in the Mediterranean with records of the worst invasive species. Mediterr Mar Sci 6(2):63–118

Zenetos A, Gofas S, Verlaque M, Çinar ME, Garcia Raso JE, Bianchi CN, Morri C, Azzurro E, Bilecenoglu M, Froglia C, Siokou I, Violanti D, Sfriso A, San Martin G, Giangrande A, Katağan T, Ballesteros E, Ramos-Esplà A, Mastrototaro F, Ocaña O, Zingone A, Gambi MC, Streftaris N (2010) Alien species in the Mediterranean Sea by 2010. A contribution to the application of European Union's Marine Strategy Framework Directive (MSFD). Part I spatial distribution. Mediterr Mar Sci 11:381–493

Zibrowius H (1992) Ongoing modification of the Mediterranean marine flora and fauna by the establishment of exotic species. Mésogée 51:83–107

Chapter 4
Scourge in the Mediterranean. Policies Against the Argentine Ant in the First Half of the 20th Century

Ana Isabel Queiroz

Abstract The Argentine ant (*Linepithema humilis*) is a Neotropical species from the Paraná River basin in central South America which has been spread unintentionally through human commerce from the nineteenth century onwards. It was soon recognised as an urban and plant pest and, more recently, as a threat to biodiversity in many of the places it has been introduced. This chapter analyses the causes of and motivations for controlling this species in Portugal, Spain, France and Italy in the first half of the twentieth century, describing and comparing legal and administrative measures among countries. Containment (in France), fight (in Italy) and eradication (in Portugal) were laid down in mandatory regulations dated from the 1920s, but none of these three aims were achieved. Despite a long history of control efforts, Argentine ants remain in Southern Europe and other invaded areas, where people still strive to control them.

Keywords Argentine ant · Regulations · Southern Europe

4.1 Introduction

Ants (order Hymenoptera, family Formicidae) are one of the most successful groups of insects in the world. Like bees, wasps and termites, ants are "eusocial" and have overlapping generations, cooperative care of offspring, and a reproductive division of labour with distinct queen and worker castes. They can be found on all continents, except Antarctica, occupying a wide range of ecological niches and exploiting many different food resources. Currently, an online database lists an impressive 14,096 described species of ants (Antbase website); however, one can assume that there are still many unidentified species in highly remote biodiversity hotspots. They are among the organisms whose travel is facilitated by humans, taking advantage of all sorts of transfer vectors and pathways. Therefore, many ant

---

A. I. Queiroz (✉)
IHC—Institute of Contemporary History, NOVA-FCSH, Lisbon, Portugal
e-mail: ai_queiroz@fcsh.unl.pt

A. I. Queiroz and S. Pooley (eds.), *Histories of Bioinvasions in the Mediterranean*, Environmental History 8, https://doi.org/10.1007/978-3-319-74986-0_4

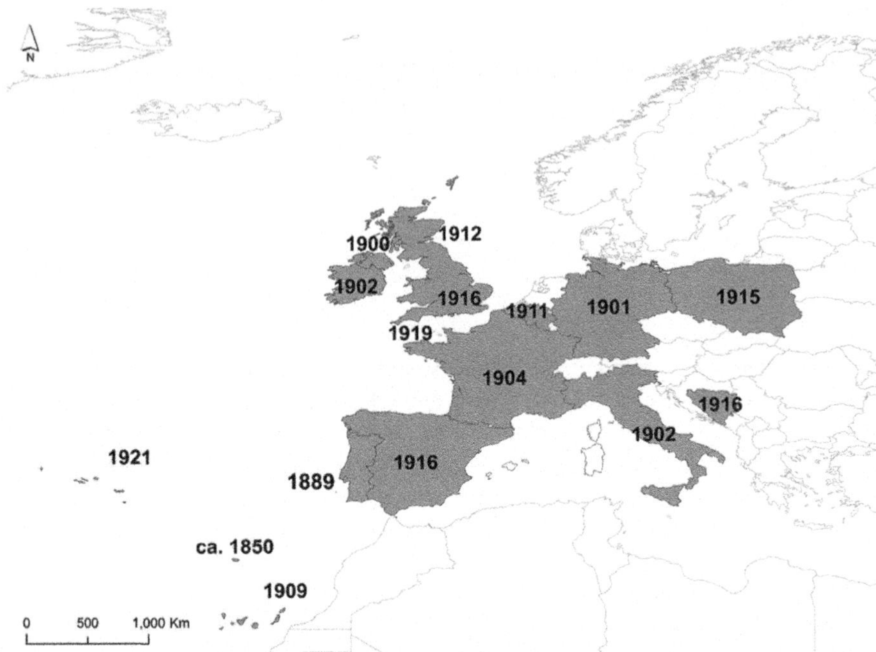

Fig. 4.1 Historical distribution of the Argentine ant in its European range until 1920 (adapted from Queiroz and Alves 2016a). Shaded areas represent the existence of at least one reference of Argentine ant' occurrence; numbers are years of its first record in each country or archipelago from the nineteenth century to 1920

species have been accidentally introduced in territories outside their native range: in Europe, for example, at least 32 alien ant species have been added to the 545 native ones (Schlick-Steiner et al. 2008). Several of them have become invasive, drawing the attention of authorities, the scientific community, the agricultural sector and the home owner.

The Argentine ant (*Linepithema humilis*) is a Neotropical species from the Paraná River basin in central South America, which has been spread unintentionally through human commerce. After its first records on Madeira island ca. 1850, the species obtained a mainland European foothold in Portugal (1890) and thereafter in a set of other European countries in the following decades, among which are Italy (1902), France (1904) and Spain (1916) (Wetterer et al. 2009; see Fig. 4.1).

In the meantime, it was also found in other Mediterranean-like regions such as California, USA (1905, Titus 1905), the Cape Peninsula, South Africa (1908, Lounsbury 1909) and Concepción, Chile (1910, Newell and Barber 1913).

Research on the Argentine ant has largely been framed by biological disciplines like genetics, chemical ecology, ethology, etc. Such studies, for example those of Suarez et al. (2001), Brandt et al. (2009) and Vogel et al. (2010), are crucial to reconstruct the history of invasion. Nevertheless, they overlook the social and economic factors that determine and explain the arrival, the invasion and the subsequent responses. Even when the focus is on the management efforts, such as Choe et al. (2010) and Boser et al. (2014), the cultural and political dimensions are scarcely elaborated. This chapter aims to fill this gap by addressing the historical policies that concern the Argentine ant in Mediterranean countries. Specifically, it analyses the causes of and motivations for controlling the pest in Portugal, Spain, France and Italy, from the 1920s to the 1950s, comparing legal and administrative measures among countries.

4.2 Impacts Before Regulations

Invasive ant species explore food resources and shelters made available in houses and agricultural fields. When in large numbers, their presence has an impact in ecology, economy and human wealth. In Macaronesean[1] and Mediterranean areas, the Argentine ant proliferates. It has been referred to as an urban pest, a plant pest and a threat to other species in many of the places it was introduced. Indoors, the ants invade warehouses, hutches, and every room in the house. In fields, they care for aphids, mealybugs and scale insects (Hemiptera), feeding on the honeydew they excrete. By protecting them from fungi and parasites, ants promote the proliferation of these insect pests, indirectly causing an adverse effect on the plants in which they are installed, namely fruit trees.

A Portuguese entomologist used the epithet "ant-thief" (*formiga-ladra*) "to distinguish [the Argentine ant] from other ants, serene inhabitants of the fields, which are far from troubling men in their villages and assaulting them for the supplies already sheltered in their homes" (Martins 1907a, pp. 111–2).[2] He was responding to the invasion of the areas surrounding the *Palácio de Cristal*, in Porto (Portugal), where the first specimens had been accidentally introduced in 1894 through live plants and objects imported for an exhibition. He also reported attacks to entomological collections in Lisbon, despite all the strategies used by scientists to carefully wrap them (Martins 1907b). During the first Luso-Spanish Scientific Congress for the Progress of Science (held in Porto, 1921), Garcia Mercet (1921) wryly stated that he "shared a room in the *Grande Hotel* with the *hormiga argentina* of Oporto" which "enters through one of the windows in the room in which he is housed; climbs furniture, and invades the delicatessen, eating the remnants of the meal" (p. 99). The same author also narrated that two Spanish entomologists hosted nearby, in the Seminar, lost their collection of chalcidoids (wasps), which they had left in their room. The destruction of entomological collections, exacerbated by the value Garcia Mercet (and Martins before him)

[1]A collective term for four archipelagos in the North Atlantic Ocean west of the continents of Africa and Europe.

[2]Wherever non-English literature is quoted, the translations are the author's. Herein, the word "men" refers to humans.

attributed to these specific losses, highlights the voracity and inconvenience these invaders caused inside the buildings.

In France, by the second decade of the twentieth century, Argentine ants had become a problem for the growing tourism business in the Cote d' Azur (Marchal 1917). They were considered "a real obsession for the inhabitants of the localities where it took up residence. The house tenants are becoming increasingly rare and the properties thus lose a large part of their value (...)" (Marchal and Poitiers 1920, pp. 318–9). Adapting furniture was an odd and costly solution presented to those living in the invaded houses: Chopard (1921) suggested a suspended pantry to store food, supported by a hook plunged into a recipient full of petroleum, where the intruder ants would die.

The most serious damages occurred in agricultural fields. In Italy, the pest was particularly harmful in the orange groves of the western part of the Gulf of Genoa, and in Rome and its surroundings (Seri 1923). Chopard (1921) noted that, in Southern France, the vineyards, some ornamental trees and leguminous crops (such as beans) were affected. The orange groves of the Algarve province (Portugal) were invaded in 1926 and the pest was considered an impediment to agricultural development (Coutinho 1929). Also in the 1920s, Spanish entomologists recorded the presence of the Argentine ant in the Valencia region (Font de Mora 1923). This species promoted the proliferation of citrus mealybugs, threatening the economy of the Valencia orange, a valuable product in foreign trade.

4.3 Phytosanitary Regulations in the Early Twentieth Century, by Country

During the first decades of the twentieth century, the impacts resulting from the Argentine ant's occurrence in Southern European countries were enough to trigger local, regional and national responses, including mandatory regulations.

4.3.1 France

In 1920, the Agricultural Society of Nice and the Maritime Alps decided to tackle the scourge, calling "farmers, gardeners and owners of buildings or crops invaded by the Argentine ant to urgently come together in a defence Union, to intensively and continuously apply the methods of destruction already recognized and to be encouraged and assisted in this task by the Municipalities, the department and the Ministry of Agriculture" (Florent 1920, p. 172). In October, the "Association for the defence against ants" (*Association de defense contre les fourmis*) asked the General Council of the Maritime Alps for a grant. This money would be used to buy sugar and paraffin bags in large quantities to destroy infested loci: "the means of struggle are known; those concerned (...) are available to implement them seriously, but the costs are excessive (...) the estimated expenditure is 100 thousand francs for three years" (Capron 1920, p. 578). An official reply appeared next year, in the form of a Regional Decree, signed by the governor of the province (Bernard 1921).

This decree emerged due to the potential "severe damage to agriculture and housing hygiene". This regulation aimed to proceed "with urgency to the methodical destruction of invasion loci", imposing a mandatory regime on the owners, tenants, overseers, and beneficial holders; they should proceed to control ants and support the treatment expenses (Article 1). The State, municipalities and official agencies supervised this process and if those in charge did not comply with the rules they would be replaced, however, they would still have to cover the expenses (articles 2 and 4).

To achieve this goal of "preventing its extension and cleaning the coast", some very precise tasks should be undertaken. Article 4 regulated the appropriate poisonous substances (No. 1), means and methods for the repellent treatments (No. 2) and winter treatments (No. 3), and also the disinfection measures that were to be applied to potted plants before their sale. These procedures had previously been suggested by the United States Department of Agriculture, as soon as the species spread in that country. *The Argentine Ant* by Newell and Barber (1913) described the species' origin and biology, and proposed methods for controlling it in the States invaded. Exactly at that time, the eminent French entomologist Paul Marchal spent three months visiting the U.S. Bureau of Entomology, where this text was being prepared (Howard 1930). After his return to Europe, his experience proved influential in the French academy and in official decisions. Arsenical insecticides seemed to be the appropriate method to destroy entire colonies. Treatments should be carefully prepared in a dilution that would delay the death of the workers while they poisoned the breeding females (queens), which they fed (Marchal and Poitiers 1920). At home, people might "enclose all food in pantries, which are protected by placing their feet in wells filled with naphthalene or petrols, or better by suspending them to a stem dipping in a funnel full of petrols" (Salagnac 1922, p. 4).

It was assumed that the Argentine ant could not survive in the climatic conditions in the north of France (Académie d'Agriculture de France 1920). The coordinated and universal control efforts would hopefully result in the confinement or the elimination of the species in Southern France.

One year after the French regulation, a newspaper reported that: "the question of the Argentine ant is not as worrisome now as it was in 1920; our winter resorts of the Côte d'Azur will not suffer, especially if the Municipalities continue to provide assistance in the fight moved by individuals against this unpleasant host of both homes and properties" (Vayssière 1922, p. 130). But, less than two decades after, in 1943, others reported the failure of the control measures carried out and the occurrence of the ant in the gardens and suburbs of Paris (Anon 1923; Guy 1927). No one knows if this resulted from a population's expansion, a secondary introduction or a combination of these phenomena.

4.3.2 Italy

Provisions against the spread of the Argentine ant were set out by the Ministerial Decree of 27 May 1922, signed by Giovani Bertini, minister of Agriculture, just before the beginning of the fascist regime. The species was added to an official list of plant pests, in force since 1919 (Article 1), and "[t]he fight against the Argentine ant" became "mandatory in all places where it is found and the treatment must be done at the expense of the interested parties" (Article 2), in other words, of the farmers and the people who had their houses and gardens invaded by the pest. The description of the methods to be used were forwarded to the regional phytopathological agencies (Article 3, in Italian, *Reggi Osservatori regionale de fitopatologia*) but the provincial governors were the ones responsible for the implementation of these measures. To this end, they were required to create local task-forces (*Consorzi obbligatori de difesa contro la formica argentina*) (Article 4).

The provinces and communes followed the regulations and issued their orders. In Trento, it was brought to the attention of the "citizens in general, and farmers in particular, the dangers that a tiny exotic ant, recently imported to Italy, could bring about if introduced in our province" (Perfetto di Trento 1926). Stakeholders were also urged to inform the prefecture if ants were detected and to send some specimens to the phytopathological regional agency if that was the case. The species would be recognized "because its prolificacy is portentous, its intrusiveness extraordinary, its mobility lightning, its voracity insatiable, and it causes damages not only in cultivated plants, fiercely protecting the aphids, but also in food, and out of the kitchens, harassing people even during sleep" (*idem*).

In 1926, Mussolini's government amended and reinforced the fight against the pest with the Decree-Law of 1st July 1926, turned into the Law No. 1266 of 17th January 1928. It was signed by Vittorio Emanuel III, King of Italy, Benito Mussolini and the ministers Giuseppe Belluzzo (Economy), Luigi Federzoni (Colonies) and Giuseppe Volpi (Finance). When it was declared that a territory had been invaded by the Argentine ant, by Provincial decree (Article 1) and Communal order (Article 2), the State would bear half of the expense necessary for the execution of the measures devised by the regional agencies of phytopathology, in accordance with the public health authorities (Article 3). The remaining costs were allocated, in equal shares, to the budgets of provinces and communes. With the strong support of the Fascist Government, the national agency controlled the policy for plant diseases in the country, assisted by regional agencies.

The Liguria's regional agency of phytopathology was an active institution at the time. Several publications by its Director reported the Argentine ant's invasion and suggested control measures (e.g. Paoli 1922a, b, 1923) following the Barber's method (Newell and Barber 1913), as in France.

The Italian writer Italo Calvino (1923–1985) lived there during his childhood. In "La formica argentina" (1952), he described the torments suffered by a family when they first moved to San Remo, a city where the houses were particularly infested by the ants, and the authorities imposed severe procedures, not always accepted or understood by the inhabitants: [the worker from the "*Argentine Ant Control Corporation*"] "was doing all he could to help the ants increase so as to not lose his job, and he was perfectly capable of doing this on purpose, because now he was always on the side of the ants and not on that of human beings" (Calvino 1991, p. 176).

In the years that followed, the ravages due to the Argentine ant never ceased. On 7 October 1948, the Commission of Agriculture and Food of the Italian Senate discussed the allocation of ITL 200 millions[3] to continue the fight against the Argentine ant in the provinces of Imperia, Savona, Rome, Latina, Napoli, Reggio Calabria, Messina and Palermo (Senato della Republica Italiana 1948). From then on, this official body recommended the old arsenical baits should be replaced by "D.D.T. and the *gammaesano*[4] powder, scattered over the anthills and in places frequented by ants" (*idem*).

4.3.3 Portugal

Since the beginning of the 1920's, the invasion was severe in a number of Portuguese regions, the worst occurrence being in the Algarve (the southernmost province), with impacts on orchards, henhouses, rabbit hutches, and inside houses in several regions (Coutinho 1929).

Four ministers of the Portuguese Government (the Prime Minister and those of the Interior, Justice, Public Instruction and Agriculture) signed the "Regulation for the extinction of the Argentine ant (*Iridomyrmex humilis*, Mayr)", "in view of the urgent complaints from the native farmers and after hearing the director of the Plant Pathology Laboratory" (from the introduction of the Decree No. 17577, 08/11/1929). The Lab was the national phytopathological agency at the time, and it belonged to the administrative structure of the Ministry of Agriculture (Decree No. 4151, 26/04/1918), which developed research, supervised inspection and implemented measures. The success of the measures depended on the "indispensable uniformity of action and management, which can only be achieved in a practical manner under a mandatory service, run by the proper authorities and constantly inspected by specialized technicians" (*idem*).

The decree urged the official entities and civil society to take action, engaging the central and regional services with authority over agriculture, the district and municipal authorities, farmers' guilds and associations, and all private individuals

[3]corresponding to $320,000 on that date, according to the official exchange rate that had been established in 1949 by the system "Breton Woods" and remained stable until 1968 (McNamara 1998, pp. 72–75); in 2016, the comparable amount is $3,148,361.83 (Measuring Worth—Relative Worth Comparators and Data Setswebsite).

[4]Trade name of the gamma isomer of the esacloro-cyclohexane, powerful synthesis insecticide that is obtained by chlorination of benzene.

who had some sort of relationship with the land. Local task-forces were created in the invaded areas (Article 2). Owners, beneficial owners, tenants, overseers, and beneficial holders, as well as associations (e.g. agricultural unions) and local authorities who managed land, were obliged to apply control measures (Article 3). The instructions on the methods and timing suitable for the application of the measures were provided in an official announcement (Article 4). The local task-forces, together with the police, court officials and city servants, could replace the addressees if they did not comply with the obligations (Article 7). Penalties for non-compliance were laid out (articles 10, 11 and 12).

As with Italy during Mussolini's regime, the military dictatorship in Portugal, which lasted from 1926 to 1933, and the authoritarian regime that followed under Salazar's rule (only ousted in 1974), turned agricultural development into a political and economic bastion, anchored in agronomic sciences, namely plant pathology. However, the difficult political period of the 1930s led to successive changes in the institutional structure of the Ministry of Agriculture (in 1931, 1932, 1933, 1936 and 1940). The fight against plant pests underwent great instability and operational difficulties. During this period, the Plant Pathology Laboratory, which initially had an operational role, became an advisory body (Decree No. 20526, 18/11/1931) and its executive and inspection tasks, including the services directed to the eradication of the Argentine ant, were adjusted to fit into a new administrative structure: the Phytopathological Inspection Services Division (Article 35(d)). Five years later, in another administrative review, this jurisdiction was transferred to the newly formed Phytopathological Services Office (Decree-Law No. 27207 of 16th November 1936).

The reports on the implementation of the legislation revealed that the means and methods were being disputed, the costs of treatment were high and were being contested, and the results obtained were a long way from the advertised eradication (Bensaúde and Neves 1937; Neves no date).

The Ordinance No. 9216 of 10th May 1939, strengthened the previous decree (although it did not mention it) by reiterating the mandatory fight against the Argentine ant in Portimão, Lagos and Silves municipalities, reinforcing, in this manner, the efforts in the Algarve region. Nevertheless, the work continued in other regions. Although the Decree No. 17577 did not foresee the State's contribution in the treatment expenses considered mandatory to private individuals, the expenditure books from the Phytopathological Services Office revealed funds, intended for the control of this ant, transferred to local services in the north, centre and south of the country, from 1937 to 1949[5].

In 1950, in the central region of the country, the principle of the "optional execution with the contribution from the Official Services to the respective expenses" presided the campaign against the Argentine ant (Gamelas Júnior 1950, p. 1). As occurred in other countries, arsenical poisons were abandoned for a "new"

[5]Arquivo Nacional da Torre do Tombo, Ministério da Agricultura, 2ª incorporação.

chlordane-based method, which was supposedly cheaper and more effective.[6] Demonstrative clusters were selected in vineyards, citrus orchards and fig trees. Experiments with oils and wettable powders, using different percentages of insecticide, were made. The local outcome was encouraging: "the results on the decrease of cochineal were remarkable (…) a big decline of these specimens was observed in citrus fruit, as a result of the fight against the Argentine ant" (*idem*, p. 16). But the treatments were not carried out enough times (once every two months until eradication) nor were they conducted at an appropriate scale (Silva Dias 1953). The economic expenditure needed for a universal and systematic control in the invaded fields was unsustainable, whether paid for by private individuals or the State.

4.3.4 Spain

Throughout the twentieth century, the legal texts on agricultural pests in Spain did not identify the Argentine ant as an insect that needed to be fought. The Spanish Plant Pest Law (*Ley de Plagas del campo*), from 21 May 1908, only highlighted specific provisions for the phylloxera and the locust outbreaks, the two major concerns at that time. This law was accompanied by the Royal Order of 3 January 1912, in which the prefects were assigned the task of assembling Local Boards of defence against pests in all of the country's municipalities and of creating a budgetary fund to take all the necessary control measures.

At the end of 1922, the greatest phytosanitary concern fell on the cottony cushion scale (*Iceria purchasii*) which was destroying the orange groves. The Royal Order of 14 October, which reinforced the implementation of the previous legislation, created provisions to monitor living plants in seaports and border railway stations, in order to prevent the accidental introduction of these new damaging organisms. It was known that the pest occurred south of Provence (France), "so close to Spain and with so many and so frequent communications". The Order authorised the director of the national phytopathological agency (Estación de Patología Vegetal del Instituto Agrícola de Alfonso II) to start the study of "any outbreak that may occur in Spain" or even to "study it abroad", to propose the measures he deemed pertinent.

A report of the National Conference on orange production explained how the orange grove's pests were composed by "several ants' species including the fearsome and misnamed *hormiga argentina* (*Iridomyrmex humilis*, Mayr), which indirectly cause very large damage associated with coccidae" (Estacion Central de Patologia Vegetal de Madrid 1926, p. 262). In response to a letter by a reader from Gibraltar whose orange groves were affected by *Icerya,* Rafael Font de Mora (1927,

[6]Chlordane is a member of a group of chemical compounds generically termed "chlorinated cyclodienes."

p.232) wanted to confirm the expansion of the species in the Southern coastline: "if you observe ants on the *Icerya*, (…) collect a few, and after killing them by introducing them into alcohol, please send them; and I will give you the treatment against the Argentine ant, I suspect it is found in that".

Between 1930 and 1950, the Ministry of Agriculture published instructions on the methods to eliminate ant pests (among which the Argentine ant). As in other countries already mentioned, the use of arsenical poisons (Anon 1935; Castro 1949) and chlordane (Cañizo 1958) stood out.

4.4 Similarities and Differences in Legal Texts

The disturbances caused by plant pests and diseases from the second half of the nineteenth century onwards (e.g. on vineyards, potato plants, olive groves) brought the European plant pathologists closer together and led them to share their knowledge with their American colleagues. They gathered around scientific societies and international governmental organisations (e.g. the International Institute of Agriculture, founded in Rome in 1905), influencing the national and international policy that regarded pest control strategies and prevention measures against the accidental introduction of new agents (Castonguay 2010). These scientific networks, strengthened at the end of World War I by the resurgence of a science diplomacy, might explain the similarity in the processes regarding the Argentine ant, proposed by the legislation that was approved in many countries during the 1920s.

Comparing these different national processes allows for an analysis of the main similarities and differences regarding the typology of legal acts, territorial scope, statement of the main goal, implementation, methods, enforcement, and accessibility to private properties and penalties for France, Italy and Portugal (as summarised in Table 4.1). The aforementioned Spanish regulations can be at some points considered pioneering but they are not included in this summary because they did not mention specific provisions to the Argentine ant.

The French Regional Decree (1921) contrasted with other legislation concerning the Argentine ant due to its sub-national geographic scope. It revealed the expectancy of a successful "methodical destruction of the invasion's loci", through the methods recommended by the American entomologists. Among other specific regulations, this was the one that came closest to scientific language and knowledge from which it carefully transcribed the means and methods proposed therein. The Italian Ministerial Decree (1922) assumed that "the fight against the Argentine ant" at a national scale depended on regionalized management, attributing to provincial governors responsibility over its implementation. The adequate measures were also defined, according to local circumstances, by the regional phytopathologic agencies. The language focused on the elimination of all specimens of this ant species as

Table 4.1 Summary of the mandatory regulations against the Argentine ant's invasion, adopted in France, Italy and Portugal in the 1920s

	France (1921) Regional Decree, 26 January	**Italy** (1922) Ministerial Decree of 27 May	**Italy** (1926) Law Decree, 1st July	**Portugal** (1929) Decree 17:577, 8th November
Territorial scope	Regional (province of the Maritime Alps)	National (all provinces declared invaded)	National (all provinces declared invaded)	National (all areas considered invaded)
Main goal	Methodical destruction of the invasion's loci	Fight against Argentine Ant	Fight against Argentine Ant	Extinction of the Argentine ant
Implementation	Public entities and private individuals who were landowners, tenants or usufructuaries supported the expenses and should proceed to control the ants; regional authorities replace private in case of non-compliance	The provincial governors were in charge of local task forces that carried out the control measures	National, regional and local power bodies supported the expenses; a national agency of phytopathology coordinated regional delegations which coordinated the local task-forces that carried out the control measures	Public entities and private individuals who were landowners, tenants or usufructuaries supported the expenses; a national agency of phytopathology coordinated local task-forces that assisted the public entities and private individuals in the mandatory control measures, and replaced them in case of non-compliance
Methods	It defined substances and treatments, and also disinfection measures applied in potted plants for sale	Adequate measures were defined by the regional agencies of phytopathology	Adequate measures were defined by the regional agencies of phytopathology	Adequate measures were defined by the national agency of phytopathology and publicized by the local authorities

(continued)

Table 4.1 (continued)

	France (1921) Regional Decree, 26 January	**Italy** (1922) Ministerial Decree of 27 May	**Italy** (1926) Law Decree, 1st July	**Portugal** (1929) Decree 17:577, 8th November
Enforcement	The responsibility fell on the public administrators, civil servants and police forces	–	–	The responsibility fell on the public administrators, civil servants and police forces.
Accessibility	The authorities had free access to all kinds of properties	–	No compensation was due for damage in plants or any private belonging due to measures	Local task-force members had free access to properties
Penalties	Fines were applied in case of non-compliance with the obligations	–	–	Fines were applied in case of non-compliance with the obligations

a reactive offensive against invasion, which contrasted with the defensive formulation found in the French decree.

The warlike terminology was maintained in the subsequent Italian acts, dated from 1926 and 1927, and also in the Portuguese decree (1929), which aimed for the "extinction of the Argentine ant". Keeping the insects that invaded the orange groves in California in mind, Woodworth (1910) also used strong language, stating that although "eradication (…) has been so often used in this State [California] for the partial destruction of an insect pest" (p. 61), the term should continue to be used due to its motivational advantages: "it would be too bad not to make at least a serious effort toward eradication (…) it is not a work that can be done in a day, but should be undertaken with the idea that final success will only come by destroying the insect step by step" (pp. 62–3).

None of these acts mentioned the ant's non-native character, nor were "the fight" or need for "extinction" justified by any explicit nativist argument, despite the nationalisms that took power in Italy (from 1922) and Portugal (from 1926). In Nazi Germany, in contrast, cases of obvious impacts of invasive species were abusively fused to a speech in favour of a "master race" some years later. According to Uekoetter (2006, p. 79), Alwin Seifert, leader of the Reich Landscape Advocates (*Reichlandschftanwalt*), "became increasingly radical in his attacks on non-native species, eventually designating the blue spruce as 'public enemy number one'". Instead, Italian and Portuguese regulations focused on the ants' impact on crops, as it was expressed in the introduction of the Portuguese decree.

From the legislation approved in the Mediterranean countries during the 1920s, the Italian one stood out, because it placed the responsibility for the control of the Argentine ant on the State, and consequently, made the State responsible for enacting and funding the required actions. Several regional plant health agencies, as well as a national one, worked on the diagnosis and control of the pests. Italy was recognized as one of the most active countries in economic entomology, at the time (Howard 1930, p. 258). This and other scientific disciplines, like genetic studies focussed on the improvement of plants, were supported by Mussolini's fascist ideology, which sought to increase food production (namely of wheat, through the *Battaglia del Grano*, launched in 1925) (Saraiva 2010) and "extolled the virtues of country life, (...) publicized legislation that encouraged 'ruralisation' and internal colonization" (Cohen 1979, p. 85).

The Portuguese dictator Salazar (Minister of Finance in 1928 and Prime Minister after 1933), who shared these political beliefs, implemented similar measures to create a rural state (see e.g. Pais et al. 1976; Rosas 1991). But the instability and financial crisis Portugal was facing at the end of the 1920s prevented the State from spending great sums of money on hiring employees or on acquiring the materials used for the "extinction" measures. For that reason, the regulation, although centralising and anti-liberal, assigned liability to the public entities and private individuals that had a relationship with the land (owners, tenants or beneficial holders). They were forced to pay for the treatments and other expenses of the local task-forces, who took action or assisted in doing so. In this respect, it differed from the French regulation (1921), which also assigned liability to the same subjects, but did not establish local task-forces ("the extinction services"), as had happened in Italy and Portugal.

4.5 Conclusion: Living with Invasive Ants

Containment (in France), fight (in Italy) and extinction (in Portugal) were laid down in the Southern European countries' regulations in the 1920s, and all three main aims failed. In the European Mediterranean territories, the Argentine ant is currently widely distributed, due to the expansion of its early non-native range, novel primary and secondary introductions throughout the twentieth and twenty-first centuries, and most of the first records of occurrence happened after the mandatory control measures (Queiroz and Alves 2016a). As a result, people and their crops must now live with the ants, controlling them locally, when needed.

With the current knowledge about the complex management of invasive species, it is questionable if these regulations, with their similarities and differences, were appropriate for the natural, social and political conditions of the countries in which they were implemented. In light of the subsequent attempts to control this and other ant species which invaded the Mediterranean region, it seems the aims were too ambitious and that, regardless of the terminology used or the efforts undertaken, they were just not feasible.

Due to the ant's resistance, reproductive capacity and ability to regenerate colonies from a low number of individuals, all kinds of efforts to eradicate the invasive Argentine ant have proved insufficient, when attempted in extensive mainland areas. Failed attempts are common worldwide for this target species and other invasive ants, even when the campaigns only cover small isolated areas (<10 ha) (Randall et al. 2011; Hoffmann et al. 2016).[7] The Argentine ant, considered one of the *100 of the World's Worst Invasive Alien Species* (Lowe et al. 2000) because of its impact on biodiversity, is still a current subject to be researched and managed in the Southern European countries and elsewhere (Berville et al. 2016; Obregon and Reyes-López 2016; Ward et al. 2006; Zina et al. 2015). Portugal and Spain still keep this ant in nature conservation legislation as an invasive species, although there are no provisions obliging private individuals or the authorities to take binding control measures.[8]

Nowadays, reasons for controlling the Argentine ants in high-value natural areas in Southern Europe justify conservation concerns in Doñana/Spain (e.g. Carpintero et al. 2003); Port-Cros and Porquerolles/France (Berville et al. 2012) and Islets of Porto Santo/Portugal (Queiroz and Alves 2016b). In urban environments and agricultural fields, people who are annoyed by the presence of these ants, often resort to the use of commercial insecticide sprays based on pyrethroids. These are only temporarily effective against ants, as everyone knows, and their toxic degradation products persist thereafter in urban waterways and aquatic sediments (Delgado-Moreno et al. 2011). Arsenicals, organochlorines and these new insecticides used in the fight against invasive insects were, and still are, a threat to human health and the environment, in general.

Recently, researchers from the Department of Entomology, University of California have proposed innovative methods to maximize the efficacy of conventional baits targeting Argentine ants. The new baits incorporated the (Z)-9-hexadecenal, a pheromone which converts them in a more palatable, non-repellent, slow-acting, transferable form, which is inaccessible to non-ants (Welzel and Choe 2016). This could prove to be a new effective way for fighting invasive ants in the first half of the twenty-first century.

Acknowledgements This work was supported by FCT—the Portuguese Foundation for Science and Technology [UID/HIS/04209/2013 and IF/00222/2013/CP1166/CT0001].

[7]The first successful eradication in the Tiritiri Island (New Zealand) was recently announced: "after many years of annually reducing the population by 99%, an incredibly successful control program but still 1% short of total eradication, scientists from the New Zealand Department of Conservation continued to refine their methodologies until it has now been three years since Argentine ants were last recorded on the Island" (Russell 2016).

[8]See the Portuguese Decree-Law 565/99 of December 21 which regulates the introduction, detention and management of Invasive Alien Species and the Spanish Royal Decree 630/2013 of August 2, which regulates the Spanish Catalog of Invasive Alien Species.

References

Antbase. http://www.antbase.org. 21 July 2017

Anon (1923) Aprés la Côte d'Azur, les fourmis d'Argentine menacent les jardins parisiens. Le Matin (23 August)

Anon (1935) Medios modernos de lucha contra las hormigas. Hojas Divulgadoras (Ministerio de Agricultura) 23:14–16

Bensaúde M, Neves M (1937) A Formiga Argentina. Métodos para a Combater. Ministério da Agricultura, Lisboa.

Bernard A (1921) Fourmi d'Argentine—Arrété Préfectorale des Alpes-Maritimes (January 26)

Berville L, Renucci M, Provost E (2012) Mise en place de protocoles de contrôle de la fourmi d'Argentine sur les îles de Port-Cros et de Porquerolles (Var, France). Sci Rep Port-Cros Natl Park 26:91–108

Berville L, Renucci M, Vidal P, Provost E (2016) Peuplement myrmécologique et évaluation de l'invasion de *Linepithema humile* sur les îles de Marseille (Bouches-du-Rhône, France). La Terre et la Vie (Revue d'Ecologie) 71(3):278–287

Boser CL, Hanna C, Faulkner KR, Cory C, Randall JM, Morrison SA (2014) Argentine ant management in conservation areas: results of a pilot study. Monogr W North Am Nat 7:518–530

Brandt M, Van Wilgenburg E, Tsutsui ND (2009) Global-Scale analyses of chemical ecology and population genetics in the Argentine ant. Mol Ecol 18:997–1005

Calvino I (1991) The Argentine ant. The watcher and other stories. HBJ, San Diego

Cañizo J (1958) Las hormigas. Hojas Divulgadoras (Ministério de Agricultura) 10:1–20

Capron A (1920) Subvention à l'Association de défense contre les fourmis de Cannes, séance du 8 Octobre. Rapports et délibérations: 578. Conseil général des Alpes-Maritimes, Nice

Carpintero S, Reyes-Lopez J, Reyna LA (2003) Impact of human dwellings on the distribution of the exotic Argentine ant: a case study in the Donana National Park, Spain. Biol Cons 115:279–289

Castonguay S (2010) Creating an agricultural world order: regional plant protection problems and international phytopathology, 1878–1939. Agric Hist 84(1):46–73

Castro AR (1949) Las Hormigas. Servicio de Capacitacion y Propaganda, Ministério da Agricultura, Madrid

Chopard L (1921) La fourmi d'Argentine Iridomyrmex humilis var. arrogans Santschi dans le midi de la France. Annales des Épiphyties 7:237–65

Choe DH, Vetter RS, Rust MK (2010) Development of virtual bait stations to control Argentine ants (Hymenoptera: Formicidae) in environmentally sensitive habitats. J Econ Entomol 103:1761–1769

Cohen JS (1979) Fascism and agriculture in Italy: policies and consequences. Econ Hist Rev 32(1):70–87

Coutinho MFP (1929) A formiga argentina Iridomyrmex humilis Mayr var. arrogans Santschi. Ministério da Agricultura, Lisboa

Delgado-Moreno L, Lin K, Veiga-Nascimento R, Gan J (2011) Occurrence and toxicity of three classes of insecticides in water and sediment in two southern California coastal watersheds. J Agric Food Chem 59:9448–9456

Estacion Central de Patologia Vegetal de Madrid (1926) Las plagas de los naranjales españoles. Informe presentado a la Conferencia Nacional Naranjera. Producción 120:263–265

Florent P (1920) La fourmi d'Argentine. Journal d'agriculture pratique, de jardinage et d'économie domestique (Nouvelle Série) 34:172

Font de Mora R (1927) Consultas. El Progreso agrícola y pecuario 1478, March 15:231–32

Font de Mora R (1923) Sobre la presencia de la hormiga argentina (Iridomyrmex humilis Mayr) en Valencia. Boletin de la Real Sociedade de História Natural 23:77–78

Gamelas J (1950) Relatório Geral da campanha de combate à formiga Argentina "Iridomyrmex humilis". Brigada técnica da IV Região, Ministério da Economia, DGSA

Garcia Mercet R (1921) La hormiga argentina de Oporto (Iridomyrmex humilis Mayr). Ass Esp Prop de las Ciencias VI:99–104

Guy (1927) Fourmis d'Argentine. L'Ouest-Eclair, 27 August

Hoffmann BD, Luque GM, Bellard C, Holmes ND, Donlan CJ (2016) Improving invasive ant eradication as a conservation tool: a review. Biol Cons 198:37–49

Howard LO (1930) A history of applied entomology (somewhat anecdotal). Smithsonian Institution, Washington

Lounsbury CP (1909) Appendix IV. Report of the government entomologist for the year 1908. Cape of Good Hope. Report of the Department of Agriculture 1908:55–70

Lowe S, Browne M, Boudjelas S, De Poorter M (2000) 100 of the world's worst invasive alien species a selection from the global invasive species database. The Invasive Species Specialist Group (ISSG) of the Species Survival Commission (SSC) of the World Conservation Union (IUCN)

Marchal P (1917) La fourmi d'Argentine (Iridomyrmex humilis Mayr). Bulletin de la Societé d'Etude et de Vulgarisation de Zoologie et d'Agriculture de Bordeaux 16:23–26

Marchal P, Poitiers R (1920) La fourmi de l'Argentine. C.R.Ac.Agric VII:315–23

Martins MN (1907a) A formiga-ladra (Iridomyrmex humilis Mayr). Broteria (Série de Vulgarização Scientífica) 6:111–19

Martins MN (1907b) Une fourmi terrible envahissant l'Europe (Iridomyrmex humilis Mayr). Broteria (Série Zoológica) 6:101–2

McNamara KR (1998) The currency of ideas: monetary politics in the European Union. Cornell University Press, Ithaca, New York

Measuring Worth—Relative worth comparators and data sets. Measuring Worth. https://www.measuringworth.com. 24 May 2016

Neves M (n/d) Experiências efectuadas no Algarve durante o ano de 1932, sobre a eficácia e custo dos meios de combate à formiga Argentina. Ministério da Economia, DGSA

Newell W, Barber TC (1913) The Argentine ant. US Department of Agriculture, Bureau of Entomology

Obregón R, Reyes-López J (2016) Primera cita de la hormiga invasora Linepithema humile (Mayr, 1868) en el País Vasco (Hymenoptera, Formicidae). Revista gaditana de Entomología 7(1):515–18

Pais JM, Valadas de Lima AM, Baptista JF, Jesus MF, Gameiro MM (1976) O fascismo nos campos em Portugal: a campanha do trigo (1a parte). Análise Social XII(46):400–74

Paoli G (1922a) Isolator, per difendere le piante contro la formiga dell' Argentina. Redia XV

Paoli G (1922b) Un modo di difendere le piante contro gli assalti della formica d'Argentina. La Costa Azzurra Agrícola—Floreale II (6)

Paoli G (1923) La formica dell' Argentina: descrizione, costumi, mezzi di lotta: istruzioni e decreti. La Costa Azzurra Agrícola—Floreale II (5 suppl)

Perfetto di Trento (1926) Circolare Lotta contro la Formica Argentina. Bollettino Uficiale Prefettura di Trento (deciembre) 51425

Queiroz AI, Alves D (2016a) People, transports and the spread of the Argentine ant in Europe, from c.1850 to present. CEM/Cultura, Espaço & Memória 7:37–62

Queiroz AI, Alves D (2016b) Local responses to a global invasive species. Varying reasons for controlling the Argentine ant in the Madeira archipelago (1850–2014). e-J Portuguese Hist 14(2):72–92

Randall JM, Faulkner KR, Boser C, Cory C, Power P, Vermeer LA, Lozier L, Morrison SA (2011) Argentine ants on Santa Cruz Island, California: conservation issues and management options. In: Veitch CR, Clout MN, Towns DR (eds) Island invasives: eradication and management, pp 108–13. IUCN, Gland, Switzerland

Rosas F (1991) Rafael Duque e a política agrária do Estado Novo (1934–44). Análise Social XXVI (112–113):771–790

Russell J (2016) Invasive ants eradicated from Tiritiri Island. National Geographic Society (blogs), 24 May. http://voices.nationalgeographic.com/2016/05/24/invasive-ants-eradicated-from-tiritiri-island/. Accessed 21 July 2017

Salagnac (1922) Les ravages de la foumi d'Argentine. Le Journal (Paris. 1892) (3 August)

Saraiva T (2010) Fascist labscapes: geneticist, wheat and landscapes of fascism in Italy and Portugal. Hist Stud Nat Sci 40(4):457–98

Schlick-Steiner B, Steiner FM, Pautasso M (2008) Ants and people: a test of two mechanisms potentially responsible for the large-scale human population–biodiversity correlation for Formicidae in Europe. J Biogeogr 35:2195–206

Senato della Republica Italiana (1948) Disegno di Legge: (Discussione e approvazione) Autorizzazione della spesa di lire 200 milioni peri il finanziamento della lotta contro la formica argentina e della spesa di lire 1,000,000 per assicurare il normale funzionamento dei servizi fitopatologici, VIII Commisione (Agricoltura e alimentazione) Riunione del 7 Ottobre 1948, 37–41

Seri L (1923) Instruzioni sulla lotta contra la Formica argentina e risultati ottenutti a Roma negli anni 1921-22-23. Boll. Menz de Inf e Not, R. St. Pat Veget- Roma

Silva Dias JC (1953) Ensaios de combate à formiga Argentina em pomares de citrinos: Setúbal. Brigada Técnica da XIII Região. Ministério da Economia, Repartição dos Serviços Fitopatológicos

Suarez AV, Holway DA, Case TJ (2001) Patterns of spread in biological invasions dominated by long-distance jump dispersal: insights from Argentine ants. Proc Natl Acad Sci 98:1095–100

Titus ESG (1905) Report on the "New Orleans" ant (Iridomyrmex humilis Mayr). US Bur Entomol Bull 52:79–84

Uekoetter F (2006) The green and the brown. A History of Conservation in Nazi Germany. Cambridge University Press, New York

Vayssière P (1922) La présence de la Fourmi d' Argentine en France. Revue scientifique 34:128–130

Vogel V, Pedersen JS, Giraud T, Krieger MJB, Keller L (2010) The worldwide expansion of the Argentine ant: global expansion of the Argentine ant. Divers Distrib 16(1):170–186

Ward DF, Anderson DP, Barron MC (2006) Using spatially explicit surveillance models to provide confidence in the eradication of an invasive ant. Scientific Reports 6

Welzel KF, Choe DH (2016) Development of a pheromone-assisted baiting technique for Argentine ants (Hymenoptera: Formicidae). J Econ Entomol 1–7

Wetterer JK, Wild AL, Suarez AV, Roura-Pascual N, Espadaler X (2009) Worldwide spread of the Argentine ant, Linepithema Humile (Hymenoptera: Formicidae). Myrmecological News 12:187–194

Woodworth CW (1910) The control of the Argentine ant. Bulletin 207:53–82. University of California Publications, Berkeley

Zina V, Garcia A, Valente C, Branco M, Franco JC (2015) Ant species (Hymenoptera, Formicidae) associated to eucalyptus plantations in Portugal. Artículos y notas 11

Chapter 5
Herpetological History of the Balearic Islands: When Aliens Conquered These Islands and What to Do Next

Iolanda Silva-Rocha, Elba Montes, Daniele Salvi, Neftalí Sillero, José A. Mateo, Enrique Ayllón, Juan M. Pleguezuelos and Miguel A. Carretero

Abstract Balearic herpetofauna represents a paradigmatic case of multiple biological invasions within the Mediterranean Basin, with a much higher number of alien amphibians (i.e. frogs and toads) and reptiles (i.e. lizards, snakes and turtles/tortoises) than native. The paleogeography of the Balearic Islands, located on Western Mediterranean between Spain and Sardinia, is complex, comprehending an ancient split from the continent during the late Miocene and Pleistocene climatic fluctuations connecting and disconnecting islands from one another (but not them to

I. Silva-Rocha (✉) · M. A. Carretero
CIBIO/InBIO—Research Centre in Biodiversity and Genetic Resources, Universidade do Porto, Porto, Portugal
e-mail: irocha@cibio.up.pt

I. Silva-Rocha
FCUP, Faculdade de Ciências da Universidade do Porto, Porto, Portugal

N. Sillero
CICGE: Centro de Investigação em Ciências Geo-Espaciais, Faculdade de Ciências da Universidade do Porto, Vila Nova de Gaia, Portugal

J. A. Mateo
Black Market, Cl. Paraires 23, 07001 Palma de Mallorca, Spain

E. Montes
Department of Zoology, Faculty of Biological Sciencies, Valencia University, c/. Dr. Moliner 50 Burjassot, 46100 Valencia, Spain

E. Montes
Environment Department, Ibiza Island Council, Av. España, 49, 07800 Ibiza, Spain

E. Ayllón
Asociacion Herpetologica Espanola, Espanola, Spain

J. M. Pleguezuelos
Department of Zoology, Faculty of Sciences, Granada University, 18071 Granada, Spain

D. Salvi
Department of Health, Life and Environmental Sciences, University of L'Aquila, Via Vetoio, 67100 Coppito, L'Aquila, Italy

A. I. Queiroz and S. Pooley (eds.), *Histories of Bioinvasions in the Mediterranean*, Environmental History 8, https://doi.org/10.1007/978-3-319-74986-0_5

the continent) that eventually re-shaped the archipelago's biota. The archipelago has been also influenced by humans since the Neolithic, being a cross-road for alien biota between North Africa and Southern Europe, which caused range regressions and extinctions in the native herpetofauna, nowadays restricted to one amphibian and two reptiles. During the last century, tourism development, the pet trade, and cargo transport of ornamental plants have produced a new wave of biological invasions. Recently introduced snakes are of particular concern, since the effect of predation may seriously threaten the remaining native reptiles in the main islands and endemic subspecies in surrounding islets. Balearic people have a negative social perception of such snakes, mainly due to the lack of familiarity with snakes among islanders but also to the herpetophobic attitude of many Mediterranean cultures. Here we review the herpetological invasions in the Balearic Archipelago and their impacts. We further discuss the on-going management actions on alien reptiles in this archipelago, namely the control of invasive snakes in Ibiza involving monitoring, trapping, environmental education and promotion of social participation.

Keywords Alien reptiles · Balearic Archipelago · Cabrera · Formentera Humans · Ibiza · Invasive species · Islands · Gimnesic Islands Lizards · Mallorca · Menorca · Snakes · Nursery trade · Pityusic Islands Tourism

5.1 The "Deep History" of the Balearics

The Mediterranean Basin is one of the richest and most complex regions on Earth on several facets, namely geological, biological and cultural. Nowadays, it is considered one of the global biodiversity hotspots (Myers et al. 2000). The region is the result of a complex history encompassing profound geological processes such as tectonic dynamics, climatic shifts, and biogeographical interchanges (Blondel et al. 2010). All these processes have contributed to the development of a particularly rich biota. The most influential geologic events were: (1) the Messinian Salinity Crisis (from 5.9 to 5.3 Myr ago) during which the Mediterranean Sea went into a nearly complete desiccation, which allowed substantial faunal interchanges between land masses; and (2) the Pleistocene climatic oscillations between glacial and interglacial episodes (especially in the last Myr), which played a key role in molding and shaping present-day biota, including species range shifts and extinctions (Blondel et al. 2010).

The Balearic Archipelago represents an ancient and isolated system within the Mediterranean. The last separation of this archipelago from the mainland dates back to the end of Messinian salinity crisis, about 5.3 Myr ago. At that time, two sub-archipelagos arose and never contacted each other again: the western group or Pityusic Islands comprising Ibiza, Formentera and nearly 60 smaller surrounding islets; and the eastern group or Gymnesic Islands comprising Mallorca, Menorca and about 30 medium and smaller surrounding islets.

The Messinian represented the last massive terrestrial exchange of fauna between Europe and Africa and among the Balearic Islands as a whole. Since then, faunal interchanges have only taken place within each sub-archipelago islands and islets. Thus, terrestrial species evolved there in isolation during the Pliocene under mild subtropical conditions (Bover et al. 2008). Subsequent climatic changes occurred in the Late Pliocene with a shift towards a Mediterranean climate regime characterized by long summer drought (Jiménez-Moreno et al. 2010), followed by substantial oscillations in temperature and humidity during the last 2.5 million years due to glaciations (Bover et al. 2008).

Pleistocene glaciations not only brought about radical changes in climate compared to the Pliocene, but also promoted temporary connections among islands within the same sub-archipelago. A number of vertebrate groups became extinct in that period (e.g. vipers; Bailon et al. 2010). The surviving fauna underwent separate evolutionary and ecological pathways to those of their continental relatives. In particular, there was a shift from a primarily insectivorous lizard fauna to a lizard fauna with a mainly plant-based diet (Pérez-Mellado and Corti 1993), while small species shifted to gigantism (Novosolov et al. 2013), namely many populations of *Podarcis lilfordi* present in small islets (Brown and Perez-Mellado 1994).

During glaciations, the Pityusic Islands (nowadays separated by a narrow and shallow channel) were connected, therefore sharing their fauna. In the Gimnesic Islands Mallorca and Menorca, a deeper sea channel prevented their connection during the Tertiary, but not during the Pleistocene glacial phase, when they were connected several times allowing interchanges of some faunal elements. Thus, current Mallorcan and Menorcan faunas are a mixture of shared and unique elements. Last but not least, the small islets that were connected to the main islands until the end of the glaciations became isolated due to the post-glacial marine transgression.

Nowadays, the Balearic Islands have a total area of 5040 km^2 and 1428 km of coastline, lying from 80 to 300 km east of the Iberian mainland (Fig. 5.1). Mallorca is the largest island, with very diverse ecosystems and landscapes, which can be divided into three distinct districts: the Serra Tramuntana, an elevated mountain range that shapes the northern coast of the island, with the highest peak at 1445 m; the Serres de Llevant, comprising small rounded hills that mark the landscape of white-sand coves and green pine forest; and a flatland area located between the two mountain ranges and characterized by rural landscapes.

Menorca Island is predominantly flat with subtle rolling hills, and is sheltered by a small mountain range along the northern coast (347 m above sea level). Therefore, in the north, the island is more rugged, with a jagged coastline, while the south is flatter with cliffs, water-carved gullies and white-sand coves nestled amid pine trees.

Cabrera is located south of Mallorca and it was officially declared a Marine and Land National Park in 1991. It is an area with a great wealth of biodiversity, featuring a multitude of endemic invertebrates, and 150 migratory bird species stop over on the island during both their spring and autumn migrations.

Ibiza has a uniform landscape and moderate relief, the highest point being Sa Talaiassa (486 m). Habitats include marshes, dunes, saltflats (*salines*), cliffs and

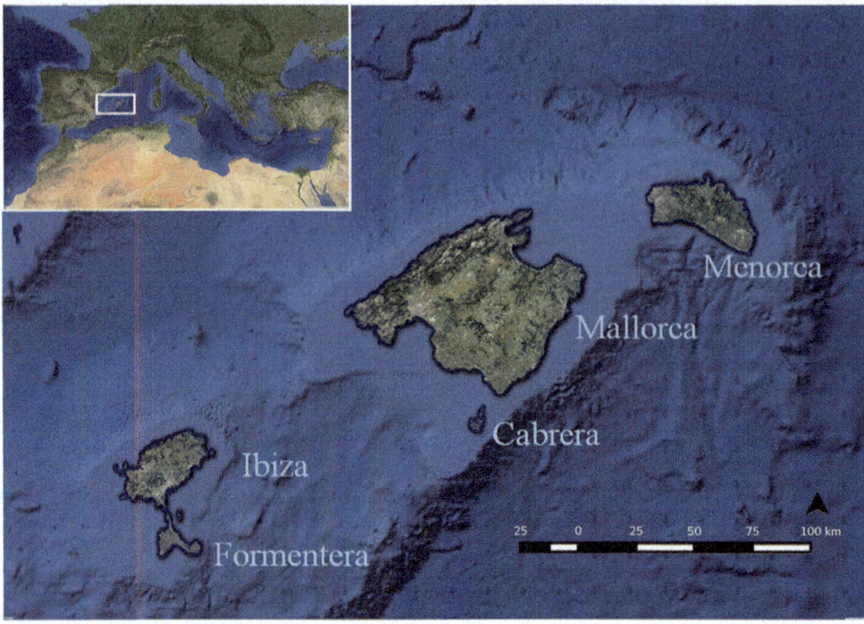

Fig. 5.1 The Balearic Archipelago and its geographical position in the Mediterranean Basin

beaches. The flat alluvial plain is formed by Quaternary calcareous deposits and two enclosing wooded hill ranges. This island is part of a Special Protection Area for Birds and The European Economic Community's Natura 2000 Network, and it is listed on the Ramsar List of Wetlands of International Importance.

Formentera is 19 km long and located 6 km south of Ibiza. It has a narrow shape that gives it a long stretch of coast (82 km) in proportion to its surface area (83 km^2). The landscape is composed of Mediterranean shrubs in dune areas as well as pine and savin (juniper) forests.

The general climate on the Balearic Islands is Mediterranean, with a hot and dry summer, mild winter and the scarce rains concentrated in autumn. During the winter the minimum temperature is 10 °C and in July temperatures can reach 35 °C (30-years standard meteorological averages; AEMET and IM 2011).

5.2 The Balearics and the Bioinvasions

5.2.1 Invasions on Islands

Islands harbour peculiar ecosystems, due to their isolation from the continents. The number and organismal taxa that can naturally reach and colonize island ecosystems is limited, the number of populations is fewer, and the total population size is smaller than in continents. Island communities are, therefore, impoverished and unbalanced in comparison with continental ones. They experience less biotic interactions and opportunities to evolve independently and differently compared to their mainland relatives, as a result of having less competitors, predators and parasites (Whittaker and Fernández-Palacios 2007). These factors create a higher sensitivity to disturbances and vulnerability to biodiversity extinction, compared to continental ecosystems (Vitousek et al. 1996; Reaser et al. 2007; Ficetola and Padoa-Schioppa 2009).

The theory of island biogeography states that species richness on islands results from the equilibrium between colonization (as a function of distance to continent or to other islands), speciation (as a function of time) and extinction (as a function of island size and habitat heterogeneity) (MacArthur and Wilson 1967). However, humans have shifted this balance on some insular ecosystems by increasing colonisation rates by alien (introduced) species; hence islands where human impact is very high may end up with more alien species.

Biological invasions on islands are even more dangerous when the presence of native predators is low, competitors are absent, and the native species are vulnerable to the aliens or to its parasites (Whittaker and Fernández-Palacios 2007; Álvarez et al. 2010). In addition to that, islands normally have a more benign environment (i.e. more stable abiotic conditions, such as milder temperatures) and lower habitat diversity, which facilitate the establishment of invasive species (Gimeno et al. 2006).

5.2.2 A Long History of Successive Herpetoinvasions

The Balearic Archipelago is one of the most isolated in the Mediterranean (i.e. farther from mainland) and because of that, its exploration and discovery by humans took place later than in other archipelagos. The first records of humans on the Balearics date from the last third of the third millennium BCE (Alcover et al. 2001; Ramis et al. 2001; Bover et al. 2008). The first settlers were hunters and farmers, also herding goats, cattle, sheep and pigs, brought from the mainland (Ayuso 2001).

A long history of migratory waves followed this period, as in other parts of the western Mediterranean with high demographic growth, the discovery of new uninhabited regions and the development of maritime contacts setting the Balearic Islands in the route of several civilizations (Iberians, Nuragics (from Sardinia), Phoenicians, Greeks, Carthaginians, Romans, Byzantines, Arabs and Berbers). All these civilizations introduced alien species passively (with cargo) or voluntarily (as food source or for religious purposes). The Romans may have introduced snakes to frighten their enemies, and—associated to Aesculapius cult—to give them good luck; tortoises were introduced by several civilizations (i.e. Phoenicians, Greeks and Arabs) to serve as food and pets. The Nuragics are thought to have introduced the Balearic green toad *Bufotes balearicus* for religious purposes (Lewthwaite 1985; Chapman 1990).

During the twentieth century, an intensification of human activity occurred in the Balearic Islands, especially on Mallorca, due to massive tourism development and to the pet trade and the nursery trade (trade of live plants for ornamental purposes). These activities have resulted in range retractions in the native herpetofauna, which is nowadays composed of only one toad, the Mallorcan midwife toad (*Alytes muletensis*) and two lizards, the Lilford's wall lizard and the Ibiza wall lizard (*P. lilfordi* and *P. pityusensis,* respectively) (Pinya and Carretero 2011).

The number of species introduced to the main islands increased strikingly, with Mallorca having the highest number of assumed aliens, with 21 species, followed by Menorca with 15, both Formentera and Ibiza with eight species, and Cabrera with two introduced species. The number of observations per year has been increasing for all the alien species in the Balearic Islands, suggesting they are expanding both in terms of range and population size.

5.2.2.1 Herpetofauna Before Invasions

On Mallorca, fossils dating from the Pliocene Epoch include a giant lacertid lizard (*Maioricalacerta refelensis*), a skink (*Chalcides* sp.), and a glass lizard (*Dopasia* sp.) (Bailon et al. 2014; Bover et al. 2014). In Menorca, two post-Messinian faunas occurred: one during the Early and Middle Pliocene (Quintana 1998; Alcover et al. 1999) and the other during the Late Pliocene or Early Pleistocene (Bover et al. 2008). The first one was composed of a giant tortoise (Bate 1914), a lizard (*Podarcis* sp., likely an ancestor of Lilfordi's wall lizard, Bailón 2004), geckos (Gekkota undetermined), an amphisbaenian (Blanidae, Garcia-Porta et al. 2002; Bailon et al. 2005), and three snakes including a colubrid and two viperids (*Vipera natientis* and *Vipera* sp., Bailon et al. 2005). The Late Pliocene/Early Pleistocene records report a lizard (*Podarcis* sp.) and two amphibians (Mallorca midwife toad *A. muletensis*, Barbadillo 1987, and *Discoglossus* sp., Alcover et al. 1981; Quintana 1998).

Regarding the Pityusic Islands, three different fossil assemblages have been recorded in three different periods: during the Late Miocene/Early Pliocene, a lizard and a tortoise (Moya-Solà et al. 1984, 1999), during the Late Pliocene, a middle sized tortoise (*Cheirogaster* sp., Bour 1985) and a lizard (*Podarcis* sp, Kotsakis 1981); and in the deposits of the Late Pleistocene, a lizard species (*P. pityusensis*, Boscá 1883).

5.2.2.2 Prehistoric and Ancient Introductions (Older Than 5th Century)

The lack of older fossils in the palaeontological record on the Balearic Islands suggests ancient introductions of one toad (Balearic green toad), two turtles (Hermann's tortoise, *Testudo hermanni*; and the European pond turtle, *Emys orbicularis*), two geckos (Mediterranean house gecko, *Hemidactylus turcicus*; and the Moorish gecko, *Tarentola mauritanica*) and one snake (Ladder snake, *Rhinechis*

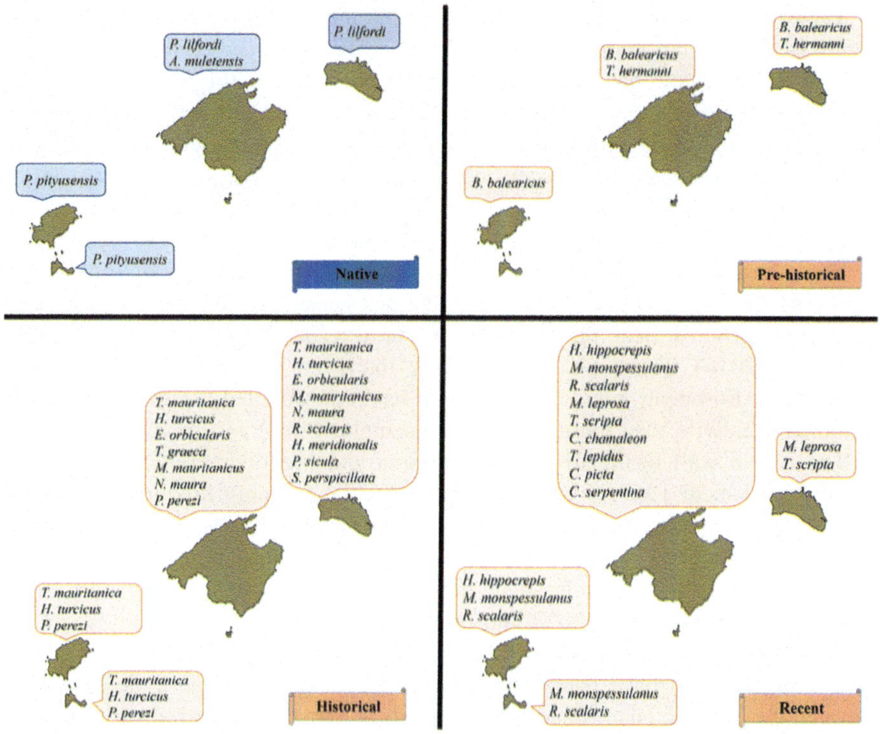

Fig. 5.2 Herpetoinvasions in the Balearic Archipelago, through time, by island

scalaris) (Fig. 5.2). The exact timing and the arrival pathway of the Balearic green toad is unknown. Both the Hermann's tortoise and the European pond turtle were introduced as a food source from the Western Mediterranean region (Fritz et al. 2006). Both gecko species were presumably introduced by passive transportation, with apparently no impact on native communities. No negative impacts on native species or ecosystems are known for these species, with the exception of the snake, which as a predator has a high risk of impact (see below).

5.2.2.3 Historical Introductions (5th Century–18th Century)

The lack of fossils combined with the low genetic differentiation observed, suggests the introduction of two frogs and nine reptiles (i.e. lizards, snakes and turtles) during historical times (Fig. 5.2): the Mediterranean tree frog *Hyla meridionalis* and Perez' frog *Pelophylax perezi*; two turtles, the European pond turtle *Emys orbicularis* and the spur-thighed tortoise *Testudo graeca*; two lizards, the Moroccan rock lizard *Scelarcis perspicillata* and the Italian wall lizard *Podarcis sicula*; and two snakes, the false smooth snake *Macroprotodon mauritanicus* and the viperine snake *Natrix maura*.

The Mediterranean tree frog could have been introduced from the south-western region of the Iberian Peninsula (Silva-Rocha 2012), either voluntarily or accidentally. It is suggested that this frog may have been unwittingly transported to the Canary Islands by the first settlers in cargos such as cattle and seeds (Recuero et al. 2007); this could also be the case in the Balearics. On the other hand, Perez's frog is suspected to have been introduced to control insects (Pinya and Carretero 2011). The latter could have had competitive interaction with the native Mallorcan midwife toad (Román 2004) and predated on this same species (Alcover et al. 1984).

Regarding the chelonians (turtles and tortoises), the pathway of introduction both for the spur-thighed tortoise and the European pond turtle is considered, similarly to pre-historical introductions, as a food resource (Velo-Antón et al. 2011; Cau et al. 2016). These two species have likely no impacts on native species, though the European pond turtle could have contributed to the decrease of Mallorcan midwife toad populations (Pleguezuelos 2004).

The Italian wall lizard arrived in Mallorca and Menorca from Sicily-Sardinia (Greca and Sacchi 1957; Silva-Rocha et al. 2012), probably involuntarily, as this species is known to live close to human populations, which increases the chances of it being introduced by passive transport. From our experience in other Mediterranean islands, this is an aggressive lizard, able to compete with native species (Silva-Rocha et al. 2012 and references within).

As to the two snakes introduced, their arrival pathways are not clear. A passive transportation could have happened in ships due to the maritime trade established between the Balearic Islands and the mainland, as may have occurred with European pond turtle in Corsica and Sardinia (Pedall et al. 2011). On the other hand, they could have been introduced for religious purposes, as referred to above. The origin of the Balearic populations of the false smooth snake is likely Tunisia, while the viperine snake populations originated from Europe, either from the Iberian Peninsula or France (Pleguezuelos 2004; Guicking et al. 2008; Silva-Rocha et al. 2015). The impacts of the false smooth snake extends beyond the native herpetofauna (i.e. Lilford's wall lizard), as they are also able to predate birds and mammals, which can cause a substantial disturbance in these insular ecological communities.

5.2.2.4 Recent Introductions (19th Century to Nowadays)

Over the past two centuries, a total of ten species were introduced in the Balearic Islands, as a consequence of increasing tourism and trade in this region (Fig. 5.2): one toad (the common midwife toad, *Alytes obstetricans*); three chelonians (the painted turtle *Chrysemys picta*, the Mediterranean turtle *Mauremys leprosa* and the pond slider *Trachemys scripta*); four lizards (the common chameleon *Chamaleo chameleon*, the occelated lizard *Timon lepidus*, the Ibiza wall lizard *Podarcis pityusensis* and the Algerian Psammodromus *Psammodromus algirus*); and three snakes (the horseshoe whip snake *Hemorrhois hippocrepis*, the Montpellier snake *Malpolon monspessulanus* and the ladder snake, the latter on islands other than Menorca where it had already been introduced).

The common midwife toad was found in Menorca in 2007, and its introduction is associated with the trade in exotic live plants for ornamental purposes (Carrera and Pons 2010); however, it has not been confirmed later than 2007. The geographical origin is unknown.

The chelonians have been introduced mostly due to the pet trade. It is important to notice that the pond slider is an invasive turtle species native to the eastern USA and adjacent areas of Mexico and introduced worldwide, which can impact on native ecosystems by predation and pathogen transmission. The common snapping turtle can also impact on local aquatic fauna, such as birds and amphibians. For the rest of the chelonians, the impacts are not known.

As to the lizards, an introduction due to the pet trade is assumed for the common chameleon, whereas for the ocellated lizard either pet trade or passive introductions through olive trees (Rivera et al. 2011) are equally possible. The origin of the former is not known, while the latter is suggested to come from southeastern Spain, based on genetic and phenotypic data (Mateo et al. 2011; Silva-Rocha 2012).

Three large snake species were introduced during the last decade: the horseshoe whip snake and the Montpellier snake during 2003 in Ibiza (the last not seen since 2010) and Mallorca; the ladder snake during 2003 in Ibiza, 2004 in Mallorca, and 2006 in Formentera (Álvarez et al. 2010; Mateo et al. 2011). In all cases, the introduction pathway is attributed to the nursery trade, through live plants for ornamental purposes (Kraus 2009; Álvarez et al. 2010).

Individuals from all the three species were found primarily in trunks of olive trees in the nursery centres of Capdepera (Mallorca) and Sant Llorenç de Balàfia (Ibiza). The origin of the first two species is thought to be the southern Iberian Peninsula. The origin of the ladder snakes is uncertain but its alien status in the Balearics is beyond doubt (Carretero and Silva-Rocha 2015). Similarly to the horseshoe whip snake and the Montpellier snake, it can be found near olive trees translocated by the importation industry (Álvarez et al. 2010). As reported by the same authors, at least one company from Spain is involved in the olive trees trade, while no importations from France and Portugal (countries where the species also ranges) are known.

The olive trade business has increased through the years in the Balearic Islands, as a consequence of a trend in garden decorations. Reptiles use the olives as a refuge and there is evidence of several species living together in old trees (Rivera et al. 2011; Graziani et al. 2006; pers. obs. of the authors), which increases the probability of introductions of multiple species in many sites. Thus, the olive trade represents an important pathway for the introduction of reptiles throughout this region, and therefore must be considered when devising prevention and conservation measures.

5.2.2.5 The Future

Ecological niche models calculated with scenarios of climate change (i.e. extreme temperature/humidity and moderate temperature/humidity), forecast an increased suitability in most of the island territory for the alien snakes (with the exception of the false smooth snake) (Silva-Rocha et al. 2015). Indeed, recent studies suggested that climate change is already increasing the activity period and benefiting the survival and reproduction of Montpellier snake and horseshoe whip snake in the Iberian Peninsula (Moreno-Rueda and Pleguezuelos 2009; Zamora-Camacho et al. 2010). Therefore, the population size and spread of some of these Mediterranean snake species is expected to increase, with an associated increase of their impact on native biota.

Unfortunately, a projection in the future was not performed for other introduced species due to lack of records. Nevertheless, in those species coming from warm regions with heterogeneous habitats, it may be expected they will also benefit from climate change.

5.3 Italian Wall Lizard: 'Let It Be' Does not Work

The Italian wall lizard *P. sicula* represents a case where nothing was done to prevent the expansion of an introduced species (Silva-Rocha et al. 2014), which leads to its wide dispersion across islands. Analysing what was (not) done with this species in other areas, allows us to forecast the trends of this introduced species in the Balearics. In addition, the analysis of what occurs on the Balearics allows us to check those trends and to understand them in an island environment.

The species is native to the Italian Peninsula and Sicily and has been introduced in several places, including islands and coastal areas of the Eastern Adriatic, Iberian Peninsula, Turkey, North Africa and the United States. This species often inhabits human modified areas and uses ornamental plants as refuges, being very prone to accidental transportation by man. This is likely the main pathway for its expansion.

Several ecological and behaviour characteristics make the Italian wall lizard a concerning invader, able to adapt easily to a new environment and to displace native lizard species. Behavioural interference was reported with native lizards such as the Dalmatian wall lizard *Podarcis melisellensis* which has gone extinct in some small islets of the Adriatic Sea after the introduction of the Italian wall lizard by competitive exclusion, namely of basking sites (Nevo et al. 1972). Hybridization between the Italian wall lizard and other *Podarcis* species has been also reported (the Tyrrhenian wall lizard *P. tiliguerta*, the Aeollian wall lizard *P. raffonei* and the Sicilian wall lizard *P. wagleriana*; (Capula 1993, 2002; Capula et al. 2002) and may be particularly harmful for native *Podarcis* populations in small islands. In California, the Italian wall lizard *Italian wall lizard* shows a diet which overlaps with native lizards (the Western fence lizard *Sceloporus occidentalis* and Southern alligator lizard *Elgaria miulticarinata*). It may also have an impact on native lizards due to predation (Kirschbaum and Pauly 2016) and competition (e.g. in Kansas with the great plains skink *Eumeces obsoletus*, Oliverio et al. 2001). Impacts on the invertebrate community are unknown, but likely.

In Menorca, the Italian wall lizard has been present at least since the Middle Ages (Alcover et al. 1981), and nowadays is all over the island. This case exemplifies the kind of population dynamics of this species following its introduction in absence of measures to prevent its spread. A species with rapid adaptation/acclimation, like the Italian wall lizards, can easily expand all over an island. The same probably happened earlier in Sardinia, where the species is now only absent at the highest altitudes.

The extinction of Lilford's wall lizard in Menorca was probably not due to this species but to the earlier introduction of alien predators (Mayol 1985; Pérez-Mellado 1998). However, a possible arrival of the Italian wall lizard to the islets surrounding the main islands represents a high risk to the survival of microinsular populations of Lilfordi's wall lizard, as the replacement of native populations following the arrival of the Italian wall lizard have already been documented in the case of the Dalmatian wall lizard in Adriatic islets (Nevo et al. 1972). The Italian Wall lizard has already reached Ses Mones, one islet off-Menorca where Lilfordi's wall lizard is still present (Speybroeck et al. 2010; van den Berg and Zawadzki 2010).

Thus, preventive measures are a priority over other actions to avoid translocations of the Italian wall lizard from Menorca to surrounding islets, where endemic Lilfordi's wall lizard is present. This could be implemented by inspections of cargo and boats, and a campaign of public awareness for both islanders and tourists.

5.4 Ibiza and Formentera: When Social Perception Helps Conservation

The Pityusic islands, Ibiza and Formentera, remained less affected by introductions until the beginning of the twenty-first century. Because of this, the recent invasion process still can be seen in progress. Therefore, these two islands offer the opportunity to study the first stages of the invasion process as well as the effectiveness of conservation measures adopted in an appropriate time frame, i.e. soon after the introduction, with the involvement of local people.

At some point in the Pleistocene, the fauna of the Pityusic Islands was affected by a crucial and determining circumstance within the Balearic Archipelago: a massive extinction happened (Bailon et al. 2010), with no correspondence on the rest of the Mediterranean Islands (Alcover and McMinn 1993). There is only evidence of two terrestrial survivors, lizards and snails, which used to live in rock fissures. This big extinction might have been caused by a catastrophic event, such as a volcanic process—for instance, the Columbretes Archipelago, 100 km to the northwest of Ibiza had volcanic activity until 300,000 years ago. Later, during Pleistocene glaciations, as a consequence of sea level regression Ibiza and Formentera formed a single large island of more than 700 km^2, the biggest island in the Mediterranean without terrestrial mammals. During that period, paleontological records show that the number of bird species inhabiting these two islands was rather high, since they are capable of overseas colonization and they had little competition for resources (i.e. food) and supposedly little predation pressure. Additionally, fossil raptor pellets show a large number of lizards, as well as fish (Alcover and McMinn 1993).

5.4.1 Human and Reptile Invasions

When humans arrived for the first time in the Pityusic Islands, around the end of the third millennium BCE, their dietary options would have been similar to those first human settlers found in the Polynesian islands; essentially terrestrial and sea birds, fishes, and sea mollusks. Lizards and terrestrial snails were likely too small to constitute a food resource. Hence, the existing fauna provided poor resources for sustaining a human population (Alcover and McMinn 1993).

Initially, humans brought only the Moorish Gecko (Rato et al. 2010; Pinya and Carretero 2011), the Mediterranean house gecko (Rato et al. 2010; Pinya and Carretero 2011) and the spur-thighed tortoise (Mateo et al. 2011) to the Pityusic islands. Remarkably no other reptiles, particularly no snakes, were introduced in the islands in that period (Pinya and Carretero 2011; Mateo 2015). The Phoenician newcomers ruled the entire island, placing it under the protection of the Canaanite god Bes (Mateo and Ayllón 2012). He was represented with a snake on one hand and a knife on the other, since he protected the family and the home from any evil, especially venomous snakes (Guirand 1960; Fig. 5.3). Consequently, Ibiza became the Bes' Island, its name probably coming from there (Ebusus; Mateo 2015). Almost all the coins minted in Ibiza during this Punic period showed the image of the snake-killer god (Fernández 1975; Fig. 5.3). There is a similar case: according to the tradition of the 4th century (CE), the absence of snakes in Ireland—an animal associated to the devil by Christianity—is attributed to the protection of Saint Patrick.

This aversion towards snakes and their absence on the island was adopted as an identity symbol by the Ibizan inhabitants. In contrast, on those Mediterranean locations where snakes were common and therefore considered as beneficial or neutral, their peoples worshiped gods related to snakes (e.g. Aesculapius, Astarté, Isis, Mitra; Fernández 1975; Velázquez Brieva 2007). Transport of snakes among those places was something usual and a high number of acclimatization and some naturalizations (i.e. species were able to reproduce) took place, some of them lasting until our days in most of the large and medium sized Mediterranean islands (Mateo et al. 2011).

In the 2nd century BCE, when Carthage was defeated, the Ibizan population surrendered to the Romans. They were able to keep the Canaanite dialect Punic, and the worship of the god Bes. Later, in the first century CE, when Phoenicians and Romans still coexisted unmixed in Ibiza, two Latin authors wrote about the island and the mysterious lack of snakes. The geographer Pomponius Mela, in his work *De Chorographia* (Mela, around 43 D.A.), stated:

Fig. 5.3 Coins and Bas-relief with god Bes holding the snake and the knife. Clapboard from the Ptolemaic necropolis of Saqqara (3rd century BCE). *Source* Allard Pierson Museum, Amsterdam

> (…) (Ebusus) lacks harmful or wild animals, to the point that it does not breed them nor tolerate those brought there. In front of it there is the Colubraria, that comes to my mind because, being very abundant in many malefic genus of snakes, it is completely uninhabitable; however, if someone enters a place previously surrounded by Ebusitan soil, it becomes a pleasant place without danger, because those same snakes that otherwise attack all they find so often, they flee away frightened by the appearance of the dust.

Thirty years later, Pliny the Elder, naturalist and historian, probably copying the previous author, included a similar reference in his work *Naturalis Historia* (Plinius the Elder, 77):

The soil of Ebusus banishes snakes, that from Colubraria begets them and it is very harmful for everyone, unless one brings with himself Ebusitan soil; ...

Both Latin authors assigned a magical character to the unusual absence of ophidians, as the Pityusic Islands were the largest Mediterranean Islands without snakes. By saying this, they were indirectly suggesting the existence of Ibizan soil trade to the city of Rome, as an element with occulted properties (Salazar de Mendonza 1770).

If we compare the number of acclimated species in Ibiza and Formentera before 1975 with other Mediterranean Islands, such as Mallorca or Menorca (Mateo et al.

2011), the Pityusic Islands have suffered much less historical introductions than the rest. It was only at the beginning of the 21st century that three snake species were introduced in Ibiza and Formentera. The change in agricultural practices occurred during this period may have indirectly prepared the ground for such introductions. After Spain became a member of the European Union, in 1986, the more competitive market forced Spanish farmers to transform their traditional agriculture into a mechanized one. Hence, between 1996 and 2005, more than 900,000 olive trees were ripped out ending up as firewood or used for gardening (Fortuny Santos 2002) in order to gain more agricultural fields. The massive trade of these olive trees between the south of the Iberian Peninsula and the Pityusic Islands provoked the first introduction of snakes in these islands.

The Montpellier snake was introduced to Ibiza, but it has not been sighted since 2010, and we therefore presume it did not establish and disappeared from the island.

The Ladder snake has a low capture rate in Ibiza, but is the most abundant snake in Formentera, having invaded around 350 ha of the island by 2016 (4.2% of the territory; *unpub.* data of the Balearic Government).

The Horseshoe whip snake was introduced to Ibiza and to Formentera, but seems to have disappeared from the latter, given that not a single specimen was captured during 2016. It is the most widespread in Ibiza, increasing the occupancy area quickly −148 km^2 occupied by the species by 2015 (27% of the island area).

Even though not all the traits of the natural history of the last two introduced snakes have been studied, preliminary analysis and observations indicate that both have successfully naturalized.

5.4.2 *Impacts of Snake Introductions on Biodiversity*

Snakes are efficient predators, and commonly absent or poorly represented in island native fauna. Therefore, their introduction on islands often dramatically impacts on the native fauna (Rodda et al. 1996). The only terrestrial vertebrate endemic to the Pityusic Islands is seriously endangered by these alien snakes.

Information on prey type and size consumed by alien predators is a key to our understanding of the success of invasive species (Herrel et al. 2008), and helps in evaluating the threat to the native organisms (Salo et al. 2007). The source populations of both the horseshoe whip snake and the ladder snake are well studied, and we compared body size, diet and reproduction between source and invasive populations.

The horseshoe whip snake in Ibiza has shifted its diet towards lizards, with the native Ibiza wall lizard, representing 55.4% of the observed prey (Hinckley et al. 2017), whilst in the diet of horseshoe whip snake in the south of the Iberian Peninsula (the source area of introduced populations) lizards represent only 24.2% of prey items (Pleguezuelos and Moreno 1990). Moreover, the high body condition (mass related to length) and the larger body size of individuals found in Ibiza offers compelling proof of the good ecological conditions experienced by the horseshoe whip snake on this island. Remarkably, the record for the maximum body length and weight for the species in the Iberian-Balearic region (1830 mm and 1440 g) was found in Ibiza, surpassing those previously described by 105% and 213%, respectively (Hinckley et al. 2017). This evidence suggests that this species grows longer and more robust in Ibiza than in its continental range (see Table 1 in Gil and Pleguezuelos 2001). This shows a rapid functional response (likely phenotypic plasticity) of this snake's size and shape to the new biotic and abiotic situation on the island.

A similar dietary shift to the one observed in the horseshoe whip snake in Ibiza is probably occurring with the ladder snake in Formentera. Continental populations of this latter species show an endotherm-based diet (i.e. small mammals and birds) with a negligible percentage of reptiles (Pleguezuelos et al. 2007). In contrast, the preliminary study of the diet in Formentera specimens shows that more than half of their prey is constituted by the endemic Ibiza wall lizard (Samuel Pinya, pers. com.).

The secondary prey type of these ladder snakes in frequency, and primary in biomass, are micromammals, which are aliens in Ibiza, abundant, and of no conservation concern. However, these alien prey of early arrival to the island provide food resources for later aliens like the horseshoe whip snake, in a process of invasional meltdown (Simberloff and Von Holle 1999). This situation will probably have negative repercussions on the Ibiza wall lizard by a hyperpredation process (Bull and Courchamp 2009; Hinckley et al. 2017).

Snakes may also be influencing the diet of other insular, native predators, with shared prey, such as the Barn owl (*Tyto alba*). Reproduction may be changing also to be more frequent (Shine 2003), with larger clutch size (Seigel and Ford 1987), larger hatchlings (Shine 1981) or earlier maturing individuals (Parker and Plummer 1987).

Populations of Ibiza wall lizard from islets surrounding Ibiza and Formentera are also potentially menaced by the arrival of snakes. A total of 39 islets hold 23 endemic subspecies (Salvador 2015; but see Rodríguez et al. 2013). Regardless their taxonomic status, most of these microinsular populations represent unique adaptations resulting from local ecological conditions and isolation, with very small sizes and undergoing strong demographic fluctuations. The arrival of snakes in these islets may produce a catastrophic and irrecoverable event for lizard populations as they would be the only available prey. On the other hand, the probability of such arrivals are not so unlikely as these islets are very close to the main islands' coast (1.17 km on average); six snakes have been observed swimming in the sea, and a snake skin was already found in one islet (unpub. data of the authors).

Besides Ibiza wall lizard, which is listed as Near Threatened (NT) (IUCN categories and criteria applied to regional scale of the Balearic Islands; Viada et al. 2006), there are 57 native birds that nest on the Pityusic islands whose adults or pulls are within the prey size range of these snakes. Among them, there are five species listed as Vulnerable (the common quail *Coturnix coturnix*, the little ringed plover *Charadrius dubius*, the Kentish plover *C. alexandrinus*, the European turtle dove *Streptopelia turtur* and the Woodchat shrike *Lanius senator*), two as

Endangered (the Cory's shearwater *Calonectris diomedea* and the barn owl *Tyto alba*), one as Critically Endangered (the Balearic shearwater *Puffinus mauritanicus*), and six as Near Threatened. Two breeding birds are Balearic endemics: the Balearic Shearwater, that nests in surrounding islets of Ibiza and in the main island of Formentera, and the Balearic Warbler (*Sylvia balearica*), occurring in both islands.

5.4.3 Impacts of Snake Invasions on Humans

Native species of Ibiza have a great significance not only from a biological point of view, but also for the inhabitants of this island. The endemic Ibiza Wall Lizard is considered a symbol of the Pityusic Islands (Dappen et al. 2013). Ibizan and Formenteran people are proud of their lizards and their images are present everywhere: on clothes, jewelry, lizard-shaped stickers on the cars, in decorative elements such as lamps, wall figures, mirrors, etc. Magazines and pamphlets frequently exhibit the lizard, and several companies use it in their logos. It is used as a tourist attraction too, and any tourist information about the island will have the symbol on it (Fig. 5.4a and b).

What does this symbol mean? Why is it so easy to relate Ibiza and Formentera with it, instead of any other animal or plant? To answer these questions, we have to remember that the first humans that arrived to the Pityusic Islands found only lizards within the community of ground dwelling vertebrates of these islands (Bover et al. 2008). The sole terrestrial vertebrate on such area was this attractive and harmless reptile. It is the only vertebrate exclusive to Ibiza and Formentera. Different civilizations introduced other vertebrates like genets (*Genetta genetta*)

Fig. 5.4 Different examples of the presence of Ibiza wall lizard, *Podarcis pityusensis,* in the everyday cultural life of Ibiza and Formentera. Photos by E. Montes

or hedgehogs (*Erinaceus europaeus*). In a way, the lizard symbolizes the naiveté and the purity of the original Ibiza and Formentera, as well as the harmless and innocuous fauna to humans that every visitor would find there. It is the best advertising anyone can make for an idyllic place like the Pityusic Islands. As a consequence, the invasive snakes are destroying the safety sense that secularly characterized Ibiza and Formentera, islands with no dangerous species until now. In fact, rural hotels are concerned and claim that the snakes are jeopardizing their business, as they have clients that come to Ibiza precisely because no dangerous animals can be found there.

Pityusic society is also having a very strong response to the presence of snakes. They are not used to snakes so their observation causes strong social alarm. Every spring and autumn, with the increasing activity of snakes, the island government receives hundreds of calls from frightened people that have seen a snake in their properties or just want to warn the authorities and give information, eager to collaborate in solving the problem. Rural inhabitants from the high-density snake zones tell us that they do not see lizards anymore, with a mix of sadness and anger. And maybe that is the main reason why they are so eager to help; the majority of these neighbors volunteer to take care of snake-traps (see below) in order to help control the invasion. The embrace of the endemic lizards as a symbol by the locals may be the key to its conservation, and the control of snakes (Bowen-Jones and Entwistle 2002).

5.4.4 Science and Management: An Inevitable Cross-Road

Eleven years after the first report of snake introduction, during 2014 and 2015, the Ibiza Island Council (island government) carried out a pilot project to control snake populations under the scientific advice and participation of the Spanish Herpetological Society (AHE). This first approach gave useful information about distribution, density, and the methods that best worked to eradicate the snakes. The main goals of this project consisted of promoting public awareness (Boudjelas 2009) and testing methods to capture snakes. Pamphlets in three different languages were distributed among the community; informative meetings were held for neighbors, as well as formative sessions for authority agents who had to deal with emergency calls related to snakes. A unified protocol was agreed between all the authority agents.

Among the capture methods tested in 2014 (plywood cover boards, PVC double funnel traps, and box-traps—baited and unbaited), lab mice baited box-traps gave the best results. Therefore, in the second year of the project, the Ibiza Island Council arranged the manufacturing of wooden box-traps, according to their guidance, with the Arts School (Secondary Level) of Ibiza. A total of 121 snakes were captured in 2015 by 200 box traps (Table 5.1, Fig. 5.5). Trained dogs were also used and proved quite effective in detecting snakes, but not very practical in the study area,

Fig. 5.5 Wooden box trap for capturing snakes on Ibiza Island. Note the access hatch on the top, a funnel in the lateral, and the small cage for restraining the lab mouse within the snake trap. Photo by E. Ayllón

Table 5.1 Individuals of the Horseshoe Whip Snake, *Hemorrhois hippocrepis*, and the Ladder Snake, *Rhinechis scalaris*, captured in Ibiza Island during the period 2011–2016, and invaded area by the Horseshoe whip snake according to years (Minimum Convex Polygon method, from the data in Fig. 5.6)

Year	Individuals		Area (km^2)
	Horseshoe whip snake	Ladder snake	Horseshoe whip snake
2011	12	7	63.2
2012	15	5	63.7
2013	38	10	109.4
2014	95	8	96.4
2015	302	10	149.3
2016	317	38	142.8
Total	779	78	

which is covered by hundreds of kilometers of stone fences where the snakes were difficult to catch (Table 5.1).

With the information gathered from the different trapping methods, we mapped the area covered by snakes each year, to evaluate the progression of the invasion.

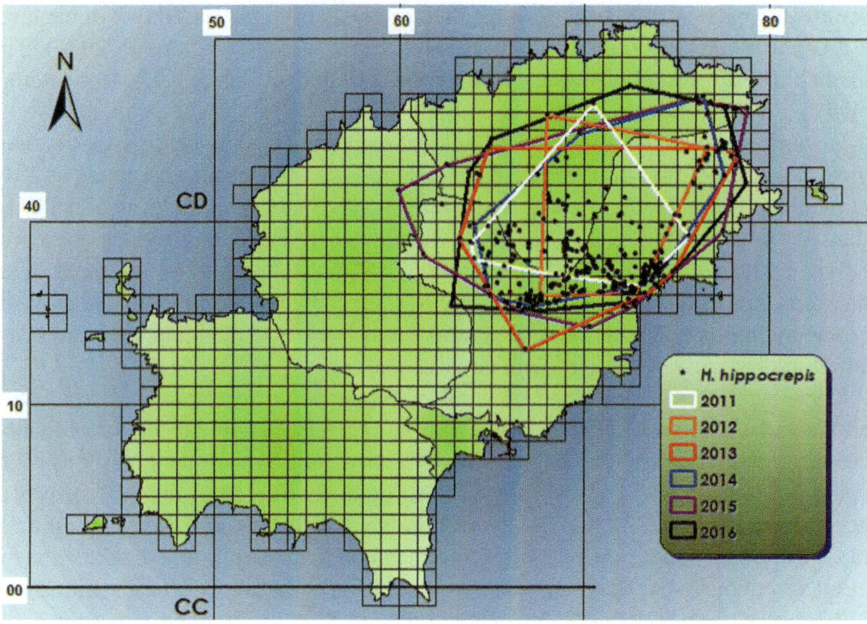

Fig. 5.6 Range of the invasive Horseshoe Whip Snake, *Hemorrhois hippocrepis*, in Ibiza Island in the period 2011–2016 (Minimum Convex Polygon method). Yearly sample size and area in Table 5.1. Figure prepared by Pedro Luis Hernandez -Sastre

As observed (Table 5.1, Fig. 5.6), the invasive population covers the northeastern sector of the island of Ibiza and the area increased in 2015, but stabilized after the intense trapping campaign of 2015–2016. These results have to be taken with caution, given that records increased inter-annually not only with the growth of the snake population within the island, but with the awareness of people and the stronger and more effective efforts to capture them (Table 5.1).

Fortunately, it seems the invasive snakes still show a contagious distribution pattern nearby the nurseries that acted as the introduction way, which is typical of the first stage of biological invasions (Williamson 1996). The apparent fast expansion observed is expected if we take into account that two of the main abiotic factors explaining the range of the reptiles, the mean annual temperature and average yearly rainfall of Ibiza, perfectly fit within the range of mean annual temperature and rainfall in their source range (Feriche 2015; Silva-Rocha et al. 2015). It is evident that the island is ideal for the spatial and demographic expansion of the species, and climate conditions will become even more favorable if predictions for climate change in this century are confirmed.

During 2015 we frequently observed scales and feces in empty traps, meaning that some snakes were able to escape. In 2016, when the Balearic Government took the control of the project, the traps were upgraded eliminating the funnels and installing just a hole of approximately 3 cm with a tilting door, only opening towards the inside of the box. This resulted in a very high increased rate of captures (2.2 captures/day/100 traps in 2015 vs. 10.5 captures/day/100 traps in 2016). The maximum result found elsewhere in trapping invasive snakes is 2.13 (Rodda et al. 1999).

Regarding the collaboration with the local community, which has given the most important contribution to the total snake captures, 400 new traps were made in 2016 by the Ibiza Island Council and two town councils (Santa Eulària des Riu, and Sant Joan de Labritja) and were sold for a subsidized price to any citizen who was willing to take care of them, after filling out and signing a commitment. This measure comes from the people themselves, since there have been repeated requests by the community for this type of action. It has been extended and more traps are being ordered, given that the demand is not falling.

These actions were not free of conflict, though: due to the lack of staff, collaborators have to sacrifice the snakes without professional guidance. This situation leads to two complex matters; animal cruelty and government responsibility regarding eradication of invasive species. Moreover, following the current procedure, researchers do not have the control on the distribution and sampling effort of traps. Common people do not know the techniques to manipulate snakes nor the regulation about what they can do or what they are not allowed to do with them. Putting such a responsibility on people's hands is risky, as they can spread the snakes by accident. Notwithstanding these potential problems, we encourage involving citizenship in invasive animal eradication programs (Boudjelas 2009; Gallo and Waitt 2011). Public participation in invasive alien species management is important, from ethical, legal, and practical standpoints (Boudjelas 2009).

From now on, the project should grow not only in staff but also in trap numbers and spread, as the priority should be avoiding the expansion of the snakes throughout the territory of the Pityusic islands. That can be done by placing traps along the edge of the currently invaded zone, before the snakes can invade further and affect populations of the endemic Ibiza Wall Lizard in snake-free zones (Grice 2009; Melero et al. 2010). The current ground perimeter of the invaded area is approximately 36 km (Fig. 5.6). Estimating one trap every 25–30 m (Rodda et al. 1999), we will require around 1400 traps to be placed around the perimeter of the invaded area just to control the expansion, plus 200–400 more to capture and reduce the population in the high-density zone of the core areas (close to the olive tree nurseries). The effectiveness of traps increases in combination with drift fences that intercept and guide snakes towards the traps (Fitzgerald 2012). However, that procedure represents a time and economic expense even greater than the cost of the traps; that is why the abundant rock walls on the island are being used as drift fences.

The study of the natural history of invasive snakes ought to be included in the project, since it may help in understanding the biological processes that make them successful and, therefore, in fighting them more accurately (Sakai et al. 2001; Stohlgren and Jarnevich 2009). This is a typical case of island invasion: a snake-free island, poorly populated in non-flying vertebrate fauna, where invasive snakes thrive. Numerous animals have been already studied after their inclusion as vouchers in scientific collections (Hinckley et al. 2017).

5.5 Conclusion

The Balearic Islands represent another striking example of the negative effect of human-mediated species introductions on native biota as well as on local people. Here, we presented two case studies which reflect two different attitudes regarding biological invasions. First, the case of the Italian wall lizard which represents the outcomes of a negligent attitude and taught us that if we ignore the problem, it will not solve itself but will simply get worse. In this particular case, a potentially invasive species was not ever considered to be dangerous in the Balearics. As a result, it spread throughout the islands where it was introduced, threatening the native biota, namely the endemic lizards which local people care about. The second case study reflects an active attitude which resulted in a strong involvement and a close collaboration with governments and general public to face the spread of invasive species, especially of introduced snakes. A project was created to understand how the invasion is progressing and how to deal with it.

Preliminary results on snakes' density control are encouraging, indicating the effectiveness of this collaboration, but we are only at the beginning of a long-term action which also includes preventing new introductions and monitoring the status of native species populations.

While capture campaigns may allow controlling the population size and spread of introduced snakes, a major priority for the conservation of native biota is to prevent new species from arriving on the island (Wittenberg and Cock 2001). The importation of big and old trees with cozy holes which can serve as shelters for alien species need to be limited, at least during critical periods (hibernation and egg-laying period, Shine et al. 1997). That limitation would leave a safe frame of two and a half months during spring time (April–June). Moreover, it is recommended that trees entering the Pityusic Islands within those periods are kept in quarantine at the entry point (any of the two Ibiza harbors); there are some who claim to have seen a snake falling from a moving trailer full of olive trees in spring. One-month quarantine in a walled enclosure (a minimum of 1.5 m height) with a half a dozen baited box-traps, should be enough to draw every hidden species in the cargo. The benefits from selling big olive trees are so high that the economic repercussions of such measures would be easily supported by the sellers, if a regulation obliged them (McNeely 2001).

Social collaboration, turning an eradication program into a publicly accepted project, has been shown to be vital for the project. It is strongly recommended to keep that work-line involving the public through a studied strategy that allows, on one hand, that Ibizan people feel they are working for their own island and symbol, contributing with their participation to achieve the maximum number of captures throughout the island. On the other hand, means the collaborators become a useful tool to technically improve the project, facilitating the collection of data in a programmed and organized way.

After all, the main concern of this invasion is, without any doubt, the same for everybody: the serious threat to the endemic fauna. Therefore, monitoring endemic populations is also crucial to determine whether and to what extent snake predation on lizards results on demographic impacts on lizard populations.

In conclusion, Balearic Islands exemplify an archipelago where aliens conquered the main islands and that has implications not only for biodiversity but also for the communities present. The future will determine if alien species will continue to conquer, or if the joint effort of science, policy and social participation will mitigate this invasion.

Ackowledgements The pilot project was partially funded by the European Regional Development Fund (FEADER), the Spanish Ministry of Agriculture and Environment, and the Balearic Islands Regional Government, through the Leader axis (Rural Development Program of the Balearic Islands). We are grateful to the Wildlife Conservation Service of the Balearic Government and the Partnership for the Fauna Recovery of the Balearic Islands (COFIB) for their total cooperation and for sharing their work with us. NS is supported by a research contract (IF/01526/2013); MAC is supported by the FCT project NORTE-01-0145-FEDER-000007, DS is supported by the FCT (Fundação para a Ciência e a Tecnologia, Portugal) post-doctoral grant SFRH/BPD/66592/2009 and IS-R by the FCT Ph.D. grant SFRH/BD/95745/2013 under the Programa Operacional Potencial Humano—Quadro de Referência Estratégico Nacional funds from the European Social Fund and Portuguese Ministério da Educação e Ciência.

References

AEMET IM (2011) Atlas Climático Ibérico - Iberian Climate Atlas. AEMET - Agencia Estatal de Meteorología, Ministerio de Medio Ambiente y Medio Rural y Marino IM, Instituto de Meteorologia de Portugal

Alcover JA, McMinn M (1993) Es Pouàs: una aventura paleontològica a Eivissa. Pap del Mus la Nat les Illes Balear 1:x–xx

Alcover J, Moyà-Solà S, Pons-Moyà J (1981) Les Quimeres Del Passat: Els Vertebrats Fossils Del Plio-Quaternari De Les Balears I Pitiuses. Palma de Mallorca

Alcover J, Mayol J, Jaume D et al (1984) Biologia i Ecologia de les poblacions de *Baleaphryne muletensis* a la muntanya mallorquina. In: Hemmer H, Alcover J (eds) Història Biològica del Ferreret. Monografies científiques 3. Editorial Moll, Palma de Mallorca, pp 129–152

Alcover J, Bover P, Seguí B (1999) Paleoecologia de les illes. In: Alcover J (ed) Ecologia de les Illes. Monografies de la Societat d'Història Natural de les Balears, Palma de Mallorca

Alcover JA, Ramis D, Coll J, Trias M (2001) Bases per al coneixement del contacte entre els primers colonitzadors humans i la naturalesa de les Balears. Endins 24:5–57

Álvarez C, Mateo JA, Oliver J, Mayol J (2010) Los ofidios ibéricos de introducción reciente en las Islas Baleares. Bol Asoc Herpetol Esp 21:126–131

Ayuso VM (2001) The Balearic islands: Prehistoric colonization of the furthest Mediterranean islands from the mainland. J Mediterr Archaeol 14:136–158

Bailón S (2004) Fossil records of Lacertidae in mediterranean islands: the state of the art. Biol lacertid lizards, Evol Ecol Perspect 37–62

Bailon S, Quintana J, Garcia-Porta J (2005) Primer registro fósil de las familias Gekkonidae (Lacertilia) y Colubridae (Serpentes) en el Plioceno de Punta Nati (Menorca, Islas Baleares). In: Alcover J, Bover P (eds) Proceedings of the international symposium insular vertebrate evolution: the palaeontological approach. Monografies de la Societat d'Història Natural de les Balears, Palma de Mallorca

Bailon S, Bover P, Quintana J, Alcover JA (2010) First fossil record of *Vipera Laurenti* 1768 "Oriental vipers complex" (Serpentes: Viperidae) from the Early Pliocene of the western Mediterranean islands. CR Palevol 9:147–154

Bailon S, Boistel R, Bover P, Alcover JA (2014) *Maioricalacerta rafelinensis*, gen. et sp. nov. (Squamata, Lacertidae), from the early Pliocene of Mallorca (Balearic Islands, western Mediterranean Sea). J Vertebr Paleontol 34:318–326

Barbadillo LJ (1987) La guía INCAFO de los anfibios y reptiles de la Península Ibérica. Islas Baleares y Canarias, INCAFO

Bate DMA (1914) On Remains of a Gigantic Land Tortoise (*Testudo gtmnesicus*, N.SP.) from the Pleistocene of Menorca. Geol Mag 1:100–107

Blondel J, Aronson J, Bodiou J-Y, Boeuf G (2010) The Mediterraneaen region—Biological diversity in space and time. Vasa 401

Boscá E (1883) Exploracion herpetológica de la Isla de Íbiza. An la Soc Española Hist Nat 12:241–250

Boudjelas S (2009) Public participation in invasive species management. In: Clout MN, Williams PA (eds) Invasive species management: a handbook of principles and techniques. Oxford University Press, Oxford, pp 93–107

Bour R (1985) Una nova tortuga terrestre del Pleistocé d'Eivissa: la tortuga de la cova de Ca Na Reia. Endis 10–11:57–62

Bover P, Quintana J, Alcover JA (2008) Three islands, three worlds: paleogeography and evolution of the vertebrate fauna from the Balearic Islands. Quat Int 182:135–144

Bover P, Rofes J, Bailon S et al (2014) Late Miocene/Early Pliocene vertebrate fauna from Mallorca (Balearic Islands, Western Mediterranean): an update. Integr Zool 9:183–196

Bowen-Jones E, Entwistle A (2002) Identifying appropriate flagship species: the importance of culture and local contexts. Oryx 36:189–195

Brown RP, Perez-Mellado V (1994) Ecological energetics and food acquisition in dense menorcan islet populations of the lizard *Podarcis lilfordi*. Funct Ecol 8:427

Bull L, Courchamp F (2009) Management of interacting invasives: ecosystem approaches. In: Clout MN, Williams PA (eds) Invasive species management: a handbook of principles and techniques. Oxford University Press, Oxford, pp 232–247

Capula M (1993) Natural hybridization in *Podarcis sicula* and *P. wagleriana* (Reptilia: Lacertidae). Biochem Syst Ecol 21:373–380

Capula M (2002) Genetic evidence of natural hybridization between *Podarcis sicula* and *Podarcis tiliguerta* (Reptilia: Lacertidae). Amphibia-Reptilia 23:313–321

Capula M, Luiselli L, Bologna MA, Ceccarelli A (2002) The decline of the Aeolian wall lizard, *Podarcis raffonei*: causes and conservation proposals. Oryx—Int J Conserv 36:66–72

Carrera D, Pons P (2010) Importància de les basses temporals per a la conservació dels amfibis Menorquins. In: Fraga P, Estaún I, Cardona E (eds) Basses temporals mediterrànies. LIFE BASSES: gestió i conservació a Menorca. ConsellInsular de Menorca, Institut Menorquí d'Estudis, Maó, pp 327–363

Carretero MA, Silva-rocha I (2015) La culebra de escalera (*Rhinechis scalaris*) en las Islas Baleares. Bol la Asoc Herpetológica Española 26:105–108

Cau MÁ, Valenzuela A, Alcover JA (2016) Archaeological evidence for the introduction of Emys orbicularis (Testudines: Emydidae) in the Balearic Islands, Western Mediterranean. Amphibia-Reptilia 37:229–236

Chapman R (1990) Emerging complexity. The later prehistory of South-East Spain, Iberia and the West Mediterranean. Cambridge University Press

Dappen N, Losin N, Pérez-Mellado V (2013) The symbol: wall lizards of Ibiza and Formentera. Day's Edge Productions

Feriche M (2015) Culebra de herradura—*Hemorrhois hippocrepis*. In: Salvador A, Marco A (eds) Enciclopedia Virtual de los Vertebrados Españoles. Museo Nacional de Ciencias Naturales, Madrid. http://www.vertebradosibericos.org/ Assessed in January 2017

Fernández JH (1975) Los dioses de la Ibiza cartaginesa. Eivissa 7:31–38

Ficetola GF, Padoa-schioppa E (2009) Human activities alter biogeographical patterns of reptiles on Mediterranean islands. Glob Ecol Biogeogr 18:214–222

Fitzgerald LA (2012) Finding and Capturing Reptiles. In: McDiarmid R, Foster M, Guyer C et al (eds) Reptile biodiversity. University of California Press, Berkeley, pp 77–88

Fortuny Santos J (2002) Metodologia del analisis sectorial en el sistema agroalimentario, aplicada al subsector oleicola catalan: evaluacion de la competitividad, el progresso tecnologico y la eficiencia economica empresarial. Universitat de Lleida

Fritz U, Auer M, Bertolero A et al (2006) A rangewide phylogeography of Hermann's tortoise, *Testudo hermanni* (Reptilia: Testudines: Testudinidae): implications for taxonomy. Zool Scr 35:531–543

Gallo T, Waitt D (2011) Creating a successful citizen science model to detect and report invasive species. Bioscience 61:459–465

Garcia-Porta J, Quintana J, Bailon S (2002) Prier hallazgo de *Blanus sp* (Amphisbaenia, Reptilia) en el neógeno balear. Rev Esp Herpet 16:19–28

Gil JM, Pleguezuelos JM (2001) Prey and prey-size selection by the short-toed eagle (*Circaetus gallicus*) during the breeding season in Granada (south-eastern Spain). J Zool 255:131–137

Gimeno I, Vilà M, Hulme PE (2006) Are islands more susceptible to plant invasion than continents? A test using *Oxalis pes-caprae* L. in the western Mediterranean. J Biogeogr 33:1559–1565

Graziani F, Berti R, Dapporto L, Corti C (2006) *Podarcis* lizards in an agro-environment in Tuscany (Central Italy): preliminary data on the role of olive tree plantations. In: Corti C, Lo Cascio P, Biaggini M (eds) Mainland and insular Lacertid lizards: a Mediterranean perspective. Firenze University Press, Firenze, pp 65–72

Greca M, Sacchi C (1957) Problemi del popolamento animale nelle piccole isole mediterranee. Annuario Museo Zoologico della Università di Napoli, 9

Grice T (2009) Principles of containment and control of invasive species. In: Clout MN, Williams PA (eds) Invasive species management: a handbook of principles and techniques. Oxford Biology, Oxford

Guicking D, Joger U, Wink M (2008) Molecular phylogeography of the viperine snake *Natrix maura* (Serpentes: Colubridae): evidence for strong intraspecific differentiation. Org Divers Evol 8:130–145

Guirand F (1960) Mythologie générale. Librairie Larousse, Paris

Herrel A, Huyghe K, Vanhooydonck B et al (2008) Rapid large-scale evolutionary divergence in morphology and performance associated with exploitation of a different dietary resource. Proc Natl Acad Sci USA 105:4792–5

Hinckley A, Montes E, Ayllón E, Pleguezuelos JM (2017) The fall of a symbol? A high predation rate by the introduced horseshoe whip snake *Hemorrhois hippocrepis* paints a bleak future for the endemic Ibiza wall lizard *Podarcis pityusensis*. Eur J Wildl Res 63:13

Jiménez-Moreno G, Fauquette S, Suc JP (2010) Miocene to Pliocene vegetation reconstruction and climate estimates in the Iberian Peninsula from pollen data. Rev Palaeobot Palynol 162:403–415

Kirschbaum HL, Pauly GB (2016) Feeding ecology and range expansion of the nonnative Italian wall lizard (*Podarcis sicula*). In: AAAS 2016 annual meeting: global science engagement. AAAS, Washington DC

Kotsakis T (1981) Le Lucertole (Lacertidae, Squamata) del Pliocene, Pleistocene e Olocene delle Baleari. Bolletí la Soc d'Història Nat les Balear 25:135–150

Kraus F (2009) Impacts of alien reptiles and amphibians. Alien Reptil Amphib A Sci Compend Anal 57–93

Lewthwaite J (1985) Social factors and economic change in Balearic prehistory, 3000–1000 B.C. In: Barker G, Gamble C (eds) Beyond domestication in prehistoric Europe. Academic Press, London, pp 205–231

MacArthur RH, Wilson O (1967) The theory of island biogeography. Princeton University Press, Princeton N.J., p 203

Mateo JA (2015) Los Anfibios y los Reptiles Introducidos en Baleares: un repaso a lo que sabemos y un ejemplo de puerta de entrada. Llibre Verd Protecció d'Espècies a les Balear 447–454
Mateo J, Ayllón E (2012) Viabilidad del Control de Ofidios en Ibiza y Formentera 72
Mateo JA, Ayres C, Lopez-Jurado LF (2011) Los anfibios y reptiles naturalizados en España. Historia y evolución de una problemática creciente. Bol la Asoc Herpetol Esp 22:2–42
Mayol J (1985) Rèptils i amfibis de les Illes Balears. Editorial Moll, Palma de Mallorca
McNeely J (2001) An introduction to human dimensions of invasive alien species. In: McNeely J (ed) The great reshuffling: human dimensions of invasive alien species. IUCN Publishers, Gland, pp 5–22
Mela P. (around 43 D.A.) Cosmografia sive de situ orbis. Retrieved from http://www.cervantesvirtual.com/obra-visor/cosmographia-sive-de-situ-orbis–0/html/ffa61dcc-82b1-11df-acc7-002185ce6064_4.htm. Accessed in Jan 2017
Melero Y, Palazón S, Bonesi L, Gosàlbez J (2010) Relative abundance of culled and not culled American mink populations in northeast Spain and their potential distribution: are culling campaigns effective? Biol Invasions 12:3877–3885
Moreno-Rueda G, Pleguezuelos J (2009) Climate warming and activity period extension in the Mediterranean snake *Malpolon monspessulanus*. Clim Change 92:235–242
Moyà-Solà S, Pons-Moyà J, Alcover JA, Agusti JS (1984) La fauna de vertebrados neógeno-cuaternario de Eivissa (Pitiuses). Nota preliminar. Acta Geol Hisp 19(1):33–35
Moyà-Solà S, Quintana J, Alcover JA, Köhler M (1999) Endemic island faunas of the Mediterranean Miocene. In: The Miocene Land Mammals of Europe, pp 435–442
Myers N, Mittermeier RA, Mittermeier CG et al (2000) Biodiversity hotspots for conservation priorities. Nature 403:853–858
Nevo E, Gorman G, Soulé M et al (1972) Competitive exclusion between insular Lacerta species (Sauria, Lacertidae)—Notes on experimental Introductions. Oecologia 10:183–190
Novosolov M, Raia P, Meiri S (2013) The island syndrome in lizards. Glob Ecol Biogeogr 22:184–191
Oliverio M, Burke R, Bologna MA et al (2001) Molecular characterization of native (Italy) and introduced (USA) *Podarcis sicula* populations (Reptilia, Lacertidae). Ital J Zool 68:121–124
Parker WS, Plummer MV (1987) Population ecology. In: Seigel RA, Collins JT, Novak SS (eds) Snakes: ecology and evolutionary biology. Macmillan, New York, pp 253–301
Pedall I, Fritz U, Stuckas H et al (2011) Gene flow across secondary contact zones of the Emys orbicularis complex in the Western Mediterranean and evidence for extinction and re-introduction of pond turtles on Corsica and Sardinia (Testudines: Emydidae). J Zool Syst Evol Res 49:44–57
Pérez-Mellado V (1998) *Podarcis sicula* (Rafinesque, 1810). In: Al. R et (ed) Fauna Ibérica. Museo Nacional de Ciencias Naturales, Madrid
Pérez-Mellado V, Corti C (1993) Dietary adaptations and herbivory in lacertid lizards of the genus *Podarcis* from western Mediterranean islands (Reptilia: Sauria). Bonn Zool Beitr 44:193–220
Pinya S, Carretero MA (2011) The Balearic herpetofauna: a species update and a review on the evidence. Acta Herpetol 6:59–80
Pleguezuelos JM (2004) Las especies introducidas de anfibios y reptiles. In: Pleguezuelos JM, Márquez R, Lizana M (eds) Atlas y libro rojo de los anfibios y reptiles de España. Dirección General de Conservación de la Naturaleza, 2nd edn. Asociación Herpetológica Española, Madrid, pp 502–532
Pleguezuelos JM, Moreno M (1990) Alimentación de *Coluber hippocrepis* en el SE de la península Ibérica. Amphibia-Reptilia 11:325–337
Pleguezuelos JM, Fernández-Cardenete JR, Honrubia S et al (2007) Correlates between morphology, diet and foraging mode in the Ladder Snake *Rhinechis scalaris* (Schinz, 1822). Contrib to Zool 76:179–186
Quintana J (1998) Aproximación a los yacimientos de vertebrados del Mio-Pleistoceno de la isla de Menorca. Bolleti la Soc d'Historia Nat les Balear 41:101–117
Ramis D, Alcover JA, Coll CJ et al (2001) The chronology of the first settlement of the Balearic islands. J Mediterr Archaeol 15:3–24

Rato C, Carranza S, Perera A et al (2010) Conflicting patterns of nucleotide diversity between mtDNA and nDNA in the Moorish gecko, *Tarentola mauritanica*. Mol Phylogenet Evol 56:962–971

Reaser JK, Meyerson LA, Cronk Q et al (2007) Ecological and socioeconomic impacts of invasive alien species in island ecosystems. Environ Conserv 34:98

Recuero E, Iraola A, Rubio X et al (2007) Mitochondrial differentiation and biogeography of *Hyla meridionalis* (Anura: Hylidae): an unusual phylogeographical pattern. J Biogeogr 34:1207–1219

Rivera X, Arribas O, Carranza S, Maluquer-margalef J (2011) An introduction of *Podarcis sicula* in Catalonia (NE Iberian Peninsula) on imported olive trees. Butlletí la Soc Catalana d'Herpetologia 19:79–85

Rodda GH, Fritts TH, Iii EWC, et al (1996) Practical concerns in the eradication of island snakes. Specialist 260–265

Rodda GH, Sawai Y, Chiszar D, Tanaka H (1999) Part V, capture and detection. In: Problem snake management: the habu and the brown treesnake. Comstock Publishing Associates

Rodríguez V, Brown RP, Terrasa B et al (2013) Multilocus genetic diversity and historical biogeography of the endemic wall lizard from Ibiza and Formentera, *Podarcis pityusensis* (Squamata: Lacertidae). Mol Ecol 22:4829–4841

Román A (2004) *Alytes muletensis* (Sanchiz & Adrover, 1977) In: Pleguezuelos J, Márquez R, Lizana M (eds) Atlas y libro rojo de los anfibios y reptiles de España, 2nd edn. Dirección General de Conservación de la Naturaleza, Asociación Herpetológica Española, Madrid, pp 79–81

Sakai AK, Allendorf FW, Holt JS et al (2001) The population biology of invasive species. Annu Rev Ecol Syst 32(305–332):7

Salazar de Mendonza DP (1770) Monarquía de España. In: Ibarra J (ed) Don Bartholome Ulloa. Mercader de Libros, Madrid

Salo P, Korpimäki E, Banks PB et al (2007) Alien predators are more dangerous than native predators to prey populations. Proc Biol Sci 274:1237–43

Salvador A (2015) Lagartija de las Pitiusas—*Podarcis pityusensis* (Boscá, 1883). Museo Nacional de Ciencias Naturales, Madrid

Seigel R, Ford N (1987) Reproductive ecology. In: Seigel R, Collins J, Novak S (eds) Snakes: ecology and evolutionary biology. Macmillan, New York, pp 210–252

Shine R (1981) Venomous snakes in cold climates ecology of the Australian genus *Drysdalia* (Serpentes Elapidae). Copeia 1981:14–25

Shine R (2003) Reproductive strategies in snakes. Proc Biol Sci 270:995–1004

Shine R, Madsen TRL, Elphick MJ, Harlow PS (1997) The influence of nest temperatures and maternal brooding on hatchling phenotypes in water pythons. Ecology 78:1713–1721

Silva-Rocha I (2012) Patterns of biological invasion in the herpetofauna of the Balearic Islands: determining the origin and predicting the expansion as conservation tools. Universidade do Porto, Porto

Silva-Rocha I, Salvi D, Carretero MA (2012) Genetic data reveal a multiple origin for the populations of the Italian wall lizard *Podarcis sicula* (Squamata: Lacertidae) introduced in the Iberian Peninsula and Balearic islands. Ital J Zool 79:502–510

Silva-Rocha I, Salvi D, Harris DJ et al (2014) Molecular assessment of *Podarcis sicula* populations in Britain, Greece and Turkey reinforces a multiple-origin invasion pattern in this species. Acta Herpetol 9:253–258

Silva-Rocha I, Salvi D, Sillero N et al (2015) Snakes on the balearic islands: an invasion tale with implications for native biodiversity conservation. PLoS ONE 10(4):e0121026

Simberloff D, Von Holle B (1999) Positive interactions of nonindigenous species: invasional meltdown? Biol Invasions 1:21–32

Speybroeck J, Beukema W, Crochet PA (2010) A tentative species list of the european herpetofauna (amphibia and reptilia)—an update. Zootaxa 27:1–27

Stohlgren TJ, Jarnevich C (2009) Risk assessment of invasive species. In: Clout M, Willians P (eds) Invasive species management: a handbook of principles and techniques. Oxford Biology, Oxford, pp 19–35 van den Berg M, Zawadzki M (2010) Wiederentdeckung einer ausgestorben geglaubten Population der Baleareneidechse, *Podarcis lilfordi* (GÜNTHER, 1874) auf der Illa de Ses Mones (Balearen, Menorca, Spanien) in Sympatrie mit der Ruineneidechse, *Podarcis siculus* (RAFINESQUE-SCHMALTZ, 1810). http://www.lacerta.de/AS/Artikel.php?Article=131. Accessed 10 Jan 2017

Velázquez Brieva F (2007) El dios Bes: de Egipto a Ibiza. Museu Arqueològic d'Eivissa i Formentera, Ibiza

Velo-Antón G, Wink M, Schneeweiß N, Fritz U (2011) Native or not? Tracing the origin of wild-caught and captive freshwater turtles in a threatened and widely distributed species (*Emys orbicularis*). Conserv Genet 12:583–588

Viada C, Mayol J, Oliver J (2006) Libro rojo de los vertebrados de las Baleares. Conselleria de Medi Ambient, 3rd edn. Govern de les Illes Balears, Mallorca

Vitousek PM, Loope LL, Aderson H, D'Antonio C (1996) Island Ecosystems: do they represent "natural experiments" in biological diversity and ecosystem function? Funct Roles Biodivers A Glob Perspect 245–259

Whittaker RJ, Fernandez-Palacios JM (2007) Island biogeography: ecology, evolution, and conservation. Oxford University Press

Williamson MH (1996) Biological invasions. Chapman & Hall

Wittenberg R, Cock MJW (2001) Invasive alien species: a toolkit of best prevention and management practices. Glob Invasive species Program 241

Zamora-Camacho FJ, Moreno-Rueda G, Pleguezuelos JM (2010) Long- and short-term impact of temperature on snake detection in the wild: further evidence from the snake *Hemorrhois hippocrepis*. Acta Herpetol 5:143–150

Chapter 6
The *Xenopus laevis* Invasion in Portugal: An Improbable Connection of Science, Mediterranean Climate and River Neglect

Mónica Sousa, Angela Maurício and Rui Rebelo

Abstract The source populations of one of the most worldwide spread invasive amphibians—the African clawed frog, *Xenopus laevis*—inhabit the Mediterranean-type climate region of South Africa. Frogs captured here were exported to many world regions during the 20th century with scientific purposes. Soon after, established invasive populations were found in several places, the majority of them subject to Mediterranean-type climates. Here we reconstruct the historical contingencies that explain the transport, site of arrival, establishment, lag phase and expansion of this invasive frog in Portugal. Improbable events were associated with practically all stages of the invasion process. Individuals were probably brought to Portugal by a single researcher and their escape from her lab was facilitated by a disastrous flood. Despite the striking differences from native frogs, this species apparently lived undetected by biologists and authorities for more than 20 years in two streams in Lisbon metropolitan area, and the discovery of the invasive populations was only prompted by the beginning of stream restoration actions. Since 2010, this frog is being subject to a control and eradication program. We provide details on the progress of this program, as well as on the role that geographic and historical contingencies may have for its final success.

M. Sousa (✉)
Departamento de Recursos Naturais e Conservação da Natureza,
Instituto da Conservação da Natureza e das Florestas, IP,
Avenida da República, 16 a 16B, 1050-191 Lisbon, Portugal
e-mail: Monica.Sousa@icnf.pt

A. Maurício
Departamento de Ambiente e Equipamento, Município de Oeiras,
Largo Marquês de Pombal, 2784-501 Oeiras, Portugal

R. Rebelo
Departamento de Biologia Animal & Centre for Ecology,
Evolution and Environmental Changes, Faculdade de Ciências
da Universidade de Lisboa, Campo Grande, 1749-016 Lisbon, Portugal

A. I. Queiroz and S. Pooley (eds.), *Histories of Bioinvasions in the Mediterranean*, Environmental History 8, https://doi.org/10.1007/978-3-319-74986-0_6

Keywords Amphibian · Laboratory escape · Silent invasion · Periurban
Historical contingency · Control program

6.1 Introduction

Reconstructing the historical process of a biological invasion through documenting contingent details on the transport, site of arrival, establishment and expansion of invasive species often helps identifying the crucial steps of the invasion process and may provide important insights on management options to control the invader. In this chapter, we report the current knowledge of the invasion history of an invasive amphibian in Portugal, the African clawed frog, *Xenopus laevis*. We highlight the natural and cultural features, either of local, national or Mediterranean scope that enabled the species' arrival, establishment and expansion, as well as their importance for the success of the control and eradication program that is currently underway.

Bioinvasions by amphibians and reptiles have been mainly unintentional all across the world and throughout human history (Kraus 2009). Most amphibians and reptiles are small-sized and can become inactive for long time periods, which facilitates their accidental transport over long distances, especially along with vegetables, fruit, or wood (Kraus 2009). Some introductions however were deliberate; for example, the species for which there is the most scientific information is the cane toad *Rhinella marina*, which was introduced in several continents as a (unfortunately inefficient) biocontrol measure against the sugar cane beetle (Crossland 2000). The negative impacts of this species on native fauna have been very high, as this toad is highly toxic and, for instance, its consumption by predators unaware of the toad's toxicity lead to a serious decline on the populations of Australian native predators (Llewelyn et al. 2009). The native amphibians may be predated (whether the adults, larvae or eggs), be contaminated with pathogens or parasites carried by the invasive (Price et al. 2014), or in some cases hybridize with the invasive, losing their genetic integrity (Arano et al. 1995).

The majority of the successful bioinvasions by amphibians and reptiles have been reported for subtropical and tropical climates, especially of southern North America, the Caribbean and Pacific Ocean Islands (Kraus 2009). However, the first invasion meltdown of herpetological fauna historically documented occurred across the Mediterranean Basin, especially in its islands, during pre-Roman and Roman times (1500–300 B.C.). Phoenician traditions included keeping snakes for pets, and apparently many other herp species, like lizards, turtles, geckos, toads or frogs were in some way carried along the Mediterranean coasts in Phoenician trading vessels (Silva-Rocha et al. 2015). Current herpetological communities in the Iberian Peninsula and Balearic islands are a mixture of native and pre-Roman invasive species, and some of these invasives are still spreading northward in the Peninsula (Richter-Boix et al. 2012).

6.2 World Invasions by the African Clawed Frog

Accurate predictions of human pregnancy became possible in the 1920s with the discovery that detection in a woman's urine or blood of Human Chorionic Gonadotrophin, a hormone produced by the placenta that maintains the ovarian corpus luteum during the beginning of pregnancy, is a reliable sign of successful uterine implantation of the embryo. The first tests involved injecting the woman's urine into the circulation of a laboratory mammal, which was then killed to facilitate inspection of its ovaries. This process was expensive and ethically questionable.

In the 1930s, after a visit to Cape Town, the English zoologist Lancelot Thomas Hogben brought back to London the knowledge on how to perform a simpler process using the females of African clawed frogs (*Xenopus laevis*) (Fig. 6.1) (Gurdon and Hopwood 2000): after being injected with the urine of a pregnant woman, the frog would ovulate within hours. The frogs could be maintained alive and reused, and yet the accuracy of the test remained high. The endocrinological work of Hogben was instrumental in the establishment of this test and responsible for the widespread use of this species as a model organism in laboratories in all continents. For decades, the Hogben Pregnancy Test became a standard and provided the basis for the creation of more accurate testing, only supplanted at the end of the 1960s.

Soon it became clear to those working in laboratories and research facilities that this frog was easy to keep in aquaria, robust and resistant to most diseases, and that

Fig. 6.1 Adult male (left) and female (right) of African clawed frogs *Xenopus laevis*. Photo by Francisco Moreira

was amenable to other kinds of experimentation. The characteristics that contributed to its establishment as the most important model amphibian in scientific research and one of the most important model vertebrates include morphological, physiological and behavioural traits (for a thorough review, see Tinsley and Kobel 1996). The African clawed frog has proven to be one of the most productive model organisms for studies in cell, molecular and developmental biology (Gurdon 1996). More recently, the ease with which these animals can be kept in aquaria has also facilitated its trade and maintenance as a pet.

From the 1930s, the species was spread to laboratories all over the world, which was soon followed by reports of invasive populations in several countries, including those located in three of the world's five Mediterranean climate-type regions (Measey et al. 2012). It should be noted that the frog's natural distribution in Sub-Saharan Africa also includes a Mediterranean-type zone in southwestern South Africa (Fig. 6.2).

African clawed frogs possess a remarkable suite of traits that turn them into very effective invaders: extraordinary longevity, up to 15 years (Flower 1936), the ability to colonize a variety of aquatic environments (Tinsley and McCoid 1996) as well as to migrate overland (Eggert and Fouquet 2006; Measey 2016), a high tolerance to salt water and anoxic conditions (Jokumsen and Weber 1980), the ability to survive drought by burrowing into drying mud (Lobos and Jaksic 2005) and to survive without feeding for up to 12 months (Tinsley et al. 1996). Furthermore, its reproductive traits enable the rapid buildup of local populations—African clawed frogs can become sexually mature six to eight months after metamorphosis, are able to breed over most of the year when environmental conditions

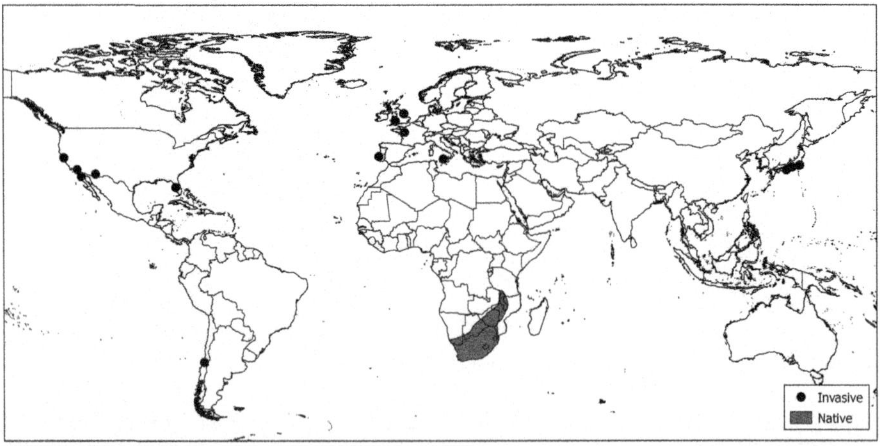

Fig. 6.2 Global distribution of African clawed frogs. The actual distribution of the species in several world areas are difficult to map at this scale, and therefore the sites of invasion marked with circles: Deux Sevres (France); Sicily (Italy); Oeiras County (Portugal); Wales and Lincolnshire (United Kingdom); Central Chile; Arizona and California (USA); Chiba, Kanagawa, Shizuoka and Wakayama Prefectures (Japan) (data from Measey et al. 2012).

related to temperature and availability of water/food are favorable, and females deposit hundreds to thousands of eggs every year (McCoid and Fritts 1989).

The discovery of the invasive populations of African clawed frogs raised concerns about their potential impacts. These were justified, as the species recently ranked second among the socio-economically most relevant invasive amphibian species of the world (Measey et al. 2016). This frog is a vector of *Batrachochytrium dendrobatidis*, a fungus implicated in the decline of many amphibian populations worldwide (Soto-Azat et al. 2010), is a voracious predator (Marques 2016; Measey et al. 2015; Courant et al. 2014; Amaral and Rebelo 2012; Faraone et al. 2008; Lobos and Jaksic 2005; Lobos and Measey 2002; Measey 1998; Tinsley et al. 1996; Schoonbee et al. 1992; McCoid and Fritts 1980) and is responsible for the reduction of reproductive success of some native amphibians, either by predating their egg masses and tadpoles, or by diminishing the quality of pond habitats so that adults avoid using them for breeding (Lillo et al. 2011).

6.3 The Portuguese Study Case

In Portugal, African clawed frogs were first spotted in the wild in 2006, on a stream running through Oeiras County, a densely urbanized area located around 20 km West of Lisbon. Although place and time of introduction are not definitely identified, there are reasons to believe that this population was the result of an introduction proceeding from research laboratories located nearby (Rebelo et al. 2010).

In 1972, a Portuguese researcher (M. G. Emílio) had started using African clawed frogs as an animal model for her Ph.D. under the supervision of Graham Shelton at the University of East Anglia, Norwich (England). In 1974, the Gulbenkian Institute for Science in Oeiras became her definitive workplace, where she worked with African clawed frogs until 1989. Most likely Portuguese specimens of African clawed frogs came from the Norwich population in the early 1970s, and these in turn originate from South Africa.

Most of the African clawed frog exports from South Africa were conducted by the Cape Provincial Administration's Department of Nature Conservation, whose Jonkershoek and Pirie fish hatcheries exported frogs collected in the south-western Cape region for over 30 years from the 1940s (Van Sittert and Measey 2016). In fact, the African clawed frogs export list of Jonkershoek fish hatchery includes during that period a single export to Norwich of 200 individuals, 120 females and 80 males, in 1968 (Van Sittert and Measey 2016). Recent studies of both mitochondrial and nuclear genes of a sample of the Portuguese exemplars showed that the genetic diversity of this population is extremely low for all the markers sampled, indicating that it results very probably from a single colonization event (De Busschere et al. 2016). The same study also found that the mitochondrial DNA of the Portuguese animals was identical to that of frogs from the South-western Cape.

The arrival of African clawed frogs at the Gulbenkian Institute for Science is undocumented, but may have occurred as early as 1974. However, several researchers recall the disastrous floods of the 1979/1980 winter, when the Laje stream overflowed and inundated the Institute basements, where the frogs were kept. This flood may have washed many individuals into the stream.

Mediterranean rivers are typically subject to stream flow variability, with dry summers and large winter floods, compounded by large inter annual rainfall variability (Boix et al. 2010). Available records from Mediterranean rivers point to an increase in the seasonal and annual hydrological variability with a higher frequency of flash floods, particularly in the 1970s and 1980s. In an historical context, such extreme events occurred over the recent past during episodes of irregular rainfall, both at seasonal and annual scales (Benito and Machado 2012).

If indeed the species' introduction in the natural environment was caused by the 1979/1980 flood, it means that the presence of an exotic large frog in Laje stream went unnoticed for more than 25 years, until its discovery in 2006. This "invisible invasion" happened in an urban area close to Lisbon and in the vicinity of research institutes hosting several generations of biologists during all that period. Furthermore, one gardener working near the Gulbenkian Institute for Science knew about the presence of those strange frogs and fed them regularly during the 1990s. Both facts are symptoms of the lack of importance devoted to the presence of exotic species by local inhabitants, and indeed by the scientific community before the increase in awareness of the impacts of biological invasions (Rebelo et al. 2010).

These frogs rarely leave water, and their drab coloration and tendency to remain at the bottom of waterbodies makes them difficult to spot. Furthermore, this is a species that only calls underwater, being inaudible to humans. However, another plausible reason why this invasive frog was overlooked for such a long time was the lack of appealing features for human use of Laje stream during the last decades of the 20th century. River neglect and extreme environmental degradation, heavy pollution by sewer discharges, combined with the difficult access to the stream, could explain the silent invasion of African clawed frogs, and may have favoured its population build-up and expansion upstream.

A new attitude towards enhancement of natural green areas, as well as concerns with water quality, led most cities in Portugal to engage in river restoration actions that eventually led to the discovery of African clawed frogs within this large urban area. The restoration of the water courses in Oeiras County began in the 1980s with a high investment in wastewater collection, transport and treatment. These actions were reinforced by the Directive 2000/60/EC of the European Parliament and of the Council of 23 October 2000, which established a framework for Community action in the field of water policy. With the implementation of the Water Framework Directive the water resources management entities were forced to move from actions mainly focused on flood risk control to a more comprehensive and global approach taking into consideration ecological and environmental systems. The main objective of the Directive is to protect, improve and restore all surface and groundwater bodies in Europe, aiming to achieve a good ecological status that may be assessed by biological, physico-chemical and hydro-morphological elements. Territorial sustainable development is required to integrate new solutions that minimize the impacts on water systems, ensuring not only the security of local human populations, but also the ecological balance of terrestrial and riverine ecosystems. At Oeiras, restoration of the watercourses and of the water infrastructure heritage was conducted also with the intent to make these places suitable for recreational purposes for the local population and to foster tourism.

In order to ensure the ecosystems' conservation, most counties started stream restoration by clearing the watercourses of invasive vegetation and planting native riparian vegetation to recover natural flood defenses. At Oeiras, large-scale hydraulic interventions in streams started in 2005 and were carried out on all streams of the county in order to promote adequate stream flow. It was within this scope that scientific biological assessments began to take place at Oeiras streams. Bioindicators have long been used in ecological assessments of surface water quality, which is subject to degradation from impacts ranging from urbanization to industrialization and agriculture (INAG 2008; Karr 1999), and a monitoring program was started as a way to assess water quality.

The first capture of a few individuals of African clawed frogs in Laje stream (Fig. 6.3) was performed in 2006 by a team of biologists from ISPA—Instituto Universitário and from Vasco da Gama Aquarium, during a freshwater fish survey (Robalo et al. 2007) near the Gulbenkian Institute for Science. The team reported their finding to the Environment Department of the municipality services which, in the same year, found the species about 2 km upstream of the first site during the stream restoration work.

In 2008 a team from ISA—Instituto Superior de Agronomia—detected African clawed frogs on a different stream in the same region. This discovery resulted from the assessment of the ecological quality of rivers at national level—Aquariport Project (Oliveira (coord.) et al. 2007)—also carried out under the Water Framework Directive. This second stream—"Barcarena stream" runs roughly parallel to Laje stream, with an average distance of approximately 5 km between both (Fig. 6.3).

The introduction into Barcarena stream may have been human-assisted, as the presence of genetically identical exemplars in both streams was an indication of a common source (De Busschere et al. 2016). However, recently Ihlow et al. (unpubl. ms.), modeled the frogs' overland dispersal ability across heterogeneous landscapes and showed that the species could cross from Laje to Barcarena basin by using the tributaries of Laje stream located near Talaíde (Fig. 6.3). This connection between basins is about 4 km upstream of the introduction site. If this was indeed the dispersion corridor, the species may have benefited from another historical episode —in 2002, a golf course was built exactly in the center of this corridor. The golf course developers used the natural landscape depressions to create four artificial lakes, and the species was detected in two of them.

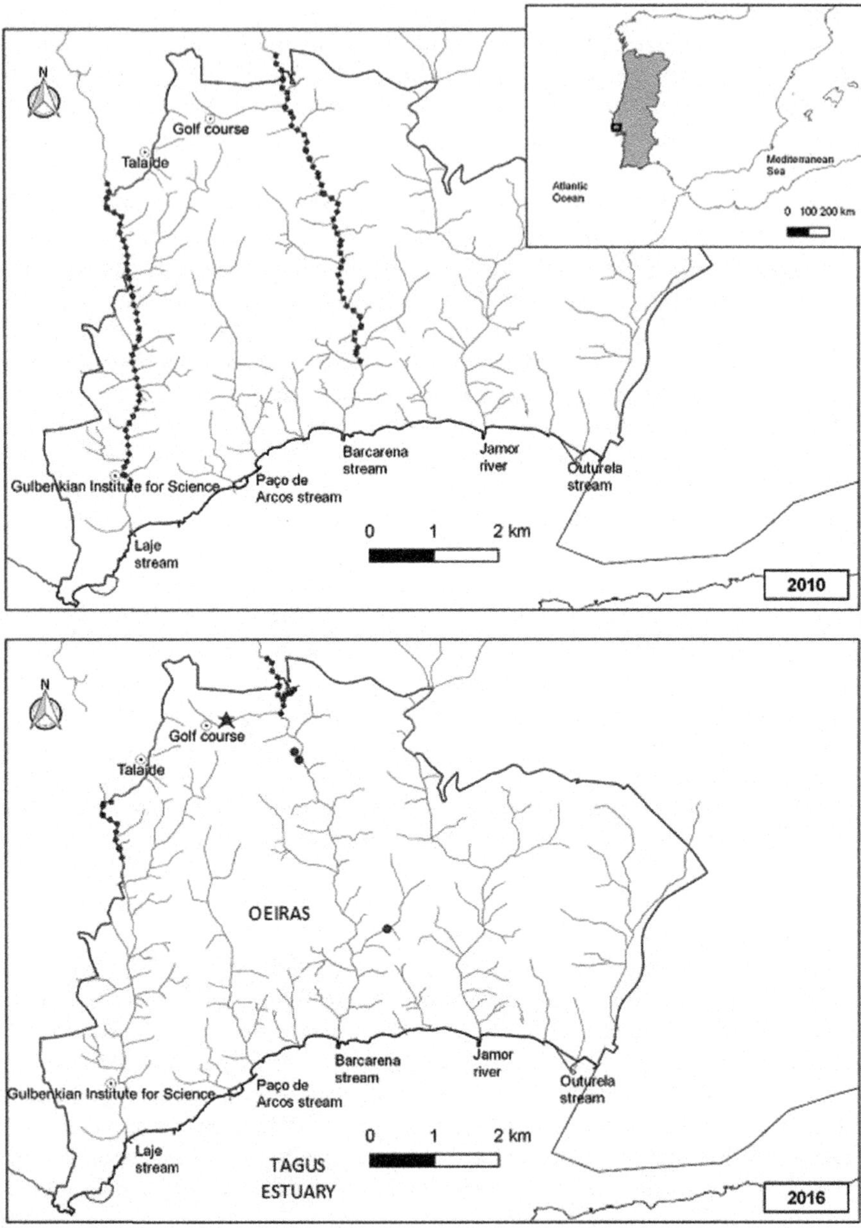

Fig. 6.3 Maximum extension of African clawed frogs known in 2010 and 2016 in Oeiras and Sintra counties near Lisbon, Portugal. Diamond icons along the river courses indicate frog presence; circles indicate cement ponds; the star indicates a golf course lake

6.4 The Portuguese Control and Eradication Program

The first sightings of African clawed frogs were all reported for the lower reaches of both streams (Laje and Barcarena), almost totally located within the Oeiras County. This discovery prompted the necessity to assess the presence of the species in the surrounding hydrographic network, such as the river Jamor and the streams of Porto Salvo and Outurela and other secondary streams, as well as in the remaining unexplored stretches of Laje and Barcarena. All these streams collect the flow of very small catchments and flow into the Tagus Estuary. The sources of the main watercourses—Jamor River and Laje and Barcarena streams—are all located east of Serra de Sintra (Sintra County).

In an attempt to evaluate the status and distribution of the species in that region the municipal services requested the technical and scientific support of the team that had conducted field surveys on the distribution and abundance of this species in the region in 2007 and 2008 (Rebelo et al. 2010). The results showed that there was a main site where the local population seemed to be more abundant, female-biased and with a more balanced body size distribution than at the other sites. This was the probable place of the initial introduction, in Laje stream near the Gulbenkian Institute for Science and a public garden—the Jardim do Marquês de Pombal.

After the initial studies and taking into account the confinement to a small catchment with apparent poor density and habitat quality for the species, and the documented impacts of this invasive, the Institute of Nature Conservation and Forestry, as the entity responsible for nature conservation, coordinated an inter institutional collaboration protocol, starting in 2010, with the aim to control and eventually eradicate the species. The African clawed frog eradication program was established between four organizations: the Institute of Nature Conservation and Forestry; the Municipality of Oeiras; the Center for Ecology, Evolution and Environmental Change of the Faculdade de Ciências, Universidade de Lisboa; and the Gulbenkian Institute of Science, and is expected to last until 2020.

Although some attempts had already been made to eradicate populations of African clawed frogs in several countries (Measey et al. 2012), only four succeeded. In all cases eradications were successful in small invaded areas or areas subject to very extreme weather events:

- in a public garden in Barcelona (Spain), where after the discovery in April 2007, the eradication was carried out through the drainage of existing ponds (Pascual et al. 2007);
- in Lily Pond, Golden Gate Park (California, USA) after persistent physical removal efforts in 2006 and 2008 and a gradual draining of the pond (Green et al. 2010);
- in North Carolina (USA) the population had not expanded in more than 20 years due to unsuitable surrounding habitat and established breeding populations, but went extinct due to freezing winter temperatures (Crayon 2005);
- in Great Britain, where the two invasive populations that settled in England (Lincolnshire) and Wales were considered extinct as a result of the combination

of eradication programs with exceptionally cold and dry winters (Tinsley et al. 2015). There have been no captures after 2010, when a single individual was caught in each of the areas.

Despite the low eradication success in other world areas, particularities in distribution and abundance of this invasive species in Portugal increase the feasibility of the control and eradication actions, and may ultimately allow its eradication in the medium term. These factors include:

- it took almost 25 years for the species be detected, suggesting a low density that may facilitate its control;
- available habitats are of low quality—the species can live in rivers but its use of this type of aquatic habitat has been mostly identified as routes for dispersion, and not as breeding sites which are mostly located in ponds or lakes (Fouquet and Measey 2006; Measey 2016);
- the two stream catchments where it was detected flow directly into the estuary, and while the species is tolerant of moderate salinity, its dispersal within the very large Tagus estuary is improbable;
- water flow in both streams has inter and intra-annual variability and there are several physical obstacles along the streams (e.g. waterfalls) that fragment population connectivity;
- a significant proportion of the stream banks have been transformed into cement walls instead of natural banks, making access to the surrounding terrestrial habitats more difficult;
- if the frogs manage to leave the streams, most of the surrounding territory is a highly urbanized area, where traffic and buildings hamper or block overland migrations.

The eradication program comprises prospecting and removing the species in all the water courses of Oeiras County: in Laje and Barcarena streams where the species has been detected, but also on all the remaining watercourses and other habitats, like garden ponds or golf course lakes close to the streams. The catch period occurs during the summer months of June to August, which is the period of higher activity of the species and water levels are low enough to allow access to the deepest reaches of the streams.

Depending on the specific context, different methods have been used for the removal of eggs, tadpoles and adults of African clawed frogs. Electrofishing[1] has been the main method for adult captures, coupled with dip-net sweeps for the tadpoles and eggs. In some artificial habitats, like irrigation and ornamental ponds, a traditional Portuguese fishing technique was applied—*xávega*, dragging a vertically-maintained fishing net along the bottom of the pond—to capture the tadpole swarms. These methodologies are safe for human beings and the

[1]The electrofishing method relies on two electrodes which deliver direct current through the water. Aquatic animals are affected by the electricity, which temporarily stuns the individuals, allowing the researchers to scoop them with nets.

environment and do not affect the recovery capacity of the native species or habitats associated.

At the beginning of the eradication program the species was captured over long sections of Laje and Barcarena streams, located almost exclusively within Oeiras County (Fig. 6.3). It was not found at the mouths of streams, subject to the saline intrusion of the Tagus Estuary. There was a higher number of captures in Laje stream in the first year of the program (84% of the total captures) than the Barcarena stream (Fig. 6.4). In this second stream the species was initially very rare and scattered along the watercourse. The most well-established population, where individuals of both sexes and of all life stages were found was located in Laje stream, namely in Jardim do Marquês de Pombal (the probable site of introduction of the species).

Skeletochronological[2] analysis of frog humerii (long limb bones) showed that the Laje population was characterized by an aged population with larger and longer-lived individuals than in Barcarena where the age structure was younger (Ferreira 2014). These results suggested that Barcarena may have better conditions for reproduction, but worse conditions for adult survival, while the opposite may happen in Laje (Fig. 6.5). These results could be explained by the environmental differences between the reproduction sites, but the aged population of Laje may also be a result of the eradication program (Ferreira 2014). The oldest animal was found in Jardim do Marquês de Pombal, and was at least 7 years old.

Throughout the duration of the control and eradication actions, the remaining population in Laje stream became residual, with isolated specimens occurring in a relatively short section that coincides with the last sites where reproduction was registered. In the last four years (2012–2016) there have been no upstream or downstream expansion records, and less than 20 adults were caught per year.

Although Barcarena stream presents poor conditions for the reproduction of the species in the main watercourse, in 2014 there was an impressive invasion episode of the stream by several thousand juveniles. The number of African clawed frogs captured in that year represents 73% of the total number of captures in that stream since the beginning of the control measures. This invasion prompted the search for potential reproduction sites that had been overlooked until then, and these sites were eventually found in small tributaries of the main river, tanks, irrigation ponds and a lake on a golf course. The majority of these sites are still-water habitats. Presently, the number of known breeding sites in Barcarena catchment is much higher than in Laje's, which reflected the higher captures at Barcarena stream in the last years. More than 15,000 tadpoles were collected at those sites in 2015 and 2016, potentially compromising successful reproduction where the species maintains its last stronghold and contributing to the prevention of mass recruitment episodes such as the one recorded in 2014. In Barcarena stream the species is now

[2]Skeletochronology is one of the histological techniques that allow the determination of age in amphibians whose annual growth is cyclical. It is based on counting the lines of arrested growth that form usually once per year, during the winter. This allows estimating the age of the individuals, their age at sexual maturation and their longevity.

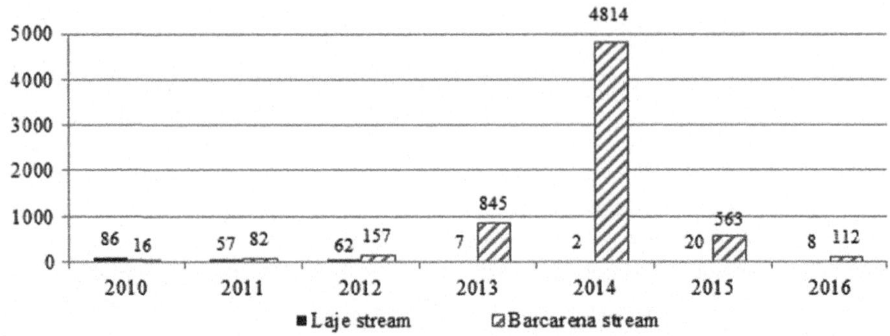

Fig. 6.4 Number of captures of African clawed frogs in Laje and Barcarena catchments over the seven years of control actions (2010–2016)

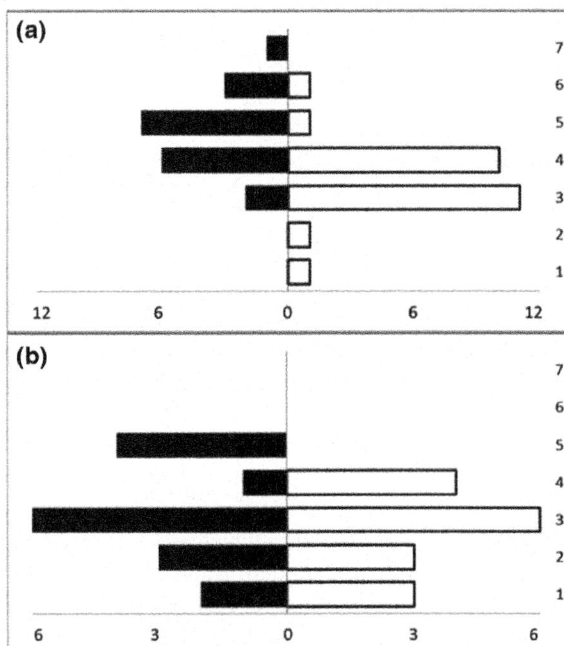

Fig. 6.5 Age structure from Laje (**a**) and Barcarena (**b**) populations of African clawed frogs in 2010 (adapted from Ferreira 2014). Black bars—females; white bars—males. X-axis—number of individuals; right Y-axis—age

restricted to a section approximately 800 meters long, and in 2016 about 60% of the captures were concentrated in two small tributaries located near the largest concentration of breeding sites in that catchment (Fig. 6.3).

After seven years of control actions the African clawed frog is restricted to upstream areas of Laje and Barcarena catchments (Fig. 5.3). The presence of still water in the immediate vicinity of the main streams appears to be one of the most important factors associated with the successful reproduction of the frog in these streams. In Laje stream the absence of these type of water bodies may have contributed to the success of the eradication program, while in Barcarena stream, its presence may be the most important deterrent of the program's success.

The reduction in the distribution extent of the species and of its abundance in both streams (Fig. 6.3) suggests efficacy of the control actions. The local population shows low recovery rates both in terms of adult immigration and recruitment of juveniles.

6.5 Role of Historical Contingencies in the Invasion by the African Clawed Frog in Portugal

The history of the introduction of African clawed frogs in Oeiras streams and of its control and eradication attempts is an example of the importance of accidents and contingencies. In fact, it has been characterized by a series of unlikely events which together have facilitated the establishment and spread of the species, as follows:

- the African clawed frog, which is the amphibian most commonly used in laboratories globally, happens to be a widely distributed species, with populations adapted to different climates, including a Mediterranean-type climate in southwestern South Africa. A large majority of the specimens exported worldwide come from that region;
- a Portuguese scientist allegedly brought some specimens to a laboratory in Portugal during a time when there were few concerns with the detention of invasive alien species;
- a disastrous flood inundated the basement where the species was being kept in aquaria;
- the expansion of the species was allowed because for years the streams were neglected;
- a golf course was built in the single potential passageway between the two basins, which allowed the expansion to a second stream catchment with more favorable conditions for reproduction and tadpole development.

Once the species was able to establish itself in the natural environment, its control and possible eradication became much more difficult. The eradication program and its success greatly depend on the commitment of institutions and organizations involved, sustained over time. Success will also be dependent on chance. In fact, the geographical proximity to urban centers and the territorial characteristics that have apparently restricted the species dispersal are also contributing for the success of the control and eradication program. Another important fact is that the species has little social or commercial value (it is still being sold as a pet in many countries), and hence there is no opposition to its eradication, which is essential to ensure an effective management.

The introduction of *Xenopus laevis* in Portugal is also an example of the role that research institutes have had in the past on the introduction of invasive species.

Indeed, deliberate releases or escapes from research laboratories were the cause of most invasive populations of this species (Measey et al. 2012). Whether intentionally or by accident, there are several known instances of introductions of other species from laboratories (Katsanevakis et al. 2015), and the Portuguese example prompted the adoption of more stringent biosafety measures in Portugal, including those that had already been proposed in the Decree-law 565/99, 21st December 1999, which, among other things, defines the safety requirements for installations intended to hold exotic species.

Acknowledgements The authors would like to acknowledge the funding of this study by BiodivERsA: "Invasive biology of *Xenopus laevis* in Europe: ecology, impact and predictive models" (BIODIVERSA/0001/2012). The authors would also like to thank Nicolau Ser, Francisco Moreira and Raquel Marques for their contributions to the field work. Alexandra Freitas provided constructive comments on the manuscript. Animals were captured under Permit from Instituto da Conservação da Natureza e das Florestas, I.P., within the scope of the "Plano de erradicação de *Xenopus laevis* nas ribeiras do Concelho de Oeiras."

References

Amaral P, Rebelo R (2012) Diet of invasive clawed frog *Xenopus laevis* at Lage stream (Oeiras, W Portugal). Herpetological J 22:187–190

Arano B, Llorente G, García-Paris M, Herrero P (1995) Species translocation menaces Iberian waterfrogs. Conserv Biol 9:196–198

Benito G, Machado MJ (2012) Floods in the Iberian Peninsula. In: Kundzewicz ZW (ed) Changes of flood risk in Europe, IAHS special publication, vol 10. IAHS Press, Wallingford, pp 372–383

Boix D, García-Berthou E, Gascón S, Benejam L, Tornés E, Sala J, Benito J, Munné A, Solà C, Sabater S (2010) Response of community structure to sustained drought in Mediterranean rivers. J Hydrol 383:135–146

Courant J, Thirion J-M, Guillon M, Grillet P, Grosselet O (2014) Le régime alimentaire de *Xenopus laevis* (Daudin, 1802) (Anura: Pipidae) introduit en France. B Soc Herpétol Fr 150:1–7

Crayon J (2005) Species account: *Xenopus laevis*. In: Lannoo MJ (ed) Amphibian declines: the conservation status of United States species. University of California Press, Berkeley, pp 522–525

Crossland M (2000) Direct and indirect effects of the introduced toad *Bufo marinus* (Anura: Bufonidae) on populations of native anuran larvae in Australia. Ecography 23:283–290

De Busschere C, Courant J, Herrel A, Rebelo R, Rödder D, Measey GJ, Backeljau T (2016) Unequal contribution of native South African phylogeographic lineages to the invasion of the African clawed frog, *Xenopus laevis*, in Europe. PeerJ 4:e1659

Eggert C, Fouquet A (2006) A preliminary biotelemetric study of a feral invasive *Xenopus laevis* population in France. Alytes 23:144–149

Faraone FP, Lillo F, Giacalone G, Lo Valvo M (2008) The large invasive population of *Xenopus laevis* in Sicily, Italy. Amphibia-Reptilia 29:405–412

Ferreira AC (2014) Estrutura etária e crescimento das populações invasoras da rã *Xenopus laevis* em Portugal: um estudo esqueletocronológico. Universidade de Évora, Tese de Mestrado

Flower SS (1936) Further notes on the duration of life in animals—II amphibians. Proc Zool Soc London 1936:369–394

Fouquet A, Measey GJ (2006) Plotting the course of an African clawed frog invasion in Western France. Anim Biol 56:95–102

Green SL, Felt S, Wilson S (2010) South African clawed frogs (*Xenopus laevis*) in Golden Gate Park, San Francisco. In: Proceedings 17th international conference on aquatic invasive species, San Diego, CA, USA

Gurdon J (1996) Introductory comments: *Xenopus* as a laboratory animal. In: Tinsley RC, Kobel HR (eds) The biology of *Xenopus*. Oxford University Press, Oxford, pp 3–6

Gurdon J, Hopwood N (2000) The introduction of *Xenopus laevis* into developmental biology: of empire, pregnancy testing and ribosomal genes. Int J Dev Biol 44:43–50

INAG (2008) Manual para a avaliação biológica da qualidade da água em sistemas fluviais segundo a Diretiva Quadro da Água: Protocolo de amostragem e análise para os macroinvertebrados bentónicos. Ministério do Ambiente, Ordenamento do Território e do Desenvolvimento Regional. Instituto Nacional da Água, I.P., Lisboa

Jokumsen A, Weber R (1980) Hemoglobin-oxygen binding properties in the blood of *Xenopus laevis*, with special reference to the influences of estivation and of temperature and salinity acclimation. J Exp Biol 86:19–37

Karr J (1999) Defining and measuring river health. Freshwater Biol 41:221–234

Katsanevakis S, Dreiu I, Nunes AL, Sanchez SP, Crocetta F, Arianoutsou M, Bazos I, Christopoulou A, Curto G, Delipetrou P, Kokkoris Y, Panov VE, Rabitsch W, Roques A, Scalera R, Shirley SM, Tricarico E, Vannini A, Zenetos A, Zervou S, Zikos A, Cardoso AC (2015) European Alien Species Information Network (EASIN)—supporting European policies and scientific research. Manag Biol Invasions 6:147–157

Kraus F (2009). Alien reptiles and amphibians. A scientific compendium and analysis. Drake, James A. (ed), Invading nature—Springer series in invasion ecology, vol 4. Springer Science + Business Media B.V., The Netherlands

Lillo F, Faraone FP, Lo Valvo M (2011) Can the introduction of *Xenopus laevis* affect native amphibian populations? Reduction of reproductive occurrence in presence of the invasive species. Biol Invasions 13:1533–1541

Llewelyn J, Schwarzkopf L, Alford R, Shine R (2009) Something different for dinner? Responses of a native Australian predator (the keelback snake) to an invasive prey species (the cane toad). Biol Invasions 12:1045–1051

Lobos G, Jaksic FM (2005) The ongoing invasion of African clawed frogs (*Xenopus laevis*) in Chile: causes of concern. Biodivers Conserv 14:429–439

Lobos G, Measey GJ (2002) Invasive populations of *Xenopus laevis* (Daudin) in Chile. Herpetol J 12:163–168

Marques RP (2016) Impactos tróficos da rã-de-unhas-africana *Xenopus laevis* na ribeira de Barcarena (Oeiras, Portugal). Mestrado em Biologia da Conservação. Faculdade de Ciências da Universidade de Lisboa, Lisboa

McCoid MJ, Fritts TH (1980) Notes on the diet of a feral population of *Xenopus laevis* (Pipidae) in California. Southwest Nat 25:272–275

McCoid MJ, Fritts TH (1989) Growth and fat body cycles in feral populations of the African clawed frog, *Xenopus laevis* (Pipidae), in California with comments on reproduction. Southwest Nat 34:499–505

Measey GJ (1998) Diet of feral *Xenopus laevis* (Daudin) in South Wales, UK. J Zool 246:287–298

Measey GJ (2016) Overland movement in African clawed frogs (*Xenopus laevis*): a systematic review. PeerJ 4:e2474

Measey GJ, Rödder D, Green SL, Kobayashi R, Lillo F, Lobos G, Rebelo R, Thirion J-M (2012) Ongoing invasions of the African clawed frog, *Xenopus laevis*: a global review. Biol Invasions 14:2255–2270

Measey GJ, Vimercati G, Villiers FA, Mokhatla MM, Davies SJ, Edwards S, Altwegg R (2015) Frog eat frog: exploring variables influencing anurophagy. PeerJ 3:e1204

Measey GJ, Vimercati G, Villiers FA, Mokhatla M, Davies SJ, Thorp CJ, Rebelo AD, Kumschick S (2016) A global assessment of alien amphibian impacts in a formal framework. Divers Distrib 22:970–981

Oliveira JM (coord.), Santos JM, Teixeira A, Ferreira MT, Pinheiro PJ, Geraldes A, Bochechas J (2007) Projeto Aquariport: Programa nacional de monitorização de recursos piscícolas e de avaliação da qualidade ecológica de rios. Direção-Geral dos Recursos Florestais. Lisboa
Pascual G, Llorente GA, Montori A, Richter-Boix A (2007) Primera localización de *Xenopus laevis* en libertad en España. Boletín de la Asociación Herpetológica Española 18:42–43
Price SJ, Garner TWJ, Nichols RA, Balloux F, Ayres C, Mora-Cabello de Alba A, Bosch J (2014) Collapse of amphibian communities due to an introduced Ranavirus. Curr Biol 24:2586–2591
Rebelo R, Amaral P, Bernardes M, Oliveira J, Pinheiro P, Leitão D (2010) *Xenopus laevis* (Daudin, 1802), a new exotic amphibian in Portugal. Biol Invasions 12:3383–3387
Richter-Boix A, Garriga N, Montori A, Franch M, San Sebastián O, Villero D, Llorente GA (2012) Effects of the non-native amphibian species *Discoglossus pictus* on the recipient amphibian community: niche overlap, competition and community organization. Biol Inv 15:799–815
Robalo JI, Doadrio I, Valente A, Almada V (2007) Identification of ESUs in the Critically Endangered Portuguese minnow *Chondrostoma lusitanicum* Collares-Pereira 1980, based on a phylogeographical analysis. Conserv Genet 8:1225–1229
Schoonbee HJ, Prinsloo JF, Nxiweni JG (1992) Observations on the feeding habits of larvae, juvenile and adult stages of the African clawed frog, *Xenopus laevis*, in impoundments in Transkei. Water SA 18:227
Silva-Rocha I, Salvi D, Sillero N, Mateo JA, Carretero MA (2015) Snakes on the Balearic islands: an invasion tale with implications for native biodiversity conservation. PLoS ONE 10:1–18
Soto-Azat C, Clarke BT, Poynton JC, Cunningham AA (2010) Widespread historical presence of *Batrachochytrium dendrobatidis* in African pipid frogs. Divers Distrib 16:126–131
Tinsley RC, Kobel HR (1996) The biology of *Xenopus*. Oxford University Press, Oxford
Tinsley RC, McCoid MJ (1996) Feral population of *Xenopus* outside Africa. In: Tinsley RC, Kobel HR (eds) The biology of *Xenopus*. Oxford University Press, Oxford, pp 81–94
Tinsley RC, Loumont C, Kobel HR (1996) Geographical distribution and ecology. In: Tinsley RC, Kobel HR (eds) The biology of *Xenopus*. Oxford University Press, Oxford, pp 33–39
Tinsley RC, Stott LC, Viney ME, Mable BK, Tinsley MC (2015) Extinction of an introduced warm-climate alien species, *Xenopus laevis*, by extreme weather events. Biol Inv 17:3183–3195
Van Sittert L, Measey GJ (2016) Historical perspectives on global exports and research of African clawed frogs (*Xenopus laevis*). Trans R Soc S Afr 71:157–166

Chapter 7
Ecologically Benign Invasions: The Invasion and Adaptation of Common Waxbills (*Estrilda astrild*) in Iberia

Gonçalo C. Cardoso and Luís Reino

Abstract Invasive species often damage the ecosystems they colonise. But non-harmful biological invasions also exist in nature, and understanding which biological invasions are ecologically benign is important for prioritizing conservation goals. As a case study, we review research on the biological invasion of the Iberian Peninsula by the common waxbill (*Estrilda astrild*), a small estrildid finch from sub-Saharan Africa, with the objectives of assessing the potential for detrimental impact on Mediterranean ecosystems, and illustrating how biological invasions provide research models for diverse sub-disciplines in ecology. Common waxbills are traded as pets and, starting from birds escaped or released from captivity, have been invading the Iberian Peninsula since the 1960s. Range expansion initially used coastal areas and river valleys, and later progressed to more inhospitable sites at higher altitude or inland. Colonising those diverse ecological environments poses adaptive challenges, and phenotypic changes have accompanied this invasion. Differences in personality among populations appear adaptive to the different climatic regimens waxbills colonised, and changes in ornamentation during the invasion are best explained by ecologically-mediated differences in sexual selection among sites. These phenotypic changes show the usefulness of invasive species for research in behavioural and evolutionary ecology. The niche waxbills occupy in Iberia is marginal in relation to the ecological space occupied by native passerines, and waxbills appear not to compete strongly with natives. The differentiated ecological niche that waxbills occupy is in part due to anthropogenic

G. C. Cardoso (✉) · L. Reino
CIBIO/InBIO—Research Centre in Biodiversity and Genetic Resources,
University of Porto, Porto, Portugal
e-mail: gcardoso@cibio.up.pt

L. Reino
e-mail: luis.reino@cibio.up.pt

G. C. Cardoso
Behavioural Ecology Group, Department of Biology, University of Copenhagen,
Copenhagen, Denmark

A. I. Queiroz and S. Pooley (eds.), *Histories of Bioinvasions in the Mediterranean*,
Environmental History 8, https://doi.org/10.1007/978-3-319-74986-0_7

influences (irrigation of agricultural areas, and food provided by exotic plants). We conclude that anthropogenic modification of the landscape creates novel niches that native species may not have filled in, and that thus can accommodate some exotic species with minimal interference on the native community.

Keywords Avian exotic species · Range expansion · Adaptation
Vacant ecological niche

7.1 Introduction

Introductions of non-native avian species in Mediterranean-type regions have had a high probability of success in establishing permanent populations and becoming invasive, especially around the Mediterranean basin (Kark and Sol 2005). Since invasive species can compete with natives or be otherwise detrimental for the ecosystems (Blackburn et al. 2014; Vilà and Hulme 2017), this raises concern about the harm that these non-native birds may cause to Mediterranean ecosystems and their avian communities. Non-harmful biological invasions are given less attention in the literature, yet most avian invasions may not be ecologically detrimental (Blackburn et al. 2009; Goodenough 2010). Knowing which invasions are ecologically damaging and which are benign is necessary to plan and prioritise conservation strategies. Investigating why ecologically benign invasions occur is also important to understand the functioning of ecological communities.

Following much of the literature on avian (e.g. Blackburn et al. 2009) and other biological invasions (e.g. Richardson et al. 2000), we define invasive species as those that spread autonomously after introduction in a non-native area, irrespective of economic or ecological impacts. Alternative definitions of invasive (e.g. taking into account economic losses or threat to biodiversity) may be useful for, for example, policy making or conservation. But here we choose a definition based solely on demographic history because it does not create expectations as to the ecological effects of non-native species, instead leaving this to empirical research.

As a case study of an avian invasion in the Mediterranean region, here we use the common waxbill (*Estrilda astrild*), a small finch from sub-Saharan Africa that is traded as a pet bird, and that became invasive in many parts of the world (Stiels et al. 2011) including, during the last 50 years, the Iberian Peninsula (Reino and Silva 1998). We review research on its invasion of the Iberian Peninsula with two objectives in mind. First, to describe the process of invasion, assess the ecological niche waxbills occupy in Iberia and the potential for competition with native birds. Second, to illustrate how studying biological invasions, irrespective of conservation implications, can contribute fundamental knowledge to several sub-disciplines in ecology. In this case, the biological invasion of waxbills proved a useful model for fundamental research in biological dispersal, behavioural and evolutionary ecology.

7.2 The Common Waxbill and the Biological Invasion of the Iberian Peninsula

7.2.1 The Common Waxbill

The common waxbill is a small, colourful finch (Fig. 7.1) that occurs natively across most sub-Saharan Africa, where it is associated mostly with open habitats, often in the proximity of water (Clement et al. 1993; Payne 2010). It is a granivorous bird that forages in low vegetation and, since it is very small (ca. 8 grams; Batalha et al. 2013), it is extremely agile, often taking small seeds upside-down or in the bent stems of fragile plants (Clement et al. 1993; Matias 2002).

Common waxbills are not territorial, and instead often nest scattered in loose colonies, with each pair building a domed nest and providing bi-parental care to nestlings (Clement et al. 1993; Payne 2010). Time of breeding is highly variable throughout their native range, and in some places waxbills are reported to breed throughout the year (Payne 2010; Sanz-Aguilar et al. 2015), reflecting an opportunistic breeding strategy to take advantage of favourable conditions for reproduction whenever they occur. This is a common strategy in tropical bird species adapted to environments with low or unpredictable climatic seasonality (Hau et al. 2008).

Fig. 7.1 A common waxbill (*Estrilda astrild*) showing the red colour ornamentation that adult males and females have in the bill and plumage

Common waxbills are highly gregarious, roosting communally and gathering in flocks year-round (Clement et al. 1993; Payne 2010). They are not migratory, but make vagrant movements, perhaps searching for suitable foraging or breeding areas (Payne 2010). Gregariousness and, when necessary, vagrant movements seem instrumental for waxbills living in habitats that provide resources irregularly, such as savannas where it is unpredictable whether water, and the associated vegetation, will last or be depleted. Similarly, the ability of waxbill flocks to make vagrant movements, in search of suitable habitat, was likely important for their establishment and invasion success in the many non-native areas where they were introduced.

As with several other colourful species in the family Estrildidae, waxbills have for long been part of the pet trade. During the nineteenth century waxbills were already well-known cage birds in Europe (Fig. 7.2), there are records of them being actively released through the XIX and XX centuries in Brazil (Sick 1997), and they were transported by humans around the world (Stiels et al. 2011). Some estrildid species invaded large non-native ranges (e.g. Stiels et al. 2015), with the common waxbill being the most successful in terms of the diversity of locations and area

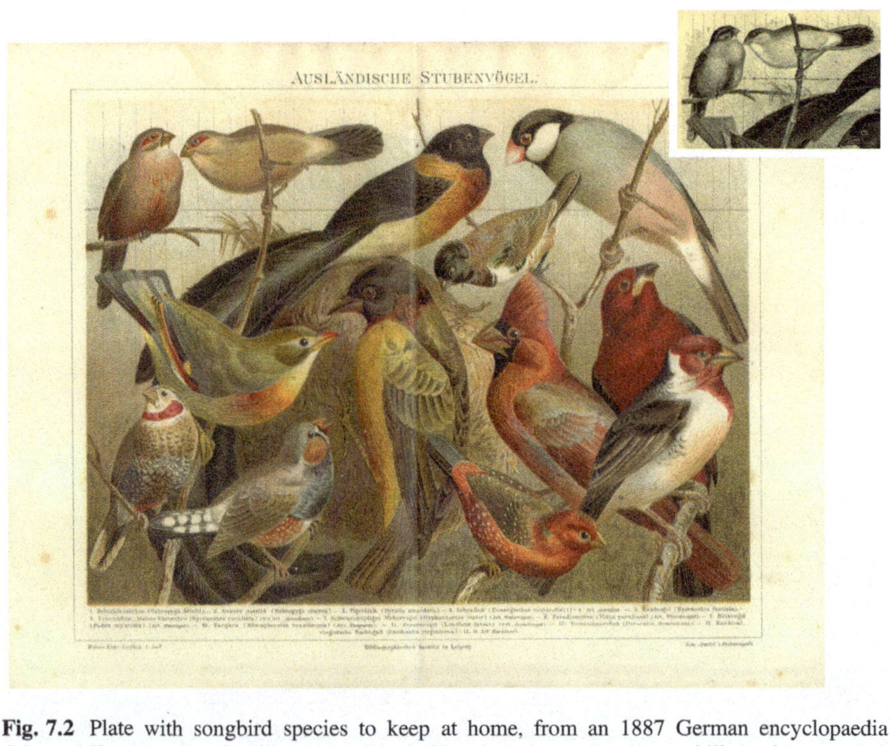

Fig. 7.2 Plate with songbird species to keep at home, from an 1887 German encyclopaedia (Meyers' Konversations Lexikon, 4th edition). The plate shows that the waxbill, on the upper left corner, was already a well-known cage bird in Europe during the nineteenth century. The inset is a detail from the 3rd edition (1874)

covered. The regions where it was introduced include many islands in the Atlantic, Indian and Pacific oceans, large continental regions of South America and, in Europe, the Iberian Peninsula (Stiels et al. 2011).

7.2.2 The Invasion of the Iberian Peninsula

There is limited information on the history of trading of common waxbills in Iberia. But Portugal had colonies in sub-Saharan Africa from the sixteenth century until the mid 1970s, the largest of which were Angola and Mozambique, where waxbills are a native species and from where they could easily be traded. Based on the evidence for waxbill commerce in another Portuguese colony (Brazil) and in Europe at least since the XIX century (Sick 1997 and Fig. 7.2), it is likely that common waxbills entered the Iberian aviculturist market long ago. By the mid-twentieth century waxbills were bred in captivity and were commercially available in different regions across Portugal (Raimundo Vicente, pers. comm.).

It was during the 1960s that the invasion of common waxbills started in Portugal. This was the decade when independence wars begun in Angola and Mozambique, and the associated human migratory movements from these colonies to Portugal might have increased waxbill commerce and breeding. Currently we do not know whether these historical events are related to the onset of the waxbill invasion in Portugal, or whether the similar timing was coincidental. The first known established waxbill populations were in Central Portugal, in locations near the coast or near the Tagus River (Fig. 7.3a; Reino and Silva 1996a, 1998), and recent reports suggest that in the late 1960s waxbills were also present more inland, near the city of Coimbra in Central Portugal (José Mourão pers. comm.). Range

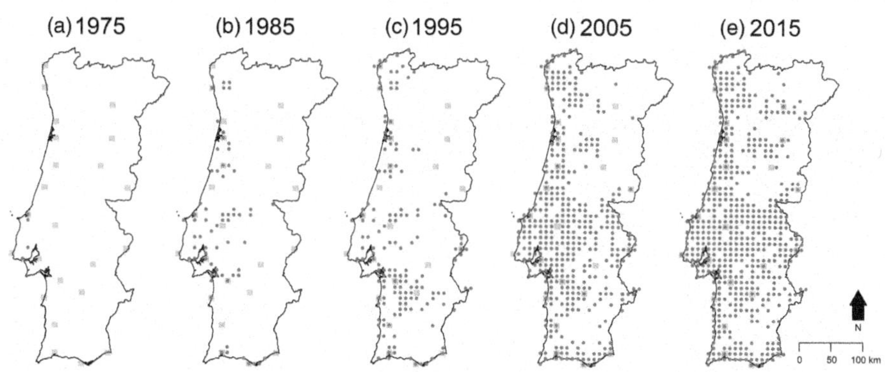

Fig. 7.3 Distribution range of the common waxbill (*Estrilda astrild*) in mainland Portugal from 1975 to 2015 (grey dots), mapped on a 10 × 10 km cell grid. Yellow squares mark grid cells where fieldwork took place, in 2010 and 2011, for the research in adaptation during the biological invasion (see main text). Distribution data was obtained mainly from breeding atlases (Rufino 1989; Equipa Atlas 2008) and also from other published and unpublished records (methods as in Sullivan et al. 2012)

expansion was quite limited until the mid-70s, but towards the end of the decade expansion accelerated, progressing mostly along coastal regions and also starting to expand inland along the Tagus valley (Reino and Silva 1996a).

The historical progression of the invasion suggests that there may have been later introductions near the coastline further north and also in the south coast. For example, in the 1980s (Fig. 7.3b) there were waxbills in the Algarve, quite far away from the remaining waxbill distribution at the time. Such discontinuous distributions were observed during the 1980s (Fig. 7.3b) and could have been due to independent introductions along the coastline or, alternatively, long-distance dispersion of individuals from central Portugal to those areas before a more continuous and sustained range expansion started to unfold (Reino and Silva 1996a, 1998).

The first habitats waxbills were associated with were closely connected with water (e.g. marshes, riverine vegetation, rice fields; Reino and Silva 1998; Sullivan and Franco 2018) and, perhaps because of this, the expansion was faster along the north coast of Portugal, eventually reaching Galicia in Spain (ca. 1988; Vidal and Reino 2002). The expansion was slower along the south coast of Portugal (Reino and Silva 1998), perhaps because there are less wet areas along the Alentejo coast south of Setúbal. In the meantime, the focus of expansion centred in the Algarve started spreading eastwards into Andalucia, in Spain, and northwards along the Alentejo coast to join the waxbills coming from the north, and then started to colonise inland as well (Reino and Silva 1998). By the 1990s the colonisation of the Portuguese coast was complete (Silva et al. 2002). Colonisation inland (eastwards) was slower, and currently expansion is still progressing in inland Spain, especially along river valleys, and there are also independent foci of invasion along the Mediterranean coast of Spain in the regions of Malaga, Valencia and Barcelona (Reino and Silva 1998; Martí and Moral 2003; Sullivan et al. 2012).

Nowadays, waxbills are naturalized across an extensive area of Iberia, mainly Portugal and adjacent parts of Spain, where they have very low rates of haemosporidian parasitism (Ventim et al. 2012; Lopes et al. 2018), lower than in their native African grounds (Lopes et al. 2018). This may have helped their successful range expansion. Waxbills in Iberia have a preference for heterogeneous habitats, usually including agricultural fields along river systems and tributaries (Reino 2005; Sullivan et al. 2012). Some areas of Portugal have not been colonised or harbour smaller waxbill populations than elsewhere, mainly in mountainous inland regions. Nonetheless, it is remarkable that waxbills eventually were able to colonise some of these relatively high-altitude and cold sites, although those sites have lower numbers of waxbills (Equipa Atlas 2008).

7.2.3 Modelling Geographic Range Expansion

If monitored closely, recent introductions of species outside their native range and subsequent expansions provide excellent data to model biological invasions and range expansion, and to test and develop new modelling approaches. Modelling attempts to synthetize the descriptive historical information on range expansions to infer patterns and causes governing those expansions. Biological invasions and range expansions have been modelled using various mathematical approaches over the last four decades (Hui and Richardson 2017; Engler et al. 2017), and more recently Species Distribution Models (SDMs) also attempt to understand which environmental factors are associated with the expansion process or may limit further colonisation to new areas (Guisan and Thuiller 2005). The introduction of waxbills in Iberia and the descriptive data on the progression of the invasion was an opportunity not just to study why this invasion was so successful, but also to use and develop innovative SDMs.

Until the 1990s, ornithological reports on the introduction and invasion of waxbills in Iberia were brief and geographically restricted (e.g. on the river Sado valley; Ferreira 1980–1981) or limited to species accounts in bird atlases (e.g. Rufino 1989). The first systematic descriptions of the spread of this exotic species were published in 1996 (Reino and Silva 1996a, b), gathering all available data on the occurrence and on the timing of first occurrence of waxbills in Iberia/Portugal (primarily using data from a Portuguese bird atlas project, Rufino 1989, and inquiries local ornithologists). Using these data, Silva et al. (2002) modelled for the first time the waxbill expansion in Portugal, using an auto-logistic model to account for spatial autocorrelation and deal with spatial heterogeneity. Auto-logistic modelling is also used in other fields of ecology, and at the time it was somewhat novel in the field of biological invasions. Results showed that spatial effects were extremely important, meaning that most range expansion progressed gradually through space rather than through leaping to distant areas, and that expansion was limited by less suitable environmental conditions, namely high altitude and high solar radiation.

Reino (2005) then used a variation partitioning method to quantify the relative importance of spatial effects and environmental conditions in the progression of the waxbill invasion. Spatial effects alone (i.e. the presence of neighbouring populations, controlling for environmental variables) explained 33% of variation when modelling this invasion, while environmental factors alone explained only 11% of the variation. An additional 25% of variation was due to cases where the presence of neighbouring colonised cells and favourable environmental conditions coincide and jointly explain the progression of the waxbill invasion. Thus, the SDMs in Reino (2005) explain over two thirds of the variation on the spatial and temporal progression of this invasion, and show that the effect of habitat differences in guiding this invasion was limited by the availability of dispersers in neighbouring areas.

Reino et al. (2009) adapted another technique, survival regression, which is used to predict the timing of discrete occurrences (e.g. death of organisms, behavioural events, disease recurrence; Moya-Laraño and Wise 2000), as a tool to model range expansions. They treated the time of colonisation of each geographical site as an event, and showed that, given information on environmental differences among sites, this technique had significant predictive capabilities. A survival regression model built with information from early waxbill expansion (1979–1981) could not accurately forecast the timing of colonisation of sites later in the range expansion, but it did show that sites colonised later on had a higher predicted probability of invasion than sites still not colonised 20 years later. Using this predictive ability, Reino et al. (2009) evaluated the potential effects of climate change on the invasion of this species in Iberia, and suggested that an increase in annual temperature increases the risk of invasion to new areas. Ongoing climate change is also predicted to have a more global impact on the future distribution of waxbills, namely poleward shifts in areas suitable for waxbill colonisation, but overall less climatically suitable areas globally (Stiels et al. 2011). For example, less suitable areas are predicted in some areas around the Mediterranean basin (e.g. Central Spain and Northern Africa). However, global-scale predictions should be interpreted with caution due to uncertainties associated with climate change scenarios and SDMs.

Finally, Sullivan et al. (2012) used the Iberian waxbill data to address the problem of modelling the distribution of non-native species when they are absent from some areas due to dispersal limitations. The problem is that ecologically suitable areas are often not yet colonised in biological invasions because of geographical barriers or insufficient time for colonisation, which can confuse SDMs and make them inaccurate (e.g. Václavík and Meentemeyer 2012). Sullivan and colleagues developed a modelling framework that estimates the spatial probability of dispersal, and uses these dispersal probabilities to make areas unlikely to receive dispersers have less influence in fitting SDMs. The waxbill data allowed estimating dispersal probabilities as a function of distance, and incorporating this dispersal probability function into the SDM substantially improved its performance. Overall, this study shows that incorporating dispersal limitations SDMs can make them more accurate and provide better predictions of range expansion during biological invasions.

The introduction and subsequent invasion of the common waxbill in Iberia was a complex ecological process, influenced both by landscape and environmental factors, and by dispersal limitations. The interplay of these factors poses challenges for ecologists attempting to predict the future expansion of biological invasions. As we noted here, the detailed waxbill data from Iberia also provided an extraordinary opportunity to understand the dynamics of an invasion, to test and improve new modelling techniques. All the studies reviewed here collected or used datasets based mainly on bird atlases and monitoring schemes, plus personal communications provided by both professional and amateur ornithologists (summarized in Fig. 7.3). This shows the importance of biodiversity monitoring schemes (e.g. bird atlas and other avian censuses) and of citizen-science data in providing the basis for fundamental research on SDMs and biological invasions.

7.3 Adaptation During the Invasion

Microevolution during biological invasions seems to be common (Cox 2004; Buswell et al. 2011) and, because of this, invasive species are recognized as useful models to study evolutionary adaptation (Sax et al. 2007; Bridle and Vines 2007). Although the ability of recently-introduced exotic species to adapt may in a few cases be limited, for example when genetic variation among individuals is insufficient for selection to act (Bridle and Vines 2007), the literature now abounds with examples of adaptive evolution during biological invasions. There are examples of evolution towards withstanding different temperatures from those in the native range (e.g. Butin et al. 2005), improving dispersal during invasions (e.g. Phillips et al. 2006), and other changes that are adaptive in the invasive ranges (Cox 2004).

We used the Iberian invasion by common waxbills to study phenotypic changes that may be adaptive when colonising novel environments, albeit in a less obvious way than the adaptations commonly reported in the invasion biology literature. Namely, we looked at animal personality (i.e. consistent differences in behaviour among individuals), sexual ornamentation, and how they changed during the waxbill invasion. Explaining why personality differences exist within animal populations is a major current question in behavioural ecology (Dall et al. 2004; Wolf and Weissing 2010, 2012; Sih et al. 2012), but there is still little information on how personality differs among different populations, especially fast changes in personality along range expansions (Canestrelli et al. 2016a). Likewise, the microevolution of sexual ornamentation towards the geographic limits of range expansions can shed light on the build-up of reproductive isolation and, ultimately, speciation. But, again, data from the wild is very scarce. Here we explain how studying a waxbill biological invasion helped progress on both these topics.

7.3.1 Ecology Influences Animal Personality

Colonising Iberia should have placed adaptive challenges to waxbill populations, as they encountered habitats with contrasting climatic regimens. The colonised areas in Portugal, where the invasion was more thoroughly studied, range from dry and hot plains to cold highlands, some of which with seasonal snow (Silva et al. 2002; Reino 2005). As explained earlier, waxbills first colonised coastal lowlands near water courses, and only later progressed inland and to higher altitudes (Silva et al. 2002; Reino 2005), which are likely less hospitable for this species and where cold winters should limit their expansion. In addition to progressing from warmer lowlands to colder highlands, during the ca. 50 years of this invasion the expansion front passed through regions that are strongly seasonal (i.e. with large differences in temperature and precipitation between the seasons) and others that are more stable (i.e. milder climatic differences between seasons). In other words, geographic differences in climate seasonality do not correlate with the direction of this waxbill invasion (Carvalho et al. 2013). As explained next, it was this colonisation of sites with stable versus seasonal climate, and not the progression from lowlands to highlands, that had lasting effects on personality traits of waxbills.

Exploratory and social behaviour often differ among individuals of the same species (Réale et al. 2007). In waxbills, behaviour in assays of exploration (the *open field test*, where a bird explores a new environment; Perals et al. 2017) and in assays of sociability (the *mirror test*, whereby a bird reacts to its mirror image) differ in a correlated manner: more explorative birds are also less attentive to a social stimulus, their mirror image, while less exploratory birds spend more time looking towards the mirror image (Carvalho et al. 2013). Behavioural differences among birds in the *mirror test* are stable through time, and are retrieved even when re-tested several weeks or months apart (Carvalho et al. 2013; Funghi et al. 2015). Therefore, exploration and social attentiveness are part of a stable behavioural syndrome, or personality axis, rather than plastic differences in behaviour that might disappear rapidly.

These personality differences of waxbills (from slow explorers that are more social, to more active individuals that are less social) are similar to, for example, the zebra finch (*Taeniopygia guttata*), another estrildid finch (Rosa et al. 2012; McCowan and Griffith 2015), and fit well with the proactive-reactive axis of coping styles commonly reported in the behavioural literature (reviewed in Koolhaas et al. 1999; Carere et al. 2010). Coping styles reflect how animals use stimuli from the environment, and can be closely related to differences in stress physiology among individuals (Koolhaas et al. 1999; Carere et al. 2010). Proactive individuals usually have bolder but less flexible behaviour, as they explore fast using relatively fixed behavioural routines from their previous experience, while reactive individuals are thought to explore more slowly because they are attentive to environmental stimuli, and are thus better at learning new routines (e.g. Verbeek et al. 1994; Exnerová et al. 2010; Guillette et al. 2011).

When looking at the geographic variation in personality across the waxbill invasion (red cells in Fig. 7.3), we find that populations in more stable climates are on average more exploratory and less attentive to social stimuli (i.e. more proactive), while those in more seasonal sites are less exploratory and more attentive to social stimuli (i.e. more reactive; Carvalho et al. 2013). These geographic differences make adaptive sense when we realize that strongly seasonal climate causes changes in the location of food and other resources, while stable climates make these resources more predictable year-round, and that reactive animals may learn and change behavioural routines more easily. Furthermore, socially-transmitted information is particularly important for gregarious animals in unstable environments, as they benefit from the experience of group members to find food sources and solve other ecological challenges (e.g. Liker and Bókony 2009; Morand-Ferron and Quinn 2011; Aplin et al. 2012).

The interest of having used a recent biological invasion to document a relationship between animal personality and climate is that this gives us a notion of the time needed for personality differences among populations to adjust to ecology. Knowing this from non-expanding, long-established populations, would require long-term research to follow populations for several years or decades. In the case of this waxbill invasion, which colonised highly seasonal and more stable climates (Carvalho et al. 2013), the correlation between behaviour and climate indicates repeated changes in personality in a low number of generations (most of the expansion took less than 40 years, and therefore less than 40 generations; see above).

Showing that the mean personality in waxbill populations adjusts fast to ecological conditions, even though personality is stable at the individual level, lends support to the hypothesis that fluctuations in ecology or other selective factors can explain the coexistence of differences in personality within natural populations (Réale et al. 2007). Fluctuations in ecology (e.g. climate) are common in nature (Halley 1996; Bell 2010), either in the form of year-to-year variation or micro-geographic differences. Therefore, by favouring different personality types from one year to the other (e.g. Nicolaus et al. 2016), or across space followed by dispersal, environmental or selective fluctuations should dynamically maintain personality differences within populations.

7.3.2 Range Expansion and Sexual Ornamentation

Range expansions provide attractive research opportunities for evolutionary biologists interested in speciation, because expansions create peripherally isolated populations where most speciation occurs (Gavrilets 2003). Of particular interest is whether sexual ornaments evolve during expansions, because changes in ornamentation can mediate reproductive isolation (Gray and Cade 2000; Boughman 2001; Seehausen et al. 2008) and are associated with speciation (Cardoso and Mota 2008; Maia et al. 2013; Gomes et al. 2016). Therefore, divergence in ornamentation along range expansions could promote speciation, but studies of whether sexual signals or ornaments change during expansions are almost non-existent (Canestrelli et al. 2016b).

Common waxbills have conspicuous red colour ornamentation (Fig. 7.1): red breast plumage, and a stroke of red in the head made up of a plumage mask and the bill that gives the species its common name (waxbill, in reference to red sealing wax). As is common among birds (Stoddard and Prum 2011; Thomas et al. 2014), red colour in estrildids uses carotenoid pigments (McGraw and Schuetz 2004) that are obtained from the diet (McGraw 2006). Therefore, expressing carotenoid-based colours often depends on the energetic and nutritional state of birds (Burley et al. 1992; Hill and Montgomerie 1994; Hill 2000; McGraw et al. 2005; Hill et al. 2009). In waxbills too, there is evidence that bill colour is redder when birds have better body condition (Marques et al. 2016) or are not subject to energetically-stressing cold nights (Funghi et al. 2018), making red colour a likely signal of condition in this species.

Male and female waxbills are ornamented, which will be very useful for interpreting geographic changes in ornamentation. Typically, in species with conventional sex roles (i.e. where females invest more in reproduction than males), males are more ornamented than females. This is because, while female reproduction is limited by their resources and physiology, males can compete to fertilize several females, resulting in higher variance in male reproductive success and, consequently, stronger sexual selection on males (Andersson 1994). As expected for a species with conventional sex roles, male waxbills are on average more ornamented than females (Cardoso et al. 2014a, b). But the average sex difference is small when compared to much larger differences in ornamentation that exist within each sex (Cardoso et al. 2014a, b). This high level of female ornamentation (close to, and often surpassing the average male) is most commonly found in social species, like the waxbill (see above), where ornamentation can have several social functions in addition to be a sexual signal (Lyon and Montgomerie 2012; Tobias et al. 2012). Accordingly, the colour ornamentation of waxbills affects not only preferences for individuals of the opposite sex, but also non-sexual social preferences among individuals of the same sex (Cardoso et al. 2014b). Therefore, if sexual selection is the main cause for changes in ornamentation along range expansions, then we expect to find more pronounced changes in male than in female ornamentation. On the contrary, if non-sexual social functions of ornamentation or other causes (e.g. environmental effects, survival selection) are the main cause for geographic changes in ornamentation, then we expect to see similar effects on both sexes.

Bill colour did not change consistently along the waxbill invasion in Iberia (Cardoso et al. 2014a). This is not surprising because their bill is living tissue and its colour can change fast, in few days, in response to short-term environmental changes (Cardoso et al. 2014b; Funghi et al. 2018). Plumage colour did change predictably along this waxbill invasion: the colour saturation of red plumage augmented, only in males, and the size of the red mask decreased, only in females (Cardoso et al. 2014a). As explained next, this increase in sexual dichromatism along the invasion (study sites marked red in Fig. 7.3), with opposed effects in males and females, indicates changes in sexual selection along the range expansions.

We noted earlier that this waxbill invasion progressed from ecologically favourable sites, which were colonised first, towards less favourable environments. This is common in range expansions, similarly to source-sink dynamics where sinks are often in the periphery of the distribution range (Sagarin and Gaines 2002; Sexton et al. 2009). When colonising ecological unfavourable areas, waxbills may experience environmental effects, such as less abundant or lower quality food, that are expected to reduce the expression of dietary, carotenoid-based red coloration. It is possible that environmental effects of this type contributed to reduce female ornamentation along the invasion, but they cannot explain the opposed changes in male ornamentation. Similarly, explanations such as differential dispersal of more ornamented birds, or differential survival based on ornamentation, cannot explain the sex-specific changes in colour or the increase in sexual dichromatism. The only hypothesis able to explain the increase in sexual dichromatism is increased strength of sexual selection along the invasion, because, as noted earlier, sexual selection targets males more strongly than females in species with conventional sex roles.

It is possible that, beyond waxbills, increased male investment in sexual signalling is common during range expansions. This is because less favourable environments (colonised later in expansions) pose survival challenges that augment the fitness differences between high- and low-quality individuals (e.g. mate quality predicts offspring quality when experimentally worsening environmental conditions; Sheldon et al. 2003; Welch 2003). Consequently, the strength of sexual selection for choosing higher quality mates increases (Hale 2008) and, since sexual signals typically mediate mate choice and indicate individual quality (Andersson 1994), the strength of sexual selection on those sexual signals is predicted to increase. Cross-species comparisons finding that species in unfavourable or less predictable environments have more elaborate sexual traits (Botero et al. 2009; Botero and Rubenstein 2012) indirectly support this hypothesis. If future studies confirm that changes in sexual ornamentation are common during expansions, this has several implications in evolutionary ecology. For example, it uncovers a novel biogeographic pattern, suggests predictably different ornaments for the young species that originated at range borders (Gavrilets 2003), and, since young species should have smaller distribution ranges, helps interpret why sexual ornamentation is associated with range size or extinction risk (McLain et al. 1995; Bro-Jørgensen 2014).

7.4 Scope for Ecological Impact on the Iberia Avifauna

One of the simplest ways invasive species can impact native communities is by competing for resources, for example food or nesting sites. While some exotic bird species were shown to compete with natives for breeding or nesting sites (e.g. Wiebe 2003; Strubbe and Matthysen 2009; Grarock et al. 2013; Charter et al. 2016), this is unlikely for waxbills. Waxbills are not territorial, they are smaller than all other granivorous passerines in Iberia (and smaller than most insectivores as well), build their own nests, and were never reported to behave aggressively towards other bird species (Catry et al. 2010; pers. obs.).

Unlike the case for nesting sites, detrimental competition for food is seldom noted in the literature on avian biological invasions (Blackburn et al. 2009). This may be because food is not the most important limiting resource for birds, or because researchers more easily detect displacement caused by interference competition than the more subtle effects of exploitation competition for food. For example, in seasonal climates food should not be a limiting factor year-round, but most likely during winter, when cold and lower food abundance place higher energetic stress on birds (Stapanian et al. 1999; Carrascal et al. 2012a, b). Therefore, assessing the potential for competition for food between exotics and natives should also consider their foraging behaviour in winter, but this has seldom been done in the literature on avian invasions.

Waxbills forage for seeds in low vegetation, in open areas not far away from water (Clement et al. 1993; Payne 2010), which in Iberia correspond mostly to agricultural landscapes (Reino and Silva 1998; Catry et al. 2010). During Spring and Summer in Iberia, waxbills forage mostly on grasses while native granivorous birds forage mostly on forbs, and waxbills use more agricultural areas compared to native birds (Sullivan et al. 2015), suggesting that waxbills use a differentiated food niche relative to those of native birds. Further evidence comes from comparing food preferences of waxbills and European serins (*Serinus serinus*) foraging on the same mosaic of agricultural fields (Cardoso et al., submitted). The European serin is the most likely competitor for food with waxbills, because it is the smallest native granivorous passerine in Iberia, it is very abundant and also forages in the low vegetation of open habitats (Clement et al. 1993). The plant species foraged upon by waxbills and serins during autumn differed substantially (probability of a feeding event by waxbills and serins coinciding on the same plant species was only 20%; Cardoso et al., submitted). In winter, when energetic stress is larger and food availability lower, and when, if food were limiting, we would expect greater similarity in the plants used by serins and waxbill, their diet similarity did not increase. On the contrary, both bird species fed on a more varied set of plants during winter and decreased their diet similarity (Cardoso et al. submitted). Thus, even comparing waxbills with its most similar native species in terms of foraging behaviour and foraging habitat, diets appear well differentiated.

Interestingly, the differences in food use between waxbills and serins were in part because both birds, but especially waxbills, included seeds of non-native plants in their diet. Without the contribution of exotic plants, the diet similarity between serins and waxbills in autumn would be three times as high (Cardoso et al. submitted). Agriculture helps the establishment of exotic, non-crop plants (e.g. Elmore et al. 2003; Bradley and Mustard 2006; Pyšek et al. 2010), and it is likely that many food resources provided by these exotic plants are not used by native faunas (Connolly et al. 2014). Especially small seeds, which tend to be less preyed upon by native faunas (e.g. Reader 1993; Howe and Brown 1999; Pearson et al. 2011), and which small birds like the waxbill are able to forage on.

Thus, human intervention appears to have doubly helped create the differentiated food niche that waxbills exploit. First, for many centuries prior to this waxbill colonisation, agriculture heavily modified the landscape and created vast open areas with or near irrigation, which are ideal foraging habitats for waxbills and are comparatively less used by native granivorous birds (Sullivan et al. 2015). Second, agriculture facilitates the establishment of non-native and other sinantropic vegetation that, again, are more used for food by waxbills than by native birds (Cardoso et al., submitted). Finding a well differentiated food niche from those of native species should be more likely in human-modified habitats, like the managed agro-ecosystems that waxbills use in much of Iberia, and perhaps also in disturbed or novel ecosystems (Morse et al. 2014). In such ecosystems, modified recently on an evolutionary timescale, novel resources may not yet be efficiently used because of insufficient time for native communities to have changed or native species adapting to those resources.

Ecological differences between waxbills and native passerines go beyond food. While the breeding phenology of most native birds is adapted to the highly seasonal Mediterranean climate, with a peak of resources around spring, waxbills are native to regions that are often less seasonal and with less predictable climate, and therefore often have more flexible and extended breeding seasons (Payne 2010; Sanz-Aguilar et al. 2015), as is common in tropical birds (Hau et al. 2008). This should allow them to efficiently exploit the extended period of abundant food resources that agricultural irrigation provides in and around cultivated fields.

Overall, evaluating different ecological traits (including diet, phenology, habitat and nesting preferences), waxbills appear to use a marginal niche relative to the ecological space occupied by native passerines in Iberia (Batalha et al. 2013). Also, the waxbill's nearest ecological distance to a native species is similar to the mean nearest distances among native species (Batalha et al. 2013), which, if we take nearest distances among natives as an indication of stable coexistence, suggests that waxbills can coexist with any native species. A possible concern is that the waxbill is the ecological nearest-neighbour of the reed bunting (*Emberiza schoeniclus*), sharing some habitat and nesting preferences (Batalha et al. 2013), and the subspecies of reed bunting breeding in Iberia (*E. s. lusitanica* and *E. s. witherbyi*) are endangered and have fragmented distributions (Kvist et al. 2011). The ecological evidence reviewed here suggests that waxbills should not pose a significant threat. Most likely, rather than managing possible effects of waxbills, habitat restoration is the key to make reed bunting populations resilient (Kvist et al. 2011).

These results suggest that some invasive species are well accommodated in their new ecosystems. We suggest that biological invasions occupying vacant niche space are more likely in human-modified habitats, where native communities may not yet be ecologically saturated. In a time when travel and commerce cause exotic species to be increasingly present in ecosystems, it is important to understand which combinations of invader traits and ecosystem traits pose the greatest or weakest threats of being ecologically harmful in order to prioritize and manage conservation efforts.

7.5 Conclusion

The common waxbill provides an example of how human history and human activities have unanticipated effects on biological communities. In this case, overseas commerce, breeding for the pet market, and release into the wild (sometimes intentionally, at least in parts of the world; Sick 1997) resulted in waxbills becoming a successful invader in many places (Stiels et al. 2011). Perhaps an unusual aspect of the waxbill example is that, at least in the Iberian Peninsula, human action had the additional effect of providing modified habitats where waxbills found an ecological niche that is distinct from those of native species (Batalha et al. 2013). This includes abundant non-native grasses in and around agricultural areas, which waxbills feed upon more than native birds (Cardoso et al. submitted). Perhaps only a minority of biological invasions use vacant ecological niche space, but the waxbill example shows that this possibility merits consideration when exotic species inhabit habitats that were already heavily modified by humans.

Our overview of recent research on the waxbill biological invasion in Iberia also shows the usefulness of historical information on biological invasions for diverse research areas within ecology. In our case, we covered research in biogeography (Species Distribution Modelling), behavioural and evolutionary ecology (changes in personality and ornamentation). When applied to biological invasions, the various subfields of ecology provide a means of synthetizing historical data and extract causes (e.g. environmental facilitators and constraints) and consequences (e.g. adaptation, microevolution) of range expansions. Historical information on biological invasions is, therefore, of wide cross-disciplinary value and, since many invasions are recent or ongoing, they are prime sources for detailed descriptions of range expansions. Cross-disciplinary analysis of those descriptive data advances both our knowledge of biological invasions, and of other research fields related to range expansion.

Acknowledgements We would like to thank Martin Sullivan for help with Fig. 7.3, and the editors and a reviewer for useful comments on an earlier version of the manuscript. We were supported by grants SFRH/BPD/110165/2015 and SFRH/BPD/93079/2013, and research projects PTDC/AAG-GLO/0463/2014–POCI-01-0145-FEDER-016583 and PTDC/BIA-EVF/4852/2014 from the Fundação para a Ciência e a Tecnologia (including the "Programa Operacional Factores de Competitividade—COMPETE").

References

Andersson M (1994) Sexual selection. Princeton University Press, Princeton

Aplin LM, Farine DR, Morand-Ferron J, Sheldon BC (2012) Social networks predict patch discovery in a wild population of songbirds. Proc R Soc B 279:4199–4205

Batalha HR, Ramos JA, Cardoso GC (2013) A successful avian invasion occupies a marginal ecological niche. Acta Oecol 49:92–98

Bell G (2010) Fluctuating selection: the perpetual renewal of adaptation in variable environments. Philos Trans R Soc B 365:87–97

Blackburn TM, Lockwood JL, Cassey P (2009) Avian invasions: the ecology and evolution of exotic birds. Oxford University Press, Oxford

Blackburn TM, Essl F, Evans T, Hulme PE, Jeschke JM, Kühn I, Kumschick S, Marková Z, Mrugała A, Nentwig W, Pergl J, Pyšek P, Rabitsch W, Ricciardi A, Richardson DM, Sendek A, Vilà M, Wilson JRU, Winter M, Genovesi P, Bacher S (2014) A unified classification of alien species based on the magnitude of their environmental impacts. PLoS Biol 12:e1001850

Botero CA, Rubenstein DR (2012) Fluctuating environments, sexual selection and the evolution of flexible mate choice in birds. PLoS ONE 7:e32311

Botero CA, Boogert NJ, Vehrencamp SL, Lovette IJ (2009) Climatic patterns predict elaboration of song displays in mockingbirds. Curr Biol 19:1151–1155

Boughman JW (2001) Divergent sexual selection enhances reproductive isolation in sticklebacks. Nature 411:944–948

Bradley BA, Mustard JF (2006) Characterizing the landscape dynamics of an invasive plant and risk of invasion using remote sensing. Ecol Appl 16:1132–1147

Bridle JR, Vines TH (2007) Limits to evolution at range margins: when and why does adaptation fail? Trends Ecol Evol 22:140–147

Bro-Jørgensen J (2014) Will their armaments be their downfall? Large horn size increases extinction risk in bovids. Anim Conserv 17:80–87

Burley NT, Price DK, Zann RA (1992) Bill color, reproduction and condition effects in wild and domesticated zebra finches. Auk 109:13–23

Buswell JM, Moles AT, Hartley S (2011) Is rapid evolution common in introduced plant species? J Ecol 99:214–224

Butin E, Porter AH, Elkinton J (2005) Adaptation during biological invasions and the case of *Adeles tugae*. Evol Ecol Res 7:887–900
Canestrelli D, Bisconti R, Carere C (2016a) Bolder takes all? The behavioural dimension of biogeography. Trends Ecol Evol 31:35–43
Canestrelli D, Porretta D, Lowe WH, Bisconti R, Carere C, Nascetti G (2016b) The tangled evolutionary legacies of range expansion and hybridization. Trends Ecol Evol 31:677–688
Cardoso GC, Mota PG (2008) Speciational evolution of coloration in the genus *Carduelis*. Evolution 62:753–762
Cardoso GC, Batalha HR, Reis S, Lopes RJ (2014a) Increasing sexual ornamentation during a biological invasion. Behav Ecol 25:916–923
Cardoso GC, Leitão AV, Funghi C, Batalha HR, Lopes RJ, Mota PG (2014b) Similar preferences for ornamentation in opposite- and same-sex choice experiments. J Evol Biol 27:2798–2806
Cardoso GC, Rodrigues G, Alves P, Vicente JR, Honrado JP (submitted) Naturalized plants decrease diet similarity between an invasive bird and its most similar native species
Carere C, Caramaschi D, Fawcett TW (2010) Covariation between personalities and individual differences in coping with stress: converging evidence and hypotheses. Curr Zool 56:728–740
Carrascal LM, Villén-Pérez S, Seoane J (2012a) Thermal, food and vegetation effects on winter bird species richness of Mediterranean oakwoods. Ecol Res 27:293–302
Carrascal LM, Seoane J, Villén-Pérez S (2012b) Temperature and food constraints in wintering birds—an experimental approach in montane Mediterranean oakwoods. Community Ecol 13:221–229
Carvalho CF, Leitão AV, Funghi C, Batalha HR, Reis S, Mota PG, Lopes RJ, Cardoso GC (2013) Personality traits are related to ecology across a biological invasion. Behav Ecol 24:1081–1091
Catry P, Costa H, Elias G, Matias R (2010) Aves de Portugal: Ornitologia do território continental. Assírio e Alvim, Lisboa
Charter M, Izhaki I, Mocha YB, Kark S (2016) Nest-site competition between invasive and native cavity nesting birds and its implication for conservation. J Environ Manage 181:129–134
Clement P, Harris A, Davis J (1993) Finches & sparrows: an identification guide. Princeton University Press, Princeton
Connolly BM, Pearson DE, Mack RN (2014) Granivory of invasive, naturalized, and native plants in communities differentially susceptible to invasion. Ecology 95:1759–1769
Cox GW (2004) Alien species and evolution: the evolutionary ecology of exotic plants, animals, microbes, and interacting native species. Island Press, Washington, DC
Dall SRX, Houston AI, McNamara JM (2004) The behavioural ecology of personality: consistent individual differences from an adaptive perspective. Ecol Lett 7:734–739
Elmore AJ, Mustard JF, Manning SJ (2003) Regional patterns of plant community response to changes in water: Owens Valley, California. Ecol Appl 13:443–460
Engler JO, Stiels D, Schidelko K, Strubbe D, Quillfeldt P, Brambilla M (2017) Avian SDMs: current state, challenges, and opportunities. J Avian Biol 48:1483–1504
Equipa Atlas (2008) Atlas das Aves Nidificantes em Portugal (1999–2005). Assírio & Alvim, Lisboa
Exnerová A, Svádová KH, Fucíková E, Drent P, Stys P (2010) Personality matters: individual variation in reactions of naive bird predators to aposematic prey. Proc R Soc B 277:723–728
Ferreira LFF (1980–1981) Contribuição para o conhecimento da distribuição de *Estrilda astrild* em Portugal. Boletim da Sociedade Portuguesa de Ciências Naturais 20:19–23
Funghi C, Leitão AV, Ferreira AC, Mota PG, Cardoso GC (2015) Social dominance in a gregarious bird is related to body size but not to standard personality assays. Ethology 121: 84–93
Funghi C, Trigo S, Gomes ACR, Soares MC, Cardoso GC (2018) Release from ecological constraint erases sex difference in social ornamentation. Behav Ecol Sociobiol72:67
Gavrilets S (2003) Models of speciation: what have we learned in 40 years? Evolution 57:2197–2215
Gomes ACR, Sorenson MD, Cardoso GC (2016) Speciation is associated with changing ornamentation rather than stronger sexual selection. Evolution 70:2823–2838

Goodenough AE (2010) Are the ecological impacts of alien species misrepresented? A review of the "native good, alien bad" philosophy. Community Ecol 11:13–21

Grarock K, Lindenmayer DB, Wood JT, Tidemann CR (2013) Does human-induced habitat modification influence the impact of introduced species? A case study on cavity-nesting by the introduced common myna (*Acridotheres tristis*) and two australian native parrots. Environ Manage 52:958–970

Gray DA, Cade WH (2000) Sexual selection and speciation in field crickets. Proc Natl Acad Sci USA 97:14449–14454

Guillette LM, Reddon AR, Hoeschele M, Sturdy CB (2011) Sometimes slower is better: slow-exploring birds are more sensitive to changes in a vocal discrimination task. Proc R Soc B 278:767–773

Guisan A, Thuiller W (2005) Predicting species distribution: offering more than simple habitat models. Ecol Lett 8:993–1009

Hale RE (2008) Evidence that context-dependent mate choice for parental care mirrors benefits to offspring. Anim Behav 75:1283–1290

Halley JM (1996) Ecology, evolution and 1/f noise. Trends Ecol Evol 11:33–37

Hau M, Perfito N, Moore IT (2008) Timing of breeding in tropical birds: mechanisms and evolutionary implications. Ornitol Neotrop 19:39–59

Hill GE (2000) Energetic constraints on expression of carotenoid-based plumage coloration. J Avian Biol 31:559–566

Hill GE, Montgomerie R (1994) Plumage colour signals nutritional condition in the house finch. Proc R Soc B 258:47–52

Hill GE, Hood WR, Huggins KA (2009) A multifactorial test of the effects of carotenoid access, food intake and parasite load on the production of ornamental feathers and bill coloration in American goldfinches. J Exp Biol 212:1225–1233

Howe HF, Brown JS (1999) Effects of birds and rodents on synthetic tallgrass communities. Ecology 80:1776–1781

Hui C, Richardson DM (2017) Invasion dynamics. Oxford University Press, Oxford

Kark S, Sol D (2005) Establishment success across convergent Mediterranean ecosystems: an analysis of bird introductions. Conserv Biol 19:1519–1527

Koolhaas JM, Korte SM, de Boer SF, van der Vegt BJ, van Reenen CG, Hopster H, de Jong IC, Ruis MAW, Blokhuis HJ (1999) Coping styles in animals: current status in behavior and stress physiology. Neurosci Biobehav Rev 23:925–935

Kvist L, Ponnikas S, Belda EJ, Encabo I, Martínez E, Onrubia A, Hernández JM, Vera P, Neto JM, Monrós JS (2011) Endangered subspecies of the reed bunting (*Emberiza schoeniclus witherbyi* and *E. s. lusitanica*) in Iberian Peninsula have different genetic structures. J Ornithol 152:681–693

Liker A, Bókony V (2009) Larger groups are more successful in innovative problem solving in house sparrows. Proc Natl Acad Sci USA 106:7893–7898

Lopes RJ, Correia J, Batalha H, Cardoso GC (2018) Haemosporidian parasites missed the boat during the introduction of common waxbills (*Estrilda astrild*) in Iberia. Parasitol http://doi.org/10.1017/S0031182018000331

Lyon BE, Montgomerie R (2012) Sexual selection is a form of social selection. Phil Trans R Soc B 367:2266–2273

Maia R, Rubenstein DR, Shawkey MD (2013) Key ornamental innovations facilitate diversification in an avian radiation. Proc Natl Acad Sci USA 110:10687–10692

Marques CIJ, Batalha HR, Cardoso GC (2016) Signalling with a cryptic trait: the regularity of barred plumage in common waxbills. Royal Soc Open Sci 3:160195

Martí R, Moral JC (2003) Atlas de las aves reproductoras de España. Dirección General de Conservación de la Naturaleza. Sociedad Española de Ornithología, Madrid

Matias R (2002) Aves exóticas que nidificam em Portugal continental. ICNB, Lisboa

McCowan LSC, Griffith SC (2015) Active but asocial: exploration and activity is linked to social behaviour in a colonially breeding finch. Behaviour 152:1145–1167

McGraw KJ (2006) Mechanisms of carotenoid-based coloration. In: Hill GE, McGraw KJ (eds) Bird coloration, vol 1. mechanisms and measurements. Harvard University Press, Boston, pp 177–242

McGraw KJ, Schuetz JB (2004) The evolution of carotenoid coloration in estrildid finches: a biochemical analysis. Comp Biochem Physiol B 139:45–51

McGraw KJ, Hill GE, Parker RS (2005) The physiological costs of being colourful: nutritional control of carotenoid utilization in the American goldfinch (*Carduelis tristis*). Anim Behav 69:653–660

McLain DK, Moulton MP, Redfearn TP (1995) Sexual selection and the risk of extinction of introduced birds on oceanic islands. Oikos 74:27–34

Morand-Ferron J, Quinn J (2011) Larger groups of passerines are more efficient problem solvers in the wild. Proc Natl Acad Sci USA 108:15898–15903

Morse NB, Pellissier PA, Cianciola EN, Brereton RL, Sullivan MM, Shonka NK, Wheeler TB, McDowell WH (2014) Novel ecosystems in the Anthropocene: a revision of the novel ecosystem concept for pragmatic applications. Ecol Soc 19:12

Moya-Laraño J, Wise DH (2000) Survival regression analysis: a powerful tool for evaluating fighting and assessment. Anim Behav 60:307–313

Nicolaus M, Tinbergen JM, Ubels R, Both C, Dingemanse NJ (2016) Density fluctuations represent a key process maintaining personality variation in a wild passerine bird. Ecol Lett 19:478–486

Payne R (2010) Family Estrildidae (Waxbills). In: del Hoyo J, Elliott A, Christie DA (eds) Handbook of the birds of the world—weavers to new world warblers. Lynx Edicions, Barcelona

Pearson DE, Callaway RM, Maron JL (2011) Biotic resistance via granivory: establishment by invasive, naturalized, and native asters reflects generalist preference. Ecology 92:1748–1757

Perals D, Griffin AS, Bartomeus I, Sol D (2017) Revisiting the open-field test: what does it really tell us about animal personality? Anim Behav 123:69–79

Phillips BL, Brown GP, Webb JK, Shine R (2006) Invasion and the evolution of speed in toads. Nature 439:803

Pyšek P, Bacher S, Chytrý M, Jarošík V, Wild J, Celesti-Grapow L, Gassó N, Kenis M, Lambdon PW, Nentwig W, Pergl J, Roques A, Sádlo J, Solarz W, Vilà M, Hulme PE (2010) Contrasting patterns in the invasions of European terrestrial and freshwater habitats by alien plants, insects and vertebrates. Glob Ecol Biogeogr 19:319–331

Reader RJ (1993) Control of seedling emergence by ground cover and seed predation in relation to seed size for some oldfield species. J Ecol 81:169–175

Réale D, Reader SM, Sol D, McDougall P, Dingemanse NJ (2007) Integrating temperament in ecology and evolutionary biology. Biol Rev 82:291–318

Reino L (2005) Variation partitioning for range expansion of an introduced species: the common waxbill *Estrilda astrild* in Portugal. J Ornithol 146:377–382

Reino LM, Silva T (1996a) Distribution and expansion of the common waxbill *Estrilda astrild* in Portugal. In: Holmes JS, Simons JR (eds) The introduction and naturalisation of birds. British Ornithologists' Union and UK Joint Nature Conservation Committee, London, pp 103–106

Reino LM, Silva T (1996b) Distribuição e expansão do Bico-de-lacre Estrilda astrild em Portugal. In: Farinha JC, Almeida J, Costa H (eds) I Congresso da SPEA, Vila Nova de Cerveira, pp 78–80

Reino L, Silva T (1998) The distribution and expansion of the Common Waxbill (*Estrilda astrild*) in the Iberian Peninsula. Biol Conserv Fauna 102:163–167

Reino L, Moya-Laraño J, Heitor AC (2009) Using survival regression to study patterns of expansion of invasive species: will the common waxbill expand with global warming? Ecography 32:237–246

Richardson DM, Pyšek P, Rejmánek M, Barbour MG, Panetta FD, West CJ (2000) Naturalization and invasion of alien plants: concepts and definitions. Divers Distrib 6:93–107

Rosa P, Nguyen V, Dubois F (2012) Individual differences in sampling behaviour predict social information use in zebra finches. Behav Ecol Sociobiol 66:1259–1265

Rufino R (coord) (1989) Atlas das aves que nidificam em Portugal continental. Serviço Nacional de Parques Reservas e Conservação da Natureza, Lisboa

Sagarin RD, Gaines SD (2002) The 'abundant center' distribution: to what extent is it a biogeographical rule? Ecol Lett 5:137–147

Sanz-Aguilar A, Carrete M, Edelaar P, Potti J, Tella JL (2015) The empty temporal niche: breeding phenology differs between coexisting native and invasive birds. Biol Invasions 17:3275–3288

Sax DF, Stachowicz JJ, Brown JH, Bruno JF, Dawson MN, Gaines SD, Grosberg RK, Hastings A, Holt RD, Mayfield MM, O'Connor MI, Rice WR (2007) Ecological and evolutionary insights from species invasions. Trends Ecol Evol 22:465–471

Seehausen O, Terai Y, Magalhães IS, Carleton KL, Mrosso HD, Miyagi R, van der Sluijs I, Schneider MV, Maan ME, Tachida H, Imai H, Okada N (2008) Speciation through sensory drive in cichlid fish. Nature 455:620–626

Sexton JP, McIntyre PJ, Angert AL, Rice KJ (2009) Evolution and ecology of species range limits. Annu Rev Ecol Evol Syst 40:415–436

Sheldon BC, Arponen H, Laurila A, Crochet PA, Merila J (2003) Sire coloration influences offspring survival under predation risk in the moorfrog. J Evol Biol 16:1288–1295

Sick H (1997) Ornitologia Brasileira, 2ed. Nova Fronteira, Rio de Janeiro

Sih A, Cote J, Evans M, Fogarty S, Pruitt J (2012) Ecological implications of behavioural syndromes. Ecol Lett 15:278–289

Silva T, Reino LM, Borralho R (2002) A model for range expansion of an introduced species: the common waxbill *Estrilda astrild* in Portugal. Divers Distrib 8:319–326

Stapanian MA, Smith CC, Finck EJ (1999) The response of a Kansas winter bird community to weather, photoperiod, and year. Wilson Bull 111:550–558

Stiels D, Schidelko K, Engler JO, van den Elzen R, Rödder D (2011) Predicting the potential distribution of the invasive common waxbill *Estrilda astrild* (Passeriformes: Estrildidae). J Ornithol 152:769–780

Stiels D, Gaißer B, Schidelko K, Engler JO, Rödder D (2015) Niche shift in four non-native estrildid finches and implications for species distribution models. Ibis 157:75–90

Stoddard MC, Prum RO (2011) How colorful are birds? Evolution of the avian plumage color gamut. Behav Ecol 22:1042–1052

Strubbe D, Matthysen E (2009) Experimental evidence for nest-site competition between invasive ring-necked parakeets (*Psittacula krameri*) and native nuthatches (*Sitta europaea*). Biol Conserv 142:1588–1594

Sullivan MJP, Davies RG, Reino L, Franco AMA (2012) Using dispersal information to model the species—environment relationship of spreading non-native species. Methods Ecol Evol 3: 870–879

Sullivan MJP, Davies RG, Mossman HL, Franco AMA (2015) An anthropogenic habitat facilitates the establishment of non-native birds by providing underexploited resources. PLoS ONE 10: e0135833

Sullivan MJP, Franco AMA (2018) Changes in habitat associations during range expansion: disentangling the effects of climate and residence time. Biol Invasions.http://doi.org/10.1007/s10530-017-1616-9

Thomas DB, McGraw KJ, Butler MW, Carrano MT, Madden O, James HF (2014) Ancient origins and multiple appearances of carotenoid-pigmented feathers in birds. Proc R Soc B 281:20140806

Tobias JA, Montgomerie R, Lyon BE (2012) The evolution of female ornaments and weaponry: social selection, sexual selection and ecological competition. Philos Trans R Soc B 367:2274–2293

Václavík T, Meentemeyer RK (2012) Equilibrium or not? Modelling potential distribution of invasive species in different stages of invasion. Divers Distrib 18:73–83

Ventim R, Mendes L, Ramos JA, Cardoso H, Pérez-Tris J (2012) Local haemoparasites in introduced wetland passerines. J Ornithol 153:1253–1259

Verbeek MEM, Drent PJ, Wiepkema PR (1994) Consistent individual differences in early exploratory behaviour of male great tits. Anim Behav 48:1113–1121

Vidal C, Reino LM (2002) Distribuição e expansão do bico-de-lacre (*Estrilda astrild*) na Galiza. In: Sánchez A (ed) Actas de las XV Jornadas Ornitológicas Españolas. SEO/BirdLife, Madrid, p 260

Vilà M, Hulme PE (eds) (2017) Impact of biological invasions on ecosystem services. Springer, Dordrecht

Welch AM (2003) Genetic benefits of a female mating preference in gray tree frogs are context-dependent. Evolution 57:883–893

Wiebe KL (2003) Delayed timing as a strategy to avoid nest-site competition: testing a model using data from starlings and flickers. Oikos 100:291–298

Wolf M, Weissing FJ (2010) An explanatory framework for adaptive personality differences. Philos Trans R Soc B 365:3959–3968

Wolf M, Weissing FJ (2012) Animal personalities: consequences for ecology and evolution. Trends Ecol Evol 27:452–461

Chapter 8
Cordgrass Invasions in Mediterranean Marshes: Past, Present and Future

Bernardo Duarte, Enrique Mateos-Naranjo, Susana Redondo Goméz, João Carlos Marques and Isabel Caçador

Abstract The invasion of natural communities by non-indigenous species (NIS) represents one of the most serious threats to biodiversity. While these invasive processes are rather well studied in river corridors and riparian communities, the invasiveness of non-indigenous aquatic plants in wetlands has received far less attention. Many NIS plants have been introduced more than 100 years ago, while others are more recent arrivals, with most of the introductions occurring at the end of the 19th and at the beginning of the 20th centuries. The *Spartina* genus (the cordgrasses) is one of the most successful among halophytes (species that can survive and complete their life cycle under saline conditions), being present in a wide range of latitudes across the globe. Typically, Mediterranean systems are inhabited by the endemic small cordgrass *Spartina maritima*, native from the Atlantic African and European Atlantic coasts. Alongside, and with very similar geographical distribution ranges, two invasive species from the *Spartina* genus have been detected in Mediterranean systems. *Spartina versicolor* was first described in the Mediterranean region during the 19th century. This taxon is considered to be originated in America, and was introduced to Europe in the 19th century. It is probable that this species was introduced as packing material in crates to various ports around the Mediterranean Sea. Denseflower cordgrass *Spartina densiflora* is an invasive grass species of South American origin that has colonized salt marshes in the Gulf of Cadiz in the southwestern Iberian Peninsula, North Africa and North America. This is a facultative halophyte species (plants that avoid the effects of high salt even though they live in a saline environment) with an

B. Duarte (✉) · I. Caçador
MARE—Marine and Environmental Sciences Centre, Faculty of Sciences, University of Lisbon, Campo Grande 1749-016, Lisbon, Portugal
e-mail: baduarte@fc.ul.pt

E. Mateos-Naranjo · S. R. Goméz
Departamento de Biología Vegetal y Ecología, Facultad de Biología, Universidad de Sevilla, 1095, 41080 Sevilla, Spain

J. C. Marques
MARE—Marine and Environmental Sciences Centre, c/o Department of Zoology, Faculty of Sciences and Technology, University of Coimbra, 3000 Coimbra, Portugal

A. I. Queiroz and S. Pooley (eds.), *Histories of Bioinvasions in the Mediterranean*, Environmental History 8, https://doi.org/10.1007/978-3-319-74986-0_8

amazing physiological and morphological flexibility, enabling it to cope with a very wide range of environment constraints (salinity, tidal submergence, soil types, drainage and nutrient availability). Having this knowledge in mind becomes important to review the history of the introduction of these NIS along with their current colonization status and physiological characteristics. In the present chapter, this approach will be integrated with future scenarios of global change and increased anthropogenic pressures to achieve a better understanding of the impact of these NIS in Mediterranean estuarine systems.

Keywords Salt marshes · Invasive Spartina · Wetland biodiversity
Mediterranean estuaries

8.1 Salt Marshes as Key Players in the Marine Ecosystem

Historically, developing human civilizations have often been concentrated in coastal areas where access to water promoted trade and commerce. Consequently, human alteration of natural ecosystems has been profound in coastal areas. Estuarine systems represent important ecosystems in the coastal landscape, providing essential ecosystems services such as water quality improvement, fisheries resources, habitat and food for migratory and resident animals, and recreational areas for human populations. In the past 40 years, estuarine conservation has been recognized as a priority at national and international levels through conventions and legislation including the Ramsar Convention (1977) and the Water Framework Directive (2000). Despite all these conservation attempts on a worldwide scale, coastal, estuarine and transitional waters have continued to be negatively affected by man's activities. This is a central concern in terms of environmental management in order to develop policies to balance socio-economic growth and environmental protection (Borja and Dauer 2008).

Salt marshes are found fringing many of the world's coasts exposed to relatively low-energy hydrodynamic environments. They are characterized by a wide variety of herbaceous, succulent and woody vascular plants. The upper limit of the occurrence of salt marshes is approximately the level of the highest astronomical tide (HAT), while the lower limit is rarely below mean high water neap (MHWN) tide level. These ecosystems appear mostly in estuarine ecosystems, but can also be found in barrier islands, spits, embayments and open shores exposed to low wave energy (Allen 2000), as well as fringing coastal lagoons. Salt marshes can be found within all tidal regime ranges, from microtidal to macrotidal, presenting a great plasticity to water-level fluctuations which can be seasonal, as in the Baltic or the estuaries of southwest Western Australia (Hodgkin and Hamilton 1998), or unpredictable, as in some lagoons where water level is determined by rainfall patterns and lagoon opening.

Salt marshes have a great ecological value for the estuarine ecosystem, namely in terms of nutrient regeneration, primary production, wildlife habitat and as shoreline stabilizers (Caçador et al. 2009; Lefeuvre et al. 2003). These ecosystems occupy the transition zone between terrestrial and marine biotopes and are characterized by a high productivity, which is considered essential in maintaining the detritus-based food chain of estuarine and coastal ecosystems (Marinucci 1982). Estuarine wetlands, as salt marshes, constitute good carbon sinks having simultaneously reduced rates of greenhouse gas (GHGs) emissions (Magenheimer et al. 1996), with a carbon sequestration capacity per unit area of about one order of magnitude higher than other wetland systems (Bridgham et al. 2006). Salt marshes are usually located in estuarine systems and their primary production allows for a greater reduction of CO_2 in the atmosphere resulting from incorporation in organic tissues through photosynthesis (Sousa et al. 2010). Wetlands represent the largest terrestrial carbon pool with a capacity of 770 Gt of carbon, overweighing the total carbon storage of farms and rainforests (Han et al. 2005).

8.2 The Native *Spartina Maritima*

The *Spartina* genus is one of the most successful among halophytes, being present in a wide range of latitudes across the globe. Some studies have described latitudinal population differentiation of *Spartina* species (Seliskar et al. 2002; Lessmann et al. 1997; Daehler et al. 1999; Otero et al. 2000; Proffitt et al. 2005; Álvarez et al. 2010; Duarte et al. 2013a). Members of the *Spartina* genus present a C_4-type photosynthesis, the most efficient photosynthetic metabolism on the planet, allowing this species to harvest carbon at higher rates than it would be observed in a C_3 organism (Hatch 1992; Álvarez et al. 2010). Despite this metabolic advantage, C_4 photosynthesis is highly dependent on the surrounding climatic environment. Differences among species and populations are expected, due to sensitivity to pressures, such as climatic variation, contaminants and grazing (Heide 2005; Stephenson et al. 2006).

The small cordgrass (*Spartina maritima*), has a very wide distribution in the northern hemisphere native to the coasts of western and northern Europe and western Africa, with also a disjunct population on the Atlantic coasts of Namibia and South Africa (Marchant and Goodman 1969). This native European cordgrass is typically found in low salt marshes with anoxic and salty sediment exposed to high tidal influence and long inundation periods (Castillo et al. 2000). In these habitats, it shows important differences between ecotypes in population structure and functionality, in terms of survivorship, shoot density and height, and biomass development (Castellanos et al. 1994, 1998; Castillo et al. 2000, 2005; Duarte et al. 2013a). Furthermore, the species rarely produces seeds in populations on the European Atlantic coasts (Yannic et al. 2004). Castellanos et al. (1994) also found the absence of germination in its Spanish populations of the Gulf of Cadiz, with all populations expanding throughout vegetative propagation.

The small cordgrass provides important ecosystem services playing a key role as a pioneer species in shoreline stabilization and salt marsh accretion (Duarte et al. 2009). It could thus be an important biotool to facilitate succession in restored areas

(Castillo and Figueroa 2009), and its high biotechnological potential in relation to the retention and accumulation of heavy metals in its tissues (Cambrollé et al. 2008) reduce the risk of heavy metal dissemination in estuary systems and to the entry of such materials into the food chain. Despite this potentially important role, its populations are not widely distributed in the Iberian Peninsula salt marshes. It occupies only approximately 12% of the salt marshes in the Portuguese transitional systems (Caçador et al. 2013), while 70% of its populations in the Spanish salt marshes show percentages of occupation lower than 5% of salt marsh surface (Espinar 2009). Together with this low relative occupied area, it has recently been withdrawing from several salt marshes either due to dieback, or to hybridization with other *Spartina* species, originating polyploid species (Ainouche et al. 2003). In sister *Spartina* species this dieback has been proposed to result from age-related decline of vigour or by competition with other species after niche elevation due to rapid expansion (Gray 2004). As a result of these circumstances, since early 1990s this species has been included in some European red lists of endangered species (England, Cooper 1993; Cantabrian coast of Spain, Bueno 1997; the southwest coast of Spain, Cabezudo et al. 2005).

8.3 First Described Occurrences by Invasive Spartina Species, from c.1848

In the western Mediterranean, damp depressions in dune habitats are colonized by *Spartina versicolor* (Fig. 8.1) that grows also in brackish marshes in the Atlantic Coast of the Southwest Iberian Peninsula. This species, also named *Spartina juncea* or *Spartina durieui* (Chevalier 1923; Saint-Yves 1932), had a controversial taxonomic status. It was initially recorded almost simultaneously in several Mediterranean places: first in Southern France near Agde (Fabre 1849), then in Italy (Parlatore 1848–1850), Algeria (Cosson and Maisonneuve 1867), and Portugal (Daveau 1897). In 1901, Neyraut detected this taxon on the Southwest French Atlantic coast, near Arcachon (Coste 1906). However, there is a severe lack of accurate information regarding the dates of discovery of each population, and thus a chronological establishment can only be made using the publication dates available. This species normally colonizes the upper and middle marsh in sandy or muddy-sandy sediments, with the potential to compete mainly with the native halophytic shrubs characteristic of these areas (for e.g. *Halimione portulacoides* and *Sarcocornia fruticosa*). These *S. versicolor* populations have been recorded with a similar distribution to that observed for *S. maritima* (Curtis) Fernald.

On the other hand, in the Mediterranean marshes, the neophyte[1] (plant species that has been relatively found in an area where it had not been recorded previously)

[1]By definition, a neophyte is a plant species species which is not native to a geographical region, and was introduced in recent history.

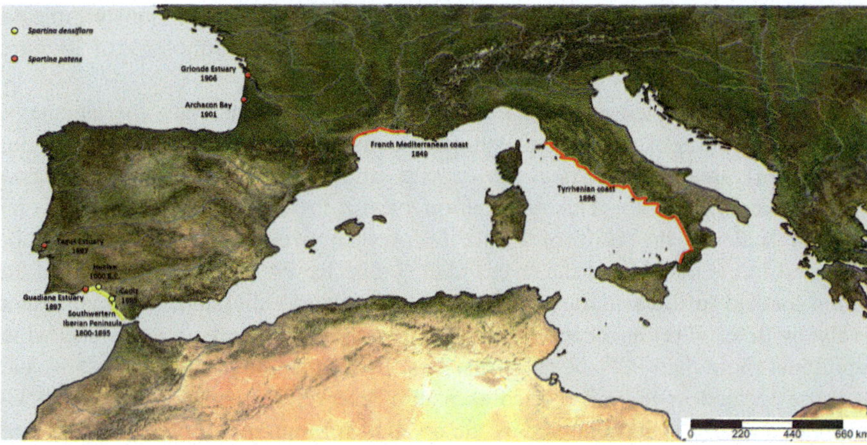

Fig. 8.1 Invasive cordgrass species historical distributions along the European coasts

Fig. 8.2 Highly dense S. patens stand (right) competing for resources with *H. portulacoides* (left) (Photo by B. Duarte)

native from the eastern South American coast *Spartina densiflora* (Fig. 8.2), has colonized mashes first in the Southwestern Iberian Peninsula (1800–1985) and then in the Northern Africa, Merja Zerga Lagoon (Morocco, first report 1978) (Nieva et al. 2001; Bortolus 2006). This species is also invading estuaries in Chile, Canada and in the United States (California late 1800s; Washington State first reported in

2001) (Bortolus 2006; Saarela 2012; Castillo et al. 2014). As previously indicated for *S. versicolor*, it is not possible to establish an exact chronology of its introductions and spread.

In terms of occupied area and the diversity of invaded habitats, the marshes of the Atlantic coasts of the Iberian Peninsula constitute the area outside the original distribution area of *S. densiflora* with the greatest degree of invasion (Mateos-Naranjo 2008). The distribution of the invasion of *S. densiflora* on a regional scale in the Gulf of Cadiz follows an irregular pattern, being more advanced in the marshes of the river Odiel Nature Reserve and of the island of San Bruno located in the estuaries of the rivers Tinto and Odiel and Guadina, respectively, with c. 4000 ha occupied by the invaded communities, a surface which represents more than 90% of these marshes. In contrast, other marsh systems such as those located in the River Piedras nature reserve and in Isla Cristina are less invaded, with c. 139 ha occupied by the invaded communities representing between 3–8% of the marsh surface of these systems (Mateos-Naranjo 2008). The rest of the marshes in the Gulf of Cadiz show intermediate degrees of invasion, and the presence of *S. densiflora* has not been observed in the Bay of Cádiz Nature Reserve, as well as in other estuary systems associated with the rivers of the Mediterranean basin, which could be ascribed to the lower tidal influence (Mateos-Naranjo 2008).

On the other hand, in relation with the heterogeneity of occupied habitats in Gulf of Cadiz, *S. densiflora* invades a wide number of habitats distributed along the whole tidal gradient, colonizing soils with very contrasting environmental characteristics, since it adapts very well to very poorly oxygenated environments, with a wide range of conductivity levels, as well as variations from brackish to saline, or temporarily hypersaline conditions, as occurs in saltpans (Nieva et al. 2001). However, the habitats that are more affected by the invasion are the areas of mid-high marsh under a clear tidal influence, where we can find the so-called 'seas of *Spartina*' (Figueroa and Castellanos 1988; Mateos-Naranjo 2008). In contrast, the habitats that are less affected are the areas of high marsh, more independent of the tide, areas of low marsh with long periods of flooding, areas of marsh dominated by glycophyte (salt intolerant) species e.g. the reed *Phragmites australis* and the southern cattail *Thypha domingensis*) which prevent the invasion of *S. densiflora* by stopping seedling establishment (Abbas et al. 2015), and salt pans (Mateos-Naranjo 2008). In addition, it should be noted that *S. densiflora* has brought it into contact with the autochthonous low-marsh dominant, *S. maritima* in some locations of the Gulf of Cadiz.

8.4 An Invasion Story Told Through Its Human History

In contrast to what happened in the case of the USA populations of *S. patens*, in Europe this cordgrass specie has commonly been disregarded, probably because its populations, which rarely flower, are easy to confuse with other common coastal grasses (Pignatti 1982). This led to some errors in native species inventories in the past, leading to varying management models according to the classification applied. On the other hand, in the USA, their home habitat, this species is under strict monitoring programs (US Forest Service).

When considered as a native species (Coste 1906; Guinochet and Vilmorin 1978; Tutin 1980; Gamisans et al. 1989; Bolòs and Vigo 2001; Fraga et al. 2003; Fenu and Bacchetta 2008), *S. patens* has been regarded as a bioindicator of a good conservation state of the ecosystem (Ferrandis Pomés 2002). On the other hand, when described as an exotic species its management has been advised accordingly, mostly with control recommendations (SanLeon et al. 1999; Duarte et al. 2015a, b, c, d).

This taxon was first described from the French Mediterranean coast (Fabre 1849), although no information is available on a specific region for this finding. Later, its presence was also recognized on the central Tyrrhenian coast of Italy (Fiori and Paleoti 1896), on the Portuguese Tagus and Guadiana estuaries (Daveau 1897) and on the French Atlantic coast, most precisely in the Gironde estuary and Aquitaine region (Coste 1906) (Fig. 8.3). Despite this taxonomical confusion that was generated across history and consequent lack of agreement in the systematics of this species, some theories arose regarding its introduction to the European coasts. The correct taxonomical designation of the NIS *S. patens* in western Iberia was only elucidated very recently, using genomic tools (Baumel et al. 2016). It is probable that the plant was introduced first to various ports around the Mediterranean, as it was used as packing material in ships' boxes and crates. So, it

Fig. 8.3 Grey plover *Pluvialis squatarola* flocking at high water on a temporary roost in the Tagus estuary. Photograph courtesy of A. Teixeira

might have been established locally from this material and eventually spread out along the coast. Saint-Yves (1932) proposed this theory, since the Gulf Stream itself would not have been enough to transport the plants from the USA to the European coasts. *Spartina patens* was also first sighted in the Arcachon Bay (Southwest France) in 1901 (Goulletquer et al. 2002). In this case, Goulletquer et al. maintain that its release was intentional, although the reasons are yet unknown.

Many aspects of *S. densiflora* have hitherto been analyzed, however most them have focused on its ecological, phenotypic physiological and genetics characteristics (Ayres et al. 2004; Bortolus 2006; Castillo et al. 2008, 2014; Mateos-Naranjo et al. 2007, 2008, 2011; Idazkin et al. 2010). There is little information about the invasion history of *S. densiflora* and in the Mediterranean framework. The first time that the presence of *S. densiflora* in the Gulf of Cadiz was discussed in a scientific paper was in the studies developed by Rubio (1985) and Figueroa and Castellanos (1988). Most authors agree that *S. densiflora* was probably introduced accidentally to the southwestern Iberian Peninsula at the end of the eighteenth century, probably due to the sea-trade in timber between South America and Spain (Fig. 8.1) (Nieva 1996; Bortolus 2006; Mateos-Naranjo et al. 2008).

The salt marshes of the joint estuary of the Tinto and Odiel rivers and Isla de San Bruno in the Guadiana estuary are the most probable initial areas of invasion of *S. densiflora* in the southwestern Iberian Peninsula. This idea is supported by two main aspects, one related to the fact that these salt marshes are where an extreme degree of invasion in terms of occupied area has been recorded, and secondly, this is where *S. densiflora* populations have colonized the greatest variety of habitats and show the highest degree of development (Mateos-Naranjo 2008). In addition, the existence of an important port in the town of Huelva (37° 26′N, 6° 94′O), just in the middle area of the Odiel and Tinto marshes, and of another port in the town of Ayamonte (Guadiana estuary; 37° 21′N, 7°40′O), only some kilometers north of the Isla de San Bruno marshes, reinforce the idea that these were the areas where the invasion could have been started. The first records of the use of Huelva as trading port go back to 1000 B.C., when it was an important Phoenician city. Nevertheless, its more expressive influence on global trading occurred after the New World discovery by Columbus. Christopher Columbus' voyage to the New World started and ended in the village of Palos de la Frontera, just across the river from the Port of Huelva, turning this harbor one of the most important in terms of global trading. And from this region, the invasions would have spread to the other marshes of the Gulf of Cadiz, of which also there are few records, although with some documented exceptions.

Moving to these exceptions, Luque et al. (2004) highlighted the first record of *S. densiflora* in 2004 in the Natural park 'Bay of Cadiz' (Guadalete River; 36° 28′N, 6° 11′O, SW Spain), where the exotic species had invaded a small area (c. 200 m^2) in a middle salt marsh dominated by *Sarcocornia fruticosa* and *Atriplex portulacoides*. But this *S. densiflora* population was eradicated from this area using mechanical and chemical methods developed early in 2005 by the Andalusia Government.

Another example is the case of *S. densiflora* invasions in specific locations of Doñana National Park (Guadalquivir River; 36° 53′N, 6° 20′O, SW Spain). Although this species is amply distributed on both banks of the Guadalquivir River, with an occupied area c. 10% of the surface with tidal influence (Mateos-Naranjo 2008), it should be noted that this invasion was indirectly stopped by human interventions in the 1980s with the construction of the dyke of 'Montaña del Río' aimed at the management and control of the supplies of fresh water in the Doñana′s marshes (Castellanos et al. 2002). However, between 2000 and 2005 it had become the dominant plant species on recent tidal Algaida marsh restorations located on the left bank of the Guadalquivir estuary (Doñana National Park; 36° 54′N, 6° 19′O, SW Spain), where it occupied tenths of hectares of the middle and low marsh, with coverage values greater than 98% (Gallego-Fernández and García-Novo 2007). Despite this demonstrated rate of expansion, it is not among the priority species targeted for control and eradication actions in Doñana National Park (Cobo et al. 2003). This is worthy of attention, since future restoration programs—such as the paralyzed 'Project Doñana 2005'—which aimed to restore the hydrological functioning of Doñana's marshes could contribute to facilitating the invasion of this exotic plant in this area, through dispersion of its prolific production of seeds of the populations located on the banks of the Guadalquivir River by tidal effect, which could reach farther inland.

8.5 *Spartina* Invasions and Changes in Ecosystem Services

The presence of NIS can be a threat for the biodiversity of salt marshes, depending on its ecophysiological fitness in this new environment. This proved to be the major constraint shaping the ecophysiological fitness of NIS, its primary productivity and consequently its spread along the Mediterranean marshes. In contrast to the native *S. maritima*, the NIS *S. patens* proved to have a lower photosynthetic fitness under the Mediterranean climate, which seems to be slowing down its spreading but not arresting it. If this is true, when comparing the ecological fitness of this species with its native congener, when it is compared to its closer neighbour species the same cannot be said. *Spartina patens* inhabits the upper-middle marshes, competing for space, nutrients and light resources with several goosefoot species (*Chenopodeaceae*), such as the sea purselane *Halimione portulacoides* and the shrubby swampfire *Sarcocornia fruticosa* (Fig. 8.2). These C_3 species have slower biomass production, thus expanding at slower rates. With the expansion of *S. patens* through the upper-middle marshes these species are being squeezed towards their altitudinal limit, reducing their colonization area.

This has several consequences for the remaining wildlife that inhabits the marshes (Teixeira et al. 2014). High-water roosts are essential features in the life-cycle of waders but other estuarine birds may also need them eventually. These roosts often occur in extensive areas of halophyte vegetation, including large man-made saltpans that remain comparatively free from human disturbance. Small islets of halophytes may be very important in the tidal cycle and they are often used as temporary roosts around high water (Fig. 8.3), while they are not yet covered by the incoming tide and can still provide safe haven to the waders perching on them above water (Teixeira et al. 2014).

Halophytes in Tagus salt marshes are habitat to some Passerine birds but these tend to be overlooked and even ignored by the non-specialist birders because they are small and unobtrusive. Some of these species are common and may be found in nearby habitats, simply moving into the areas colonized by halophytes to collect small invertebrates and maturing seeds. These food items are often picked up from hedges of the tree purslane *Atriplex halimus* in the upper marsh, and from tall bushes of the perennial saltwort *Sarcocornia fruticosa* and the golden samphire *Inula crithmoides*. The reduction of the colonization area of these areas caused by the expansion of *S. patens* inevitably reduces the feeding areas of these bird species.

If from the ecological point of view this is already a matter of concern, if salt marshes are considered as cultural, aesthetic and educational areas, then the seriousness of the problem acquires an increased concern. In the last two decades, marshes have been regarded as privileged sites for bird watching both for scientific and educational proposes, being used as examples of diverse ecosystems where plants provide unique habitats for wildlife (Teixeira et al. 2014). Attitudes of people living in urban areas around the estuary have changed dramatically in recent years. Compared to only a few decades ago (ca. 1970–1980) there is now widespread public awareness for environmental issues and many citizens praise activities that put lasting value to healthy estuarine ecosystems. These include many different uses, like kayaking, bird watching, and jogging on the water front. Shooting small birds or otherwise molesting protected species has been illegal for many years, but nowadays these practices are virtually banned around the estuary. Birds of prey also become respected by the public and they are no longer seen as mere targets for extermination. Hunting waterfowl lost terrain to other activities. All these results did come from sustained investments at environmental awareness and education that were deployed by the public services and by dedicated organizations for many years. The biodiversity reduction promoted by *S. patens* can eventually turn ecologically rich marshes into monospecific stands, removing their role as privileged biodiversity sites.

Marshes also provided other important ecosystems services to humans, especially to the nearby populations. These ecosystems act like barriers against storm surge and flooding, protecting houses and cities from extreme climatic events (Duarte et al. 2013b). This comes from the marshes' ability to growth in height due to their sedimentation function. Morphologically, marshes are very dynamic with flats, roosts and creeks that are opened and closed according to the marsh hydrodynamics and sedimentation. Again, this is highly dependent on the species dynamics (Salgueiro and Caçador 2007), with different species promoting different sedimentation rates shaping the marsh elevational gradient, while also acting as facilitators for neighboring species (Caçador et al. 2007; Salgueiro and Caçador 2007). Non-indigenous species have undeniable effects on the subjective and relative characteristics of cultural ecosystem services, such as aesthetics. Inherent to these biological invasions is that the provision of these services changes due to the effects of non-native species on the biology and ecology of the system (Kueffer and Kull 2017).

The high tolerance of *S. patens* to salinity gradients provides it with the ability to colonize the entire marsh elevational gradient, acting as a competitor with the native flora (Duarte et al. 2015a, b, c, d). In the worst-case scenario, Mediterranean marshes can lose their elevational gradients due to the retreat of the creek stabilizer *H. portulacoides* or of sedimentation enhancers like *Sarcocornia perennis*, turning a morphologically dynamic landscape into a flat marsh resembling the monospecific marshes formed by *S. patens* in its native habitats. This would make all the human structures located in the vicinity more prone to flooding events and storm surges.

Another important role widely recognized for marshes may also be at risk with the invasion of *S. patens*, namely the ability to remove contaminants from polluted sediments. Different halophytes have different abilities and behaviors towards what is normally known as sediment natural phytoremediation (Caçador et al. 2009). For example, *H. portulacoides* has a phyto-extractor character, removing contaminants from sediments and allocating them into its biomass (Duarte et al. 2010). On the other hand, the members from *Spartina* genus are known to act as phyto-stabilizers, maintaining contaminants in the sediments surrounding its roots at high concentrations (Duarte et al. 2011). The spreading of the non-indigenous *S. patens* will reduce the marshes' ability to remove contaminants, leaving these pollutants more available for the remaining biota, including fishes and bivalves, on which large fishing communities rely (Caçador et al. 2012). A study conducted in northern Spain in 2008 reinforced this fact, showing evidence that detritivore and insect herbivore food web pathways were significantly affected by this invasion (Page et al. 2010).

The invasive potential of *S. densiflora* derives from its high ecological, phenotypic and physiological versatility (Bortolus 2006; Castillo et al. 2008; Mateos-Naranjo et al. 2007, 2008, 2011), which allows it to tolerate a wide range of environmental conditions. This together with its prolific seed production and characteristic clonal growth (Kittelson and Boyd 1997; Bortolus 2006; Nieva et al. 2001; Mateos-Naranjo et al. 2008), have made the potential of this invasive species to occupy a great surface area the primary concern for conservation in the salt marshes of the Gulf of Cadiz. The creation of vast and dense meadows of cordgrass would greatly simplify habitat diversity, with serious impacts on ecosystem services and on socio-economical aspects linked with the estuary system. On a large scale, the ongoing spread of *S. densiflora* is resulting in intense occupation of these spaces, causing a significant hydrologic alteration, with physiographic changes to the salt marshes decreasing the flood control functionality of tidal channels (Castellanos et al. 1994; Castillo et al. 2000; Nieva et al. 2001; Mateos-Naranjo et al. 2008).

These structural and functional changes in the drainage system alter the sediment and nutrient cycles, as well as the sense of the fluxes of energy. This issue is exacerbated, since *S. densiflora* accumulates a great amount of dead structures (leaves, tillers, rhizomes) which remain there for years (Figueroa and Castellanos 1988) hindering the availability of nutrients and energy necessary for other levels of the trophic chain. Likewise, these structural changes are also creating a barrier which hinders the dispersal of organisms and their propagules, altering their biological cycles, and generating serious economic consequences when affecting species with fishery and shellfishery potential.

The massive occupation of space in salt marshes by *S. densiflora* is reducing species richness in less than a decade; particularly in the middle and middle-high marshes (Mateos-Naranjo 2008). Furthermore, the presence of this species is contributing to reducing the genotypic diversity in the chenopodiaceous[2] marshes (Figueroa et al. 2003). In addition, the environmental changes related with its presence also alter the species richness and diversity of the benthic macroinvertebrate community (composed mainly of annelids, crustaceans and molluscs; Curado et al. 2014), and also other animal species which also have important ecological and socio-economical values, such as bird species (Curado et al. 2013).

Moving to the effect of the presence of *S. densiflora* for the conservation of *S. maritima*, it should be highlighted that, invasion may be limited by the presence of this plant at lower elevations (Castillo et al. 2008). The alien altered the native zonation/succession pattern through the colonization of the center of *S. maritima* stands (Castellanos et al. 1994; Figueroa et al. 2003). Furthermore, it has been demonstrated that the *S. densiflora* × *S. maritima* hybrids exhibit ecological amplitudes greater than their parental species (Castillo et al. 2010), and could therefore displace the species, as well as jeopardize the resilience of the ecosystems services related with its presence (described previously in Sect. 2).

8.6 The Future of Invaded Marshes in a World of Change

The latest report of the Intergovernmental Panel on Climate Change (IPCC 2014) predicts, along with increasing atmospheric CO_2, increases in temperature, in the duration and intensity of drought, in soil salinization and sea level rise (SLR). Consequently, climate change emerges as one of the biggest challenges for ecosystems conservation and for sustaining its important services. Salt marshes of the Mediterranean area will be no exception, in fact due to their location and characteristics, these will be among the estuarine ecosystems most vulnerable to the effects of climatic change. Loss of salt-marsh habitat due to lateral erosion is already a major problem across Europe, with some reports pointing out to losses around 40 ha $year^{-1}$ due to this phenomenon (Hughes and Paramor 2004). Additionally, climatic change effects will also have its impacts on the invasion ecology of some non-indigenous species, such as *Spartina* species, altering is patterns of invasion through its effect on the indigenous species and on the invasive potential of the alien species (Dukes and Mooney 1999).

---

[2]Annual or perennial succulent halophytic or xerophytic herbs or shrubs that include most of the species present in Mediterranean and Atlantic marshes.

Studies that have assessed the influence of climatic change and rising atmospheric CO_2 concentrations on the native *S. maritima*, and on both its relative exotics are scarce (Duarte et al. 2015a, b; Mateos-Naranjo et al. 2010a, b). Mateos-Naranjo et al. (2010a) indicated that elevated atmospheric CO_2 concentration could stimulate the growth of *S. maritima* in a range of under high salinity ranges up to 65%. These results suggested that the *S. maritima* productivity and the ecosystem services it provides will likely increase in future climatic scenarios. Other studies, suggest that *S. maritima* tightens its distribution to low marshes, characterized by more stressful conditions with long submergence periods or higher salinity thus avoiding contact with the exotic species *S. densiflora* (Castillo et al. 2008). We could have hypothesized that SLR could increase the degree of stress in the inhabited area of *S. maritima*, which would lead to a displacement of its actual ecological niche to high areas regarding the tidal level, with a great occupation of the space by *S. patens* and *S. densiflora*, endangering the conservation of *S. martima* populations.

On the other hand, studies focused on the invasive potential of the alien species suggest that rising atmospheric CO_2 and the interaction with other environmental factor related with climatic change (e.g. such as temperature or salinity), could enhance the invasive success of both exotic *Spartina* species in the salt marshes of the Mediterranean. Thus, Mateos-Naranjo et al. (2010b) found that the relative growth rate of *S. densiflora* increased in plants grown at elevated CO_2 at low and moderate salinity scenarios. Nevertheless, net photosynthetic rate is highly temperature dependent, being its increment one of the predicted effects related with the climatic change. Thus, an increment of temperature could exacerbate the invasive success of *S. densiflora* in the salt marshes of the Mediterranean area, as indicated by the increment of net photosynthetic rate to future warming conditions (Fig. 8.4). In this future context, *S. densiflora* invasion would be favoured and could contribute to the current regression of *S. maritima* and its related ecosystem services in many salt marshes in the Mediterranean framework.

Increased temperature and evapotranspiration, will increase the salinity in the upper and middle marsh affecting the habitat. This has consequences at the primary productivity level of namely the succulent shrubs, reducing it beyond its physiological optimum (Duarte et al. 2013c, 2014). Additionally, this opens space to the spreading of other species, namely invasive ones, like *S. patens,* highly resistant to salinity changes (Duarte et al. 2015b). Nowadays the expansion of this species is constrained by a physiological misfit to the present climatic conditions observed in the Mediterranean marshes (Duarte et al. 2015a), preventing its fastest expansion. Thus, the on-going climatic changes can be favouring drivers for a faster expansion of this species, taking part of less productive Mediterranean species in a climate change scenario, by direct competition. In the areas where salt marsh is converted to vegetated mudflat or open water, the provision of ecosystem services is expected to decrease (Craft et al. 2009). Additionally, punctual and abrupt and extreme thermal events can induce severe damage on plant productivity on a shorter term (IPCC 2012). At this level, shrub halophytes will be more severely damaged by heat waves (Duarte et al. 2015c, d), events which are expected to be more intense and more

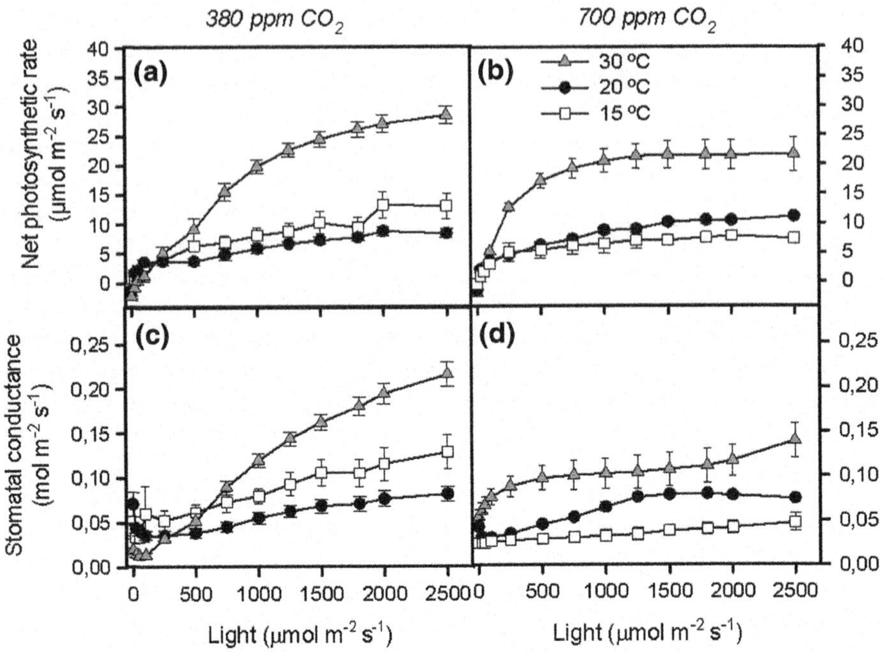

Fig. 8.4 Net photosynthetic rate **a**, **b** and stomatal conductance **c**, **d** in randomly selected fully expanded leaves of *S. densiflora* in response to increments in light intensity and temperature, at atmospheric CO_2 concentration of 380 and 700 ppm. Values represent means ± standard error, n = 5

frequent in a near future (IPCC 2012), with higher impacts at the pioneer species level (for e.g. *Aster tripolium* and. *S. maritima*). Moreover, the reduction of the primary productivity of the most abundant native species, will also open space for the colonization by NIS, changing completely the marsh landscape and, of course, its ecosystem services to the estuarine community (Duarte et al. 2016).

8.7 Invasive *Spartina* Removal and Marsh Restoration

Prevention of NIS spreading is normally more cost-effective than post-entry eradication or removal (Taylor and Hastings 2004). Nevertheless, in most cases we are already too late for this option, as several *Spartina* NIS establishments are far away from the prevention phase. This way some restorations approaches have been developed in China and in the USA that can be used as case studies for the Mediterranean marshes (An et al. 2007; Callaway 2005). While in terrestrial ecosystems these approaches are rather simple and well-studied, the aquatic environment and specially the intertidal ecosystems pose several problems (Anderson 2005). The discussion of the control options and associated constraints and the objective function for control has come under closer study more recently (Liebhold and Tobin 2008). In China, *S. alterniflora* was introduced intentionally for land reclamation proposes (Zhang et al. 2004). Recently, scientists concluded that it was rapidly out-competing the native marsh species and thus measures started to be implemented to control the expansion of this weed until a complete eradication (Chen et al. 2004).

Physical control methods are normally based in devices to remove, burn, prune, excavate, dredge, cover, flood and drain the plants of *S. alterniflora*, restricting their growth, hinder their respiration and photosynthesis and finally suppress or kill them (Hedge et al. 2003). Chemical controls use herbicides with inevitable contamination of the estuarine water body, while biological controls inevitably lead to another NIS introduction, bringing a further potential invasion and need a long time to validate their efficiency (Li and Zhang 2008). These last referred impacts led to the choice of physical measures in most of the NIS *Spartina* removal programs (Major et al. 2003; Hammond and Cooper 2002; Kriwoken and Hedge 2000). Early detection of small infestation areas still under expansion seems to be essential for the success of the eradication and native plant recolonization (Riddin et al. 2016). In South Africa, the chemical treatment of small patches (about one ha) in the Great Brak Estuary allowed to completely eradicate an invasive *Spartina* species and the regrowth of native salt marsh within one year after chemical treatment (Riddin et al. 2016).

While in the USA, China and Australia there are several records regarding the pre-invasion environment and exotic *Spartina* expansion, in Europe, several countries lack these data. For these reasons, most of the exotic *Spartina* control and management programs are based on the complete removal of NIS patches followed by natural or assisted restoration by the present native vegetation. The lack of data regarding the pristine condition of the indigenous vegetation may be imposing another serious threat to the marshes. By restoring the marshes using what is now the native vegetation and gradient, the ecological evolution of the marshes may suffer also some shifts. Nevertheless, the focus of these programs is the eradication and management of the invasive *Spartina* species expansion, and thus this has been achieved in some countries, serving nowadays as example for other regions that only now are identifying these invasive *Spartina* as problems.

The San Francisco Estuary Invasive *Spartina* Project (www.spartina.org) led by the California State Coastal Conservancy is one of the successful cases of NIS Spartina removal ad revegetation with native species. Established in 2000, the overall goal of the project is to develop a regionally coordinated project to address the rapid spread of four introduced and highly invasive *Spartina* species in the San Francisco Estuary. NIS were removed by Imazapyr herbicide treatment followed by manual removal (any seedlings or small plants in soft substrate) and covering of the terrain with mud flat substrate (Fig. 8.5). After the chemical treatment of *S. densiflora*, the dead stocks and seedlings were removed by hand and the marsh left to recover naturally, accompanied by proper monitoring programs. The absence of the

Fig. 8.5 *Spartina densiflora* chemical removal project evolution in San Francisco salt marshes. *Source* www.spartina.org

NIS alleviated the competitive stress towards the succulent halophytes allowing its spreading into their typical zonation, restoring the marsh structure to a condition close to what is considered to be pristine, by comparison with non-invaded marshes.

8.8 Conclusion

Salt marshes, and in particular the communities of halophytes which compose them, provide essential ecosystem services in terms of socio-economic growth and environmental protection. In the Mediterranean salt marshes framework, we have highlighted those communities integrated by the native European cordgrass *S. maritima* due to its key role as a pioneer species involved in the maintenance of salt marsh structure and function, and as consequence in its ecosystems services. Despite this relevant importance, the relative presence of *S. maritima* in the salt marshes is low in terms of occupied area, showing percentages of c. 12% in the Portuguese salt marshes, and lower than 5% for the 70% of its populations in the Spanish transitional systems, due to dieback or due to hybridization with other invasive *Spartina* species.

Furthermore, we could hypothesize that sea level rise related to climate change endangers the conservation of *S. martima* populations, since tidal level increment would lead to ecological niche displacement from low marshes, with long submergence periods or higher salinities, to high areas, with a greater occupation of the space by *S. patens* and *S. densiflora*. The presence of non-indigenous *Spartina* species along the Mediterranean marshes can be regarded as a threat for salt marsh biodiversity and ecosystem services. Among these species, the spread of *S. densiflora* has meant an intense occupation of the spaces, causing significant structural and functional changes in the drainage system of the marshes systems, which alters the sediment and nutrients cycles, as well as the fluxes of energy necessary for other levels of the trophic chain. These deleterious effects for normal functioning of salt marshes could be exacerbated by future environmental conditions related to climatic change, since increased atmospheric CO_2 concentrations and temperatures could exacerbate *S. densiflora* invasion, contributing to loss of ecosystem services in many salt marshes in the Mediterranean framework.

On the other hand, *S. patens* expansion is reducing the habitat of the typical succulent species that characterize typical Mediterranean marshes. Besides the obvious impacts on neighbour plant species, the consequences of this invasion expand to the whole ecosystem. Landscape changes introduced by the inevitable habitat displacement promoted by NIS, impact all the animals (birds, fishes and invertebrates) that depend on different species to shelter, feed and reproduce. Additionally, the human populations are either directly affected by reduction of the abundance of economically important seafood and fish species and indirectly by a severe reduction of the services provided by marshes such as coastal protection against storm surge and erosion or reduction of the cultural and aesthetic value of the marshes as tools for environmental education.

Humanity and the marshes face a new challenge in the twenty-first century. Climate change is undeniable, but its ecological effects are exacerbated by the expansion of NIS. Management efforts must highlight desired marsh attributes and ecosystem services in the face of human activities that threaten salt marsh ecosystems. This requires abandoning old-fashioned restoration goals, aiming to achieve "pristine" or "reference" conditions, replacing them by objectives aiming for enduring ecosystem subsistence and preservation of essential ecosystem services (contaminant sink, nursery habitat, nutrient regulation, and shoreline protection), passing inevitably by the management and control of aggressive non-indigenous *Spartina* species. The need to urgently change management and monitoring practices is undeniable, by adding measures to prevent, control and monitor this ongoing new threat, as it was made in the past in terms of pollution control and erosion mitigation. This should also be followed by the implementation of environmental education programs that attract the attention of the general public to wetlands as sources of aesthetic and cultural value and privileged seascapes, highlighting always their important ecosystems services both for wildlife and nearby human populations.

Acknowledgements This work has been co-funded by Oficina de Cooperación Universidad de Sevilla (Conv. Ay. Act. y Proy. Coop. Des. Mod. 2, 2014/15-2015/2016) and Ministerio de Economía y Competitividad (MINECO Project CGL2016-75550-R cofunded by FEDER). The authors would also like to thank to the "Fundação para a Ciência e Tecnologia (FCT)" for funding the research in the Marine and Environmental Sciences Centre (MARE) throughout the project UID/MAR/04292/2013. The authors would also like thank to the MAR 2020 program through the project RESTAURA2020 (16-01-04-FMP-0014). B. Duarte investigation was supported by FCT throughout a Posdoctoral grant (SFRH/BPD/115162/2016).

References

Abbas AM, Lambert AM, Rubio-Casal AE, De Cires A, Figueroa ME, Castillo JM (2015) Competition from hydrophytes reduces establishment and growth of invasive dense flowered cordgrass (*Spartina densiflora*). Peer J 3:e1260

Ainouche M, Baumel A, Salmon A, Yannic G (2003) Hybridization, polyploidy and speciation in *Spartina* (Poaceae). New Phytol 161:165–172

Allen JRL (2000) Morphodynamics of Holocene salt marshes: a review sketch from the Atlantic and Southern North Sea coasts of Europe. Quat Sci Rev 19:1155–1231

Álvarez R, Castillo JM, Mateos-Naranjo E, Gandullo J, Rubio-Casal AE, Moreno FJ, Figuero ME (2010) Ecotypic variations in phosphoenolpyruvate carboxylase activity of the cordgrass *Spartina densiflora* throughout its latitudinal distribution range. Plant Biol 12:154–160

An SQ, Gu BH, Zhou CF, Wang ZS, Deng ZF, Zhi YB, Li HL, Chen L, Yu DH, Liu YH (2007) Spartina invasion in China: implications for invasive species management and future research Weed Res 47:183–191

Anderson LWJ (2005) Potential for sediment-applied acetic acid for control of invasive *Spartina alterniflora*. J Aquat Plant Manage 45:100–105

Ayres DR, Smith DL, Zaremba K, Klohr S, Strong DR (2004) Spread of exotic cordgrasses and hybrids (*Spartina* sp.) in the tidal marshes of San Francisco Bay, California, USA. Biol Invasions 6:221–231

Baumel A, Rousseau-Gueutin M, Sapienza-Bianchi C, Gareil A, Duong N, Rousseau H, Coriton O, Amirouche R, Sciandrello S, Duarte B, Caçador I, Castillo JM, Ainouche M (2016) *Spartina versicolor* Fabre: another case of *Spartina* trans-Atlantic introduction? Biol Invasions 18:2123

Bolòs O, Vigo J (2001) Flora dels Pa¨ısos Catalans 4 (Monocotiled`onies). Editorial Barcino, Barcelona

Borja A, Dauer DM (2008) Assessing the environmental quality status in estuarine and coastal systems: comparing methodologies and indices. Ecol Ind 8:331–337

Bortolus A (2006) The austral cordgrass *Spartina densiflora* Brong.: its taxonomy, biogeography and natural history. J Biogeogr 33:158–168

Bridgham SD, Megonigal JP, Keller JK, Bliss NB, Trettin C (2006) The carbon balance of North American wetlands. Wetlands 26:889–916

Bueno A (1997) Flora y vegetación de los estuarios asturianos. Cuadernos de Medio Ambiente. Naturaleza 3. Consejería de Agricultura, Principado de Asturias

Cabezudo B, Talavera S, Blanca G, Salazar C, Cueto M, Valdés B, Hernández-Bermejo JE, Herrera C, Rodríguez-Hiraldo C, Nava D (2005) Lista roja de la flora vascular de Andalucía. Consejería de Medio Ambiente, Junta de Andalucía, Sevilla

Caçador I, Tibério S, Cabral H (2007) Species zonation in Corroios salt marsh in the Tagus estuary (Portugal) and its dynamics in the past fifty years. Hydrobiologia 587:205–211

Caçador I, Caetano M, Duarte B, Vale C (2009) Stock and losses of trace metals from salt marsh plants. Mar Environ Res 67:75–82

Caçador I, Costa JL, Duarte B, Silva G, Medeiros JP, Azeda C, Castro N, Freitas J, Cabral H, Costa MJ (2012) Macroinvertebrates and fishes as biomonitors of heavy metal concentration in the Seixal Bay (Tagus estuary): which species perform better? Ecol Ind 19:184–190

Caçador I, Neto JM, Duarte B, Barroso DV, Pinto M, Marques JC (2013) Development of an Angiosperm Quality Assessment Index (AQuA—Index) for ecological quality evaluation of Portuguese water bodies—a multi-metric approach. Ecol Ind 25:141–148

Callaway JC (2005) The challenge of restoring functioning salt marsh ecosystems. J Coast Res 40:24–36

Cambrollé J, Redondo-Gómez S, Mateos-Naranjo E, Figueroa ME (2008) Comparison of the role of two *Spartina* species in terms of phytostabilization and bioaccumulation of metals in the estuarine sediment. Mar Poll Bull 56:2037–2042

Castellanos EM, Figueroa ME, Davy AJ (1994) Nucleation and facilitation in saltmarsh succession: interactions between *Spartina matitima* and *Arthrocnemum perenne*. J Ecol 82:239–248

Castellanos EM, Heredia C, Figueroa M, Davy AJ (1998) Tiller dynamics of *Spartina maritima* in successional and non-successional mediterranean salt marsh. Plant Ecol 137:213–225

Castellanos EM, Luque CJ, Mateos-Naranjo E, Redondo-Gómez S, Figueroa ME (2002) Ecological implications of the invasion of the alien cordgrass *Spartina densiflora* in the tidal marshes of the Gulf of Cádiz. Neobiota, pp 192–200

Castillo JM, Figueroa ME (2009) Effect of abiotic factors on the life span of the invasive cordgrass *Spartina densiflora* and the native *Spartina maritima* at low salt marshes. Aquat Ecol 43:1–60

Castillo JM, Fernández-Baco L, Castellanos EM, Luque CJ, Figueroa ME, Davy AJ (2000) Lower limits of *Spartina densiflora* and *S. maritima* in a Mediterranean salt marsh determined by different ecophysiological tolerances. J Ecol 88:801–812

Castillo JM, Rubio-Casal AE, Redondo S, Álvarez-López AA, Luque T, Luque CJ, Nieva FJ, Castellanos EM, Figueroa ME (2005) Short-term response to salinity of an invasive cordgrass. Biol Inv 7:29–35

Castillo JM, Leira-Doce P, Rubio-Casal AE, Figueroa ME (2008) Spatial and temporal variations in aboveground and belowground biomass of *Spartina maritima* (small cordgrass) in created and natural marshes. Estuar Coast Mar Sci 78:819–826

Castillo JM, Ayres DR, Leira-Doce P, Bailey J, Blum M, Strong DR, Luque T, Figueroa ME (2010) The production of hybrids with high ecological amplitude between exotic *Spartina densiflora* and native *S. maritima* in the Iberian Peninsula. Divers Distrib 16:547–558

Castillo JM, Grewell BJ, Pickart A, Bortolus A, Pena C, Figueroa ME, Systma M (2014) Phenotypic plasticity of invasive *Spartina densiflora* (Poaceae) along a broad latitudinal gradient on the pacific coast of North America. Am J Bot 101:448–458

Chen ZY, Li B, Chen ZY, Chen JK (2004) Local competitive effects of introduced *Spartina alterniflora* on *Scirpus mariqueter* at Dongtan of Chongming Island, the Yangtze River estuary and their potential ecological consequences. Hydrobiologia 528:99–106

Chevalier A (1923) Note sur les *Spartina* de la flore française. Bulletin de la Société botanique de France 70:54–63

Cobo MD, Sánchez-Gullón E, García Murillo, P (2003) Datos acerca de la presencia y gestión de especies invasoras y xenófitas en un espacio protegido europeo paradigmático. En Contribuciones al conocimiento de las especies invasoras en España G.E.I., Serie Técnica

Cooper MA (1993) The status of *Spartina maritima* in Suffolk. Suffolk Naturalist′ Society. Transactions 29:48–54

Cosson E, de Maisonneuve MCD (1867) Introduction á la flore d'Algérie. Phanérogamie. Groupe des Glumacées (seu descriptio glumacearum in Algeria nascentium). Exploration Scientifique de l'Algérie, publiée par ordre du Gouvernement. Sciences Naturelles, Botanique. Imprimerie Impériale, Paris

Coste H-J (1906) Flore descriptive et illustrée de la France 3. Librairie des Sciences naturelles Paul Klincksieck, Paris

Craft C, Clough J, Ehman J, Joye S, Park R, Pennings S, Guo H, Machmuller M (2009) Forecasting the effects of accelerated sea-level rise on tidal marsh ecosystem services. Front Ecol Environ 7:73–78

Curado G, Figueroa E, Sanchez MI, Castillo JM (2013) Avian communities in *Spartina maritima* restored and non-restored salt marshes. Bird Study 60:185–194

Curado G, Sanchez-Moyano JE, Figueroa ME, Castillo JM (2014) Do *Spartina maritima* plantations enhance the macroinvertebrate community in European salt marshes? Estuaries Coasts 37:589–601

Daehler C, Anttila C, Ayres D, Strong D (1999) Evolution of a new ecotype of *Spartina alterniflora* (Poaceae) in San Francisco Bay, California, USA. Am J Bot 86:543–546

Daveau J (1897) La flore littorale du Portugal. Boletim da Sociedade Broteriana 14:4–54

Duarte B, Raposo P, Caçador I (2009) *Spartina maritima* (cordgrass) rhizosediment extracellular enzymatic activity and its role on organic matter dec omposition and metal speciation processes. Mar Ecol 30:65–73

Duarte B, Caetano M, Almeida P, Vale C, Caçador I (2010) Accumulation and biological cycling of heavy metal in the root-sediment system of four salt marsh species, from Tagus estuary (Portugal). Environ Poll 158:1661–1668

Duarte B, Freitas J, Caçador I (2011) The role of organic acids in assisted phytoremediation processes of salt marsh sediments. Hydrobiologia 674:169–177

Duarte B, Couto T, Freitas J, Valentim J, Silva H, Marques JC, Dias JM, Caçador I (2013a) Abiotic modulation of *Spartina maritima* photosynthetic ecotypic variations in different latitudinal populations. Estuar Coast Mar Sci 130:127–137

Duarte B, Caçador I, Marques JC, Croudace I (2013b) Tagus Estuary salt marshes feedback to sea level rise over a 40-year period: insights from the application of geochemical indices. Ecol Ind 34:268–276

Duarte B, Santos D, Marques JC, Caçador I (2013c) Ecophysiological adaptations of two halophytes to salt stress: photosynthesis, PS II photochemistry and anti-oxidant feedback—implications for resilience in climate change. Plant Physiol Biochem 67:178–188

Duarte B, Santos D, Silva H, Marques JC, Caçador I (2014) Photochemical and Biophysical feedbacks of C3 and C4 Mediterranean halophytes to atmospheric CO_2 enrichment confirmed by their stable isotope signatures. Plant Physiol Biochem 80:10–22

Duarte B, Baeta A, Rousseau-Gueutin M, Ainouche M, Marques JC, Caçador I (2015a) A tale of two *Spartinas*: Climatic, photobiological and isotopic insights on the fitness of non-indigenous versus native species. Estuar Coast Mar Sci 167:178–190

Duarte B, Santos D, Marques JC, Caçador I (2015b) Ecophysiological constraints of two invasive plant species under a saline gradient: halophytes versus glycophytes. Estuar Coast Mar Sci 167:154–165

Duarte B, Marques JC, Caçador I (2015c) Impact of extreme heat and cold events on the energetic metabolism of the C3 halophyte *Halimione portulacoides*. Estuar Coast Mar Sci 167:166–177

Duarte B, Goessling JW, Marques JC, Caçador I (2015d) Ecophysiological constrains of *Aster tripolium* under extreme thermal events impacts: merging biophysical, biochemical and genetic insights. Plant Physiol Biochem 97:217–228

Duarte B, Marques JC, Caçador I (2016) Ecophysiological responses of native and invasive *Spartina* species to extreme temperature events in Mediterranean marshes. Biol Inv 18:2189–2205

Dukes JS, Mooney HA (1999) Does global change increase the success of biological invaders? Trends Ecol Evolut 14:135–139

Espinar JL (2009) Pastizales de *Spartina* (Spartinion maritime). Ministerio de Medio Ambiente, y Medio Rural y Marino, Madrid

Fabre ME (1849) Description d'une nouvelle espèce de *Spartina*, abondante sur une portion du littoral méditerranéen. Annales des Sciencies Naturelles 3:22–125

Fenu G, Bacchetta G (2008) La flora vascolare della Penisola del Sinis (Sardegna Occidentale). Acta Bot Malacitana 33:1–34

Ferrandis Pomés C (2002) Ampliación del aeropuerto de Barcelona. Medidas de integración ambiental en la 3S¸ pista. I Congreso de Ingeniería Civil, Territorio y Medio Ambiente, Madrid, 13–15 Feb 2002, pp 1349–1362

Figueroa ME, Castellanos EM (1988) Vertical structure of *Spartina maritima* and *Spartina densiflora* in mediterranean marshes. *Plant form and vegetation structure*. In: Werger MJA, van der Aart PJM, During HJ, Verhoeven JTA (eds) SPB Academic Publishing, The Hague, The Netherlands, pp 105–108

Figueroa ME, Castillo JM, Redondo S, Luque T, Castellanos EM, Nieva FJ, Luque CJ, Rubio-Casal AE, Davy AJ (2003) Facilitated invasion by hybridization of *Sarcocornia* species in a salt-marsh succession. J Ecol 91:616–626

Fiori A, Paoletti G (1896–1908) Flora analitica d'Italia 1. Tipografia del Seminario. Padua

Fraga P, García O, Pons M (2003) Notes i contribucions al coneixement de la flora de Menorca (V). Bolletí de la Societat d'Història Natural de les Balears 46:51–66

Gallego-Fernández JB, García-Novo F (2007) High-intensity versus low-intensity restoration alternatives of a tidal marsh in Guadalquivir estuary, SW Spain. Ecol Eng 30:112–121

Gamisans J, Deschâtres R, Paradis G, Lambinon J (1989) *Spartina versicolor* Fabre. In: Jeanmonod D, Burdet HM (eds) Notes et contributions à la flore de Corse, IV, vol 44. Candollea, pp 367–368

Goulletquer P, Bachelet G, Sauriau PG, Noel P (2002) Open Atlantic coast of Europe—A century of introduced species into French waters. In: Leppakoski E et al (eds) Invasive aquatic species of Europe. Kluwer Academie Publishers, Netherlands

Gray A (2004) Will *Spartina anglica* invade northwards with changing climate? Third international conference on invasive Spartina, San Francisco, CA, 8–10 Nov 2004

Guinochet M, Vilmorin R (1978) Flore de France, 3rd edn. Éditions du Centre National de la Recherche Scientifique, Paris

Hammond MER, Cooper A (2002) Spartina anglica eradication and inter-tidal recovery in Northern Ireland estuaries. Turning the tide: the Eeradication of invasive species. Proceedings of the International Conference On Eradication of Island Invasives. Veitch CR, Clout MN (eds), IUCN SSC Invasive Species Specialist Group. IUCN, Gland, Switzerland and Cambridge (UK)

Han B, Wang XK, Ouyang ZY (2005) Saturation levels and carbon sequestration potentials of soil carbon pools in farmland ecosystems of China. Rural Eco-Environment 21:6–11

Hatch M (1992) C4 photosynthesis: an unlikely process full of surprise. Plant Cell Physiol 33:333–342

Hedge P, Kriwoken LK, Patten K (2003) A review of *Spartina* management in Washington State US. J Aquat Plant Manage 41:82–90

Heide O (2005) Ecotypic variation among European arctic and alpine populations of *Oxyda digyna*. Arctic Antarctic Alpine Res 37:233–238

Hodgkin E, Hamilton B (1998) Changing estuarine wetlands: a long term perspective for management. In A. J. McComb and J. A. Davis (eds), Wetlands for the Future. 243–255 Adelaide, Gleneagles

Hughes RG, Paramor OLA (2004) The effects of bioturbation and herbivory by the polychaete *Nereis diversicolor* on loss of saltmarsh in south-east England. J Appl Ecol 41:440–448

Idazkin YL, Bortolus A, Bouza P (2010) Ecological processes shaping Central Patagonian salt marsh landscapes. Austral Ecology 36(1):59–67

Intergovernmental Panel on Climate Change (IPCC) (2012) Managing the risks of extreme events and disasters to advance climate change adaptation. A special report of Working Groups I and II of the intergovernmental panel on climate change. In: Field CB, Barros V, Stocker TF, Qin D, Dokken DJ, Ebi KL, Mastrandrea MD, Mach KJ, Plattner G.-K, Allen SK., Tignor M and Midgley PM (eds) Cambridge University Press, Cambridge

Intergovernmental Panel on Climate Change (IPCC) (2014) Climate change 2014: impacts, adaptation and vulnerability. Contribution of working group II to the fifth assessment report of the IPCC

Kittelson PM, Boyd MJ (1997) Mechanisms of expansion for an introduced species of cordgrass, *Spartina densiflora*, in Humboldt Bay, California. Estuaries 20:770–778

Kriwoken LK, Hedge P (2000) Exotic species and estuaries: managing Spartina anglica in Tasmania, Australia. Ocean Coast Manag 43:573–584

Kueffer C, Kull CA (2017) Non-native species and the Aesthetics of nature. In: Vilà M, Hulme PE (eds) Impact of biological invasions on ecosystem services, invading nature—Springer series in invasion ecology, vol 12. Springer International Publishing Switzerland

Lefeuvre JC, Laffaille P, Feunteun E, Bouchard V, Radureau A (2003) Biodiversity in salt marshes: from patrimonial value to ecosystem functioning. The case study of the Mont-Saint-Michel bay. C R Biol 326:125–131

Lessmann J, Mendelssohn I, Hester M, McKee K (1997) Population variation in growth response to flooding of three marsh grasses. Ecol Eng 8:31–47

Li H, Zhang L (2008) An experimental study on physical controls of an exotic plant Spartina alterniflora in Shanghai, China. Ecol Eng 32:11–21

Liebhold AM, Tobin PC (2008) Population ecology of insect invasions and their management. Annu Rev Entomol 53:387–408

Luque CJ, Alvarez AA, Rubio AE, Muñoz J, Redondo S, Castillo J, Castellanos EM, Figueroa ME (2004) Memoria de Vegetación: Parque Natural Bahía de Cádiz. Informe para la Consejería de Medio Ambiente, Junta de Andalucía, Sevilla

Magenheimer JF, Moore TR, Chmura GL, Daoust RJ (1996) Methane and carbon dioxide flux from a macrotidal salt marsh, Bay of Fundy, New Brunswick. Estuaries 19:139–145

Major III WW, Crue CE, Grassley JM, Conquest LL, (2003) Mechanical and chemical control of smooth cordgrass in Willapa Bay, Washington. J Aquat Plant Manag 41:6–12

Marchant C, Goodman P (1969) *Spartina maritima* (Curtis) Fernald. J Ecol 57:287–291

Marinucci AC (1982) Trophic importance of *Spartina alterniflora* production and decomposition to the marsh estuarine ecosystem. Biol Conserv 22:35–58

Mateos-Naranjo E (2008) Hacia una metodología de gestión de la especie invasora *Spartina densiflora* Brongn.: Estado de invasión, avances ecofisiológicos y control. Universidad de Sevilla, Sevilla

Mateos-Naranjo E, Redondo-Gómez S, Silva J, Santos R, Figueroa ME (2007) Effect of prolonged flooding on the invader *Spartina densiflora* Brong. J Aquat Plant Manage 45:121–123

Mateos-Naranjo E, Redondo-Gomez S, Cambrolle J, Luque T, Figueroa ME (2008) Growth and photosynthetic responses to zinc stress of an invasive cordgrass, Spartina densiflora. Plant Biol 10:754–762

Mateos-Naranjo E, Redondo-Gómez S, Andrades-Moreno L Davy A (2010a) Growth and photosynthetic responses of the cordgrass Spartina maritima to CO2 enrichment and salinity. Chemosphere 81:725–731

Mateos-Naranjo E, Redondo-Gómez S, Alvarez R, Cambrollé J, Gandullo J, Figueroa E (2010b) Synergic effect of salinity and CO2 enrichment on growth and photosynthetic responses of the invasive cordgrass Spartina densiflora. J Exp Bot 61:1643–1654

Mateos-Naranjo E, Andrades-Moreno L, Redondo-Gómez S (2011) Comparison of germination, growth, photosynthetic responses and metal uptake between three populations of *Spartina densiflora* under different soil pollution conditions. Ecotox Environ Safety 74:2040–2049

Nieva FJ (1996) Aspectos ecológicos en *Spartina densiflora* Brongn. Tesis Doctoral. Universidad de Sevilla, España

Nieva FJ, Díaz-Espejo A, Castellanos EM, Figueroa ME (2001) Field variability of invading populations of *Spartina densiflora* Brongn. In different habitats of the Odiel Marshes (SW Spain). Estuar Coast Mar Sci 52:515–527

Otero X, Sanchez J, Macías F (2000) Nutrient status in tall and short forms of *Spartina maritima* in the salt marshes of Ortigueira (NW Iberian Peninsula) as related to physicochemical properties of the soils. Wetlands 20:461–469

Page HM, Lastra M, Rodil IF, Briones MJI MJI, Garrido J (2010) Effects of non-native *Spartina patens* on plant and sediment organic matter carbon incorporation into the local invertebrate community. Biol Inv 12:3825–3838

Parlatore F (1848–1850) Flora italiana I. Tipografia Le Monnier. Firenze

Pignatti S (1982) Flora d'Italia 3. Edagricole, Bologna

Proffitt C, Chiasson R, Owens A, Edwards K, Travis S (2005) *Spartina alterniflora* genotype influences facilitation and suppression of high marsh species colonizing an early successional salt marsh. J Ecol 93:404–416

Riddin T, van Wyk E, Adams J (2016) The rise and fall of an invasive estuarine grass. S Afr J Bot 107:74–79

Rubio JC (1985) Ecología de las Marismas del Odiel. Tesis doctoral. Departamento de Biología Vegetal y Ecología, Universidad de Sevilla, España

Saarela JM (2012) Taxonomic synopsis of invasive and native Spartina (Poaceae, Chloridoideae) in the Pacific Northwest (British Columbia, Washington and Oregon), including the first report of Spartina x townsendii for British Columbia, Canada. Phytokeys 10:25–82

Saint-Yves A (1932) Monographia Spartinarum. Candollea 5:19–100

Salgueiro N, Caçador I (2007) Short-term sedimentation in Tagus estuary, Portugal: influence of salt marsh plants. Hydrobiologia 587:187–193

SanLeon DG, Izco J, Sanchez JM (1999) *Spartina patens*: a weed in Galician saltmarshes (NW Iberian Peninsula). Hydrobiologia 15:213–222

Seliskar D, Gallagher J, Burdick D, Mutz L (2002) The regulation of ecosystem functions by ecotypic variation in the dominant plant: a *Spartina alterniflora* salt-marsh case study. J Ecol 90:1–11

Sousa AI, Lillebo AI, Pardal MA, Caçador I (2010) The influence of *S. maritima* on carbon retention capacity in salt marshes from warm-temperate estuaries. Mar Poll Bull 61:215–223

Stephenson D, Oliver L, Burgos N, Gbur E (2006) Identification and characterization of pitted morning glory (*Ipomoea lacunose*) ecotypes. Weed Sci 54:78–86

Taylor CM, Hastings A (2004) Finding optimal control strategies for invasive species: a density-structured model for Spartina alterniflora. J Appl Ecol 41:1049–1057

Teixeira A, Duarte B, Caçador I (2014) Salt marshes and Biodiversity. In: Khan MA, Böer B, Öztürk M, Al Abdessalaam TZ, Clüsener-Godt M, Gul B (eds) Tasks for vegetation science, vol 47. Sabkha ecosystems: volume IV: cash crop halophytes and biodiversity conservation. Springer, Berlin

Tutin TG (1980) *Spartina* Schreber. In: Tutin TG et al (eds) Flora Europaea vol 5. Cambridge University Press, Cambridge, pp 259–260

Yannic G, Baumel A, Ainouche M (2004) Uniformity of the nuclear and chloroplast genomes of *Spartina maritima* (Poaceae), a salt-marsh species in decline along the Western European Coast. Heredity 93:182–188

Zhang RS, Shen YM, Lu LY, Yan SG, Wang YH, Li JL, Zhang ZL (2004) Formation of *Spartina alterniflora* salt marshes on the coast of Jiangsu Province, China. Ecol Eng 23:95–105

Chapter 9
The Forest on the Peninsula: Impacts, Uses and Perceptions of a Colonial Legacy in Cyprus

O. L. Pescott, S. E. Harris, J. M. Peyton, M. Onete, A. F. Martinou and J. O. Mountford

Abstract Throughout history the values and meanings attached to habitats and species in particular places have seen considerable change. Such shifts in perspective are of particular relevance to the biology of invasions, with human attention and values often determining both the initial movement of species around the world, and the decision that subsequent independent spread should be considered damaging to the environment. This chapter examines such a case for the Akrotiri peninsula, Cyprus, where a particular colonial story about the degraded state of the environment, and the need to combat malaria, led to the introduction of various Australian trees for sanitation and other purposes. Today, some of these non-native species are considered invasive, and are having impacts on valued wetland habitats on the peninsula. We use archival research to investigate the changes in policy towards these habitats and the non-native species that affect them, and field research to describe the ecological context. Our study illustrates the complex interactions between ideas, practical aims, and values that lie behind the planned and invaded habitats at Akrotiri.

Keywords Invasive non-native species · Alien species · Colonialism *Acacia saligna* · *Eucalyptus*

O. L. Pescott (✉) · J. M. Peyton · J. O. Mountford
NERC Centre for Ecology & Hydrology (CEH) Wallingford, Benson Lane, Wallingford, Oxfordshire OX10 8BB, UK
e-mail: olipes@nerc.ac.uk

S. E. Harris
Eastern Michigan University, 900 Oakwood St., Ypsilanti, MI 48197, USA

M. Onete
Institute of Biology Bucharest, Romanian Academy, Splaiul Independentei Nr. 296, 060031 Bucharest, Romania

A. F. Martinou
Joint Services Health Unit (JSHU), Akrotiri, Cyprus

A. I. Queiroz and S. Pooley (eds.), *Histories of Bioinvasions in the Mediterranean*, Environmental History 8, https://doi.org/10.1007/978-3-319-74986-0_9

9.1 Introduction

Over and above the fundamental, discoverable truths of ecology and the environment, human value systems exert their influence, a no doubt inevitable consequence of the multiple temporal and spatial scales, the many and varied species, and the complex biotic and abiotic interactions (including those involving humans) that make up ecological systems (Castree 2014; Allen and Hoekstra 2015). It should not surprise us then, that, throughout history, the values and meanings attached to habitats and species in particular places have seen considerable change. Such shifts in perspective are of particular relevance to the biology of invasions, with human attention and values often determining both the initial movement of species around the world, and the decision that subsequent independent spread should be considered damaging to the environment. We examine such a scenario in Cyprus, where shifts in policy and approaches to the habitats of the Akrotiri peninsula over the past 150 years provide a fascinating, and well-documented, case of changing environmental priorities. We first provide an overview of these semi-natural and anthropogenic habitats from the point of view of the ecologist, assessing the intrinsic diversity and richness of the plant communities of the peninsula in a broad, and, we would hope, objective fashion. We then present the historical context and actions that created one of the most significant anthropogenic habitats of the peninsula, a forest of non-native Australian trees. The forest provides us with our non-native 'bioinvasions', albeit to differing degrees for different species. Drawing on our own fieldwork in the area and on previous studies, we examine some of the current impacts of these non-native forest species on native Cypriot wetland communities and, finally, review some of the current discussions about, and uses of, the Akrotiri forest, in order to demonstrate the continuing diversity of responses and meanings attached to the area.

9.2 The Habitats of the Akrotiri Peninsula

Wetland habitats are relatively scarce in the Mediterranean, with most of the extensive areas of wetland plant communities being found in the coastal lowlands (Britton and Crivelli 1993; Mediterranean Wetlands Observatory [MWO] 2012; Markogianni et al. 2014). Indeed, wetland sites on a similar scale to those found in Continental, Atlantic or Steppic biogeographic regions are very few in number in the Mediterranean (http://medwet.org). Wetland communities may also occur along river valleys and gorges, in inland freshwater lakes, and around springs and artificial water bodies such as dams or drainage channels (Gucel et al. 2012). The exact distribution of these features varies between countries, and existing inventories must certainly underestimate the areas of smaller features such as freshwater marshes, ponds, and wooded wetlands (Britton and Crivelli 1993; Perennou et al. 2012), many of which may be refuges for formerly more widespread plants—note, for example, some of the marginal wetland habitats described by Bergmeier and Abrahamczyk (2008) for Crete. Although wetlands are not the richest habitat for endemic plant diversity in the Mediterranean (Hobohm and Bruchmann 2011), they may still be of high local or regional value, containing many plant and animal species of restricted distribution (e.g. Smith et al. 2014). Globally, freshwater habitats are disproportionally important for the conservation of biodiversity given their area (Strayer and Dudgeon 2010), and, in recognition of this, recent efforts have been made to identify new areas of freshwater habitat important for biodiversity in the eastern Mediterranean (Barrios et al. 2014).

Being of restricted occurrence, and by virtue of their association with the regionally scarce resource of fresh water, Mediterranean wetland habitats also face ongoing threats from habitat conversion and water extraction (Perennou et al. 2012; MWO 2012, 2014). These threats appear to be increasing across the region (MWO 2014), although wetland habitat destruction is not a new phenomenon: Rackham (1990), for example, notes the extensive destruction of fen habitats in Greece during the 20th century. Across Europe, wetland drainage for agricultural intensification, or through attempts to combat malaria, has been widespread for hundreds of years (Joosten 1999), motives that have also influenced the development of these habitats in the eastern Mediterranean. In the context of Cyprus, the salt lake and marshes of the Akrotiri peninsula represent one of the most significant wetland complexes on the island (Delipetrou et al. 2008). The salt lake is itself a relatively uncommon feature in the region, and the importance of the extensive complex of permanent and seasonal saline water bodies, salt and freshwater marshes, and sand flats found at Akrotiri was recognised by its designation as a Ramsar site in 2003 (JNCC 2008). Criterion 1 of the Ramsar Sites Criteria, stating that "[a] wetland should be considered internationally important if it contains a representative, rare, or unique example of a natural or near-natural wetland type found within the appropriate biogeographic region", was cited as a key factor in its designation. More locally, Christodoulou (2003) has estimated that coastal halophytic wetlands account for only 0.3% of the land area of Cyprus.

The main wetland habitats around the lake include communities of conservation importance from the syntaxonomic classes[1] (Mucina et al. 2016) *Juncetea maritimi*, *Therosalicornietea* and *Salicornietea fruticosae*—essentially equivalent to habitats 1410 (Mediterranean salt meadows), 1310 (*Salicornia* and other annuals colonising mud and sand) and 1420 (Mediterranean and thermo-Atlantic halophilous scrubs) of Annex 1 of the European Commission Habitats Directive. Reedbed and sedge

[1]Vegetation scientists often use this hierarchical classification system for the description and naming of plant communities. Although these 'syntaxonomic' categories are "less crisply defined" (Mucina et al. 2016) than the taxonomic categories of genus, species etc., in Europe at least, the use of such systems is well established, and categories higher than the class (such as the alliance mentioned here) can provide a convenient means of indicating biogeographic, taxonomic and environmental similarities between communities. The Latinised names of these communities typically indicate key species and characteristics, as well as the level at which the category sits in the hierarchy (Kent 2012).

communities of the class *Phragmito-Magnocaricetea* are also important around the lake, and both abut and enter a forest of introduced woody species to the north. Many species of regional importance also occur on the peninsula. The most recent vascular plant Red Data Book (RDB) for Cyprus (Tsintides et al. 2007) includes the following plants that can still be found on the Akrotiri peninsula in moist habitats: saw sedge (*Cladium mariscus*), crypsis (*Crypsis factorovskyi*), slender club-rush (*Isolepis cernua*), sea rushes (*Juncus maritimus*, *J. littoralis*), sea flax (*Linum maritimum*), water mint (*Mentha aquatica*), and frog fruit (*Phyla nodiflora*) (Peyton and Mountford 2015; see Tsintides et al. 2007 for the conservation statuses of these plants). We suspect that many other wetland plants found on the peninsula, for example, long-bracted sedge (*Carex extensa*), hispid sedge (*C. hispida*), elymus (*Elymus elongatus* ssp. *haifensis*), sharp rush (*Juncus acutus*), rigid rush (*J. rigidus*), subulate rush (*J. subulatus*), sea plantain (*Plantago maritima* ssp. *crassifolia*), and water germander (*Teucrium scordium* ssp. *scordioides*), could be more threatened than currently considered, given the potential scarcity of, and threats to, suitable habitat on Cyprus. Current assessments of rarity (Tsintides et al. 2007) and broad distribution mapping (Hand et al. 2011–2017) appear largely to be based on the Flora of Meikle (1977, 1985), itself partly based on historic literature. Although much new information has accumulated since Meikle's *Flora* (see Hand et al. 2011–2017), more fine-scale information on species' distributions would be enlightening and invaluable for assessing threat.

Non-wetland habitats on the peninsula include the scrub forest, or maquis, south of the salt lake. In broad phytogeographical terms the entire southern coast of Cyprus has been classified as within the *Oleo-Ceratonion siliquae* alliance (Noirfalise 1987). Finer classifications focused on Cyprus alone have classified the Akrotiri maquis as a coastal subtype of the *Ceratonio-Pistacion lentisci* characterised by Phoenicean juniper (*Juniperus phoenicea*) (Thirgood 1987). The main forest north of the lake is, however, primarily composed of planted non-native species,[2] mainly Port Jackson wattle (*Acacia saligna*), river red gum (*Eucalyptus camaldulensis*), tuart (*E. gomphocephala*), and casuarina (*Casuarina cunninghamiana*), all originating in Australia. Wattle and casuarina are both currently invasive on the peninsula (Hadjikyriakou and Hadjisterkotis 2002), whereas the eucalypt species are not, although casual[3] occurrences can be found in the forest and elsewhere on the island. In drier areas of the Akrotiri forest, where the original marshlands may have originally graded into open woodland or maquis, Aleppo pine (*Pinus halepensis*) is present as a planted non-native, and native woody species such as mastic (*Pistacia lentiscus*) and Phoenicean juniper appear where the

[2]For more on the botanical networks that led to such species becoming part of the British colonial landscape see, amongst others, Barton (2002), Doughty (2000), and Grove (1995).

[3]We use the word 'casual' here in the ecological sense of a plant that has successfully reproduced (either sexually or vegetatively), but that has not been observed to maintain its populations over longer periods (Richardson et al. 2000). That is to say, the species is not considered to be established (or naturalised, an oft-used synonym) in the area under consideration.

plantation woodland thins out. The origins and impacts of this plantation are considered further in this chapter.

9.3 From Pestilence to Productivity: Colonial Forestry in the Lowlands

That the contemporary conservationist values diversity, whether of species within a site or at the larger scale of habitats with a landscape, is now taken for granted; whether or not diversity itself should alone be promoted as the target of biological conservation, or whether this should be linked to ecosystem functions and services to potentially create wider support for conservation in society, is still debated (Silvertown 2015). The biology of invasions and non-native species has come to feature in these discussions, with some ecologists and other commentators arguing for the recognition of the potential benefits of non-native species (Pearce 2015), or promoting a *laissez-faire* and very long-term view of the issues surrounding impacts (cf. Hulme et al. 2015a, b; Thomas and Palmer 2015; Thomas 2015). Before the 20th century such a discourse on non-native species did not exist in quite the same form, and the environmental aims of colonial administrators were mainly focused on issues of production and sanitation (Thirgood 1987; Barton 2002; Davis 2016), with little or no regard for impacts on native biodiversity, or indeed on its cultural associations and uses for local people. Although it may be anachronistic to use the term itself, in considering the environment of Cyprus the colonial British focused on a subset of ecosystem services, but, unlike today, achieving the potential 'provisioning' or 'regulating' services of forests was unfettered by any consideration of the intrinsic value of native habitats or species, nor apparently of the potential future weediness of species introduced for forestry or other purposes.

Harris (2007, 2012) documented the approach of the British administration (1878–1960) to the forests of Cyprus, as well as reviewing the physical environmental history of the island. Importantly, Harris also critically reviewed what *has been said* through the ages about the Cypriot environment—a narrative that appears to have distracted and misled the British (Harris 2007). Harris (2007) argued that the 'ruined landscape' narrative, or 'pseudo-history' (Grove and Rackham 2001), of the Mediterranean was the main driving force behind British forestry policy, and, indeed, all of the reports on the forests of Cyprus from that time emphasise the 'degraded' nature of the island's landscape (e.g. House of Commons 1880; Wild 1879; Madon 1881; Hutchins 1909), a narrative that is still encountered today (Delipetrou et al. 2008; cf. Rackham 2017). The most memorable piece of rhetoric is perhaps that of the French forester P. G. Madon (1881) who urged immediate action for the island's forests, lest the Mediterranean "soon count one island less and one rock more".

This emphasis on restocking and repairing the forests of Cyprus was largely focused on the mountains: the Southern Range of Paphos and the Troodos and the

Northern Range of Kyrenia (e.g. Thirgood 1987). Lowland forestry, with its extensive use of non-native species, has been written about to a far lesser extent, although the (re-)afforestation of parts of the Mesaoria plain in particular was at least of initial interest to British forestry personnel (House of Commons 1881). No doubt this is partly due to their relative areas, with the state-owned lowland forests and plantations being in the minority; it may also be a result of the ruined landscape theory linking with the pervasive idea of the 'Green Island' of Cyprus, so admired and utilised, we are told, in Classical times (Thirgood 1987; cf. Papacostas 2001; Rackham 2017). The emphasis on the relationship between deforestation and erosion in mountainous areas, which has been so instrumental in colouring the thoughts and theories of geographers and ecologists working in arid lands, links to this idea (Grove and Rackham 2001; Harris 2012; Davis 2016). An additional reason is simply that the forests are largely restricted to the cooler, moister hills that dominate a large part of the island (Delipetrou et al. 2008; Rackham 2017).

Compared to the mountain forests, it has proven challenging to locate travellers who have specifically commented on the Akrotiri salt lake and its surrounding marshes, or indeed similar habitats elsewhere on the island. Most commentaries are indirect, focusing on the 'unhealthiness' of the villages and towns near wetlands. For example, the first High Commissioner's Report on Cyprus (House of Commons 1880, p. 203), written soon after the arrival of the British in 1878, noted that the "unenviable notoriety which Larnaca [has] obtained for unhealthiness [is] principally owing to the close proximity of undrained marshes". Amongst the historical travellers' accounts of Cyprus reviewed by Harris (2007), de Mas Latrie (1870) and Agnes Smith (1887) also note the three main lowland wetland areas near to Limassol, Larnaca and Famagusta as centres of general 'pestilence'. More specifically for our purposes, the British naturalist Francis Guillemard (1888, pp. 108–109), who stayed near Akrotiri at Episkopi in the spring of 1887, reported that "[t] he neighbourhood of the salt lakes has a bad name, and at the village of Akrotiri, which is situated close to them, the infant mortality is said to be such that there is always one (but only one) baby in the place"; Hutchins (1909) repeats a similar sentiment over 20 years later.

As for the British commentary on the state of Cyprus's forests (Harris 2012), the state of the marshlands was also blamed on the previous administration of the Ottomans, albeit indirectly. Indeed, the link between forest degradation, erosion, and the resultant siltation and creation of fens has been noted by Grove and Rackham (2001) to be a part of the ruined landscape theory of the Mediterranean. Savile (1878, p. 78) provides an example of this for Cyprus, stating that "when the forests are cut down [...] torrents of rain [...] wash the vegetable mould off the slopes into the plains, leaving [...] miasma-breeding marshes".

Additional evidence for the potentially increased problems with malaria in the lowlands comes from a map supporting a successful 1941 application to the Colonial Office for a Development & Welfare grant by the then Conservator of Forests, R. R. Waterer (UK National Archive, Appendix D of document CO 67/328/8). This map clearly delimits the distributions of the two main malarial mosquito species in Cyprus at that time, *Anopheles superpictus* and *An. sacharovi* (given under the synonym *An. elutus*). *Anopheles sacharovi*, still one of the main vectors for malaria in other parts of the Levant (ECDC 2005–2017), is clearly shown as being present upstream of Famagusta in temporary riverside marshes, south of Larnaca around its salt lake, and north of the Akrotiri lake. The area shaded for *An. sacharovi* on the Akrotiri peninsula approximately matches the forested land under eucalyptus, casuarina and wattle today (CO 67/328/8). Both *An. superpictus* and *An. sacharovi* are vectors for the malarial parasite *Plasmodium vivax*, although current reviews suggest that *An. sacharovi* may be more problematic for human health, with some evidence suggesting that *An. superpictus* is more zoophilic and *An. sacharovi* more anthropophilic (Sinka et al. 2010; ECDC 2005–2017).[4] The drainage of the lowland marshes, then, was a priority for Cyprus's new British rulers, and one on which all the latest techniques of 'scientific forestry', including the use of non-native species, would be brought to bear.

Both Thirgood (1987) and Harris (2007) have thoroughly reviewed British forestry policy in Cyprus, albeit from different sides of the 'ruined landscape' argument, and we do not intend to repeat this material here. However, many of the reports written in the years following the British occupation contain valuable information on the introduction, establishment, and spread of the non-native woody species used for the lowland plantations, including the Akrotiri State Forest.[5] The Akrotiri State Forest was also, at one time (Harris 2007, p. 34), divided into the Main State Forest north of the lake (the plantation on the marshlands), and the Minor State Forest to the south (the coastal maquis, or 'scrub forest', but with planted additions such as the non-native Aleppo pine and mosaics of phrygana and steppe communities). Historical accounts of the habitats surrounding the salt lake, or photographic evidence, may help us to understand the likely impacts of the afforestation and natural spread of the species concerned, on both native biodiversity and ecosystem services.

A. E. Wild, an experienced forester briefly seconded from India, was the first person engaged by the British to report on the forests of Cyprus. In relation to the lowlands, Wild's report of 1879 (p. 13) suggests distributing southern blue gum (*Eucalyptus globulus*) seedlings to locals "as a sanitary measure, but also with regard to the improvement of the general appearance of such towns", although nothing specific is mentioned concerning the main areas of marsh noted above. It is not clear whether this suggestion for disseminating eucalypts was followed through at the time,[6] although, later, non-native species were apparently distributed on a small scale (Bovill 1920).

In 1881 Sir Ralph Biddulph, the High Commissioner, and the Principal Forest Officer (PFO) Madon corresponded with the British Government and the botanist

---

[4]Note the potential for historic shifts between pastoralism and agriculture to have altered the epidemiology of the disease on the island.

[5]The forest has also been known as the Limassol Salt Lake Plantation or Limassol Marsh Plantation, depending on the report.

[6]If it was, the species was not a great success—it does not feature in Meikle (1977).

William Thiselton Dyer, Assistant Director of the Royal Botanic Gardens at Kew, regarding the afforestation of the lowland marshes using eucalypts (House of Commons 1881).[7] The correspondence throws some light on the challenges of establishing eucalypts on the island, both because of the particularly cold winter of 1879–1880 that killed many saplings, damaging two trial plantations, and because of the requirement of the eucalypts for the better soils, most of which were already under the plough. Their fire-promoting properties, or the potential invasiveness of eucalyptus plantations, were apparently not appreciated at this time; indeed, Pyne (1982, p. 188) suggests that it was not until the second decade of the 20th century that the fire risks presented by eucalypts in arid environments began to concern foresters. By the end of 1881, Madon reportedly had seven additional eucalyptus species with which to experiment, including "*Tereticornis*", a name which may have covered both forest red gum (*Eucalyptus tereticornis*) and the variety *E. tereticornis* var. *rostrata*, a synonym used for river red gum (*E. camaldulensis*); given that he recommends "*Tereticornis*" for marshland planting, river red gum is the most likely (Doughty 2000). Madon had also secured limited funds to establish trial plantations with the "1st, and most important [aim of the] sanitation of towns" (House of Commons 1881, pp. 93–94), although fuel for villages is also given as a subsidiary benefit.

The reports of D. E. Hutchins (1909), a "forester-at-large" who acted as a consultant in many parts of the British Empire (Barton 2002), and of A. K. Bovill (1915, 1920), permanent PFO on Cyprus between 1896 and 1920, provide detail on the subsequent development of the Akrotiri plantations. Planting on the peninsula had modest beginnings: in 1893 the Commissioner of Limassol, Roland L. N. Michell, had "ventured [...] to suggest the formation of a plantation near the Akrotiri salt marsh, with a view to sanitary results [...]. A beginning has just been made,[8] on a very small scale. About four thousand cuttings (of plane, willow, poplar, and olive) have been planted in a diminutive enclosure [...] and I am promised [...] that provision will be made for carrying on the project, by the

[7]Note that by this date the reputation of certain eucalypts for malaria-related drainage schemes was well established, partly due to the apparent success of Trappist monks at Tre Fontane near Rome (Doughty 2000). Some sceptics remained, however (Doughty 2000), Dyer among them. In the letter cited here, Dyer also quotes Sir Joseph Hooker's account of his recent visit to Tre Fontane, which is worth repeating here in full: "I made a point of going to 'Tre Fontana,' [sic.] the Trappist convent in the Campagna cured of fever by Eucalyptus! The fact is, it was inhabited by a few monks, who died off like flies. These were succeeded by the Trappists, who anon brought 500 acres under cultivation of all sorts, including a few Eucalyptus, now multiplied by thousands of mere seedlings. Of course the sanitary state of the locality improved at once. Nevertheless, no fewer than three died of fever there last year. The Government allows a gang of 80 convicts and a guard of soldiers to aid in cultivation; and this together with a subsidy depends for its continuance on the fever being kept down and Eucalyptus planted. So of course the monks attribute all to the gum trees, which seem to me to have grown wretchedly compared to what one sees in the Riviera. I saw very little Eucalyptus elsewhere and no good ones anywhere; the Campagna winters are too cold." (House of Commons 1881, p. 107). Gaming bureaucratic systems is apparently not a mortal sin.

[8]Presumably in early 1895, Michell's report is dated May 15th, 1895.

planting of eucalyptus and other suitable trees" (House of Commons 1896, p. 61). Bovill (1915) recounts that the plantation was expanded using planting techniques recommended by Hutchins during his visit in 1909. During the early 20th century then, the plantation must have been a fairly minor affair: a memorandum on plantation work in Cyprus written by Bovill in 1902 makes no mention of it. Similarly, a plan of "Forests and mines" prepared in 1928 as a part of a set of maps of the island (a piece of propaganda associated with the 50th anniversary of the British occupation) does not depict a plantation in the area of the current Akrotiri forest, although a linear plantation of wattle between the peninsula and Limassol, started in 1900 to keep the road clear of drifting sand (Bovill 1915), is shown (Fig. 9.1). Bovill (1915, p. 20) also notes that the Limassol Salt Lake plantation "has been very difficult to deal with", and that "[m]any experiments have been made with all kinds of trees, the most successful being the wattles and pines". Hutchins' (1909) recommended technique of digging trenches and raising banks to increase establishment success is then mentioned as "fairly successful", with "the only drawback being the cost", and the common refrain that "[f]urther experiments must be carried out" (Bovill 1915).

Hutchins (1909) describes the salt lake plantation as being 40 acres planted between 1902 and 1908 (about a fifth of the area of the current forest), although

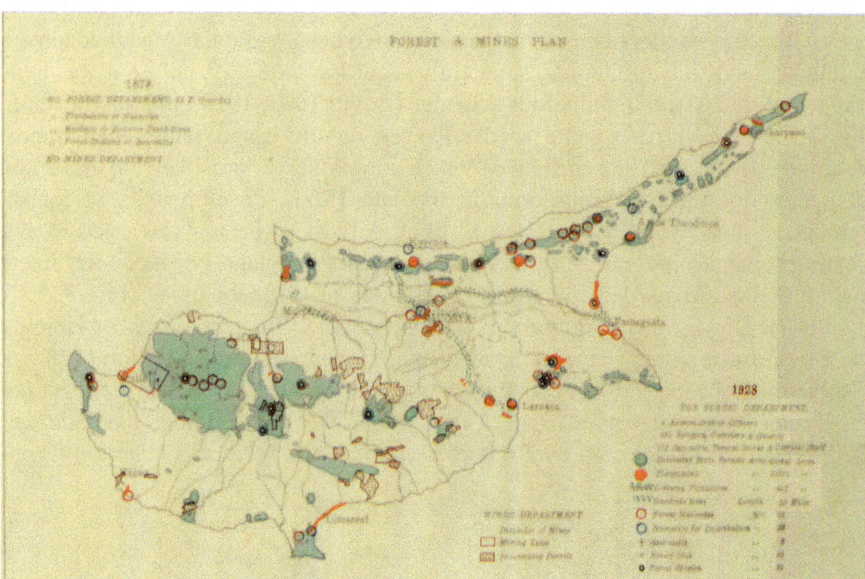

Fig. 9.1 The "Forests and mines" subpanel from a larger set of maps of Cyprus highlighting the contributions of the British administration to the island in 1928. Each panel also contains a summary of the situation in 1878 (the year of the British occupation), presumably intended to highlight the achievements of the administration on its 50th anniversary (© UK National Archives, CO 1047/364)

with only 10 acres having "succeeded" (p. 75). His comments, however, reflect Bovill's, noting "[m]arsh water close to the surface. Soil cracks badly in summer. Planted with *Pinus pinea*, *Pinus halepensis*,[9] Cypress (common), [and] *Acacia cyanophylla* [syn. *A. saligna*]" (Hutchins 1909, p. 39). Bovill (1915) concludes by optimistically noting the "great extent of land [by the salt lake] to be taken in hand" and the remunerative potential of the Limassol firewood market. Jens Holmboe (1914), a Norwegian botanist who visited Cyprus in the spring and summer of 1905, spending two days around Akrotiri in May, makes no obvious mention of the site, despite discussing the woods and plantations of the island in some detail, and having been guided for part of his tour by Bovill. Holmboe (1914, p. 243) does refer to a stand of eucalypts near Limassol, planted in 1899, but it is not clear that this was at the salt lake. What does seem clear is that the Akrotiri plantation at this time was modest, failing to thrive, and that further efforts and, indeed, experimentation with new methods and species would be needed.

The extent of the woody non-natives on Akrotiri today, and their ability to spread naturally, is not then just a simple matter of what was planted in the late 19th and early 20th centuries. It is clear from the reports of Madon, Wild, Hutchins and Bovill that numerous species were imported and trialled, with very few of these becoming widely used, and even fewer becoming naturalised. Even today the eucalypts are not considered to be naturalised (Hand et al. 2011–2017), although river red gum (*E. camaldulensis*) has become invasive elsewhere in the world (Tererai et al. 2013).

Port Jackson wattle, however, appears to have been particularly well-adapted to the climate and soils of Cyprus, apparently requiring little time before it was being used across the island in numerous schemes (Bovill 1902, 1915), although even for this species the shift to using seed collected on the island rather than imports (Holmboe 1914), and the area over which it was sown, are likely to have contributed to its rapid naturalisation. In the late 1920s, developments in forestry techniques under Bovill's successor as PFO, A. H. Unwin, may have led to even greater propagule pressure[10] from wattle on native habitats. For example, tractor sowing of the species led to carpets of saplings in some locations (Fig. 9.2; CO 1069/697). Unwin also continued to use and develop the techniques of Hutchins and Bovill for draining the marshlands; Fig. 9.3 provides an illustration of the use of these methods to establish eucalypts alongside the freshwater lake to the west of Famagusta, a project begun in 1922 (CO 1069/697). That similar methods were used at Akrotiri is indicated by a report from the Chief Medical Officer on Cyprus

[9]At this point the British did not appear to distinguish the native *P. brutia* from the non-native *halepensis*.

[10]'Propagule pressure' is the term used by ecologists to denote the influence of the number of individuals on the likelihood of the successful establishment of a species in an area (Lockwood et al. 2009).

Fig. 9.2 "Tractor sowing of Acacia Cyanophylla [sic.] after 8 months, Xylotimbo Forest". From an album of photographs prepared by A. H. Unwin, 1934, as a commemorative gift to the visiting Financial Commissioner for Cyprus, Sir Ralph Oakden (© UK National Archives, CO 1069/697)

written in 1926 (CO 67/220/15), where both wattle and eucalyptus are mentioned in this context.

Such methods were undoubtedly successful in changing the character of the larger marshlands of Cyprus, and were used for many years. For example, in the 1940s, in Waterer's previously mentioned application for Colonial Office development funds, photos very similar to Unwin's are included to support the case. Two in particular highlight the desired transition from rush (*Juncus*)-dominated salt meadows to eucalyptus plantation (Fig. 9.4; CO 1069/698 and CO 67/310/16)—these photos were taken at the Akrotiri salt lake. The caption to another photo emphasises the "unproductive marshland at Limassol swamp capable of forest reclamation which is at present an acute malarial problem and only supports rough grazing and reed production" (photo number 30, CO 67/310/16).

It is worth noting that it is not clear whether the use of non-native trees in drainage schemes significantly reduced the incidence of malaria: the Chief Medical Officer in 1944 considered *An. sacharovi* "easy to eliminate" through drainage, although the figures provided in that document and elsewhere do not indicate any decrease in malaria over the preceding decade (CO 67/328/8; Aziz 1947). Hutchins (1909) himself noted that the tree-driven, bank-and-trench methods of drainage might actually increase malaria, presumably due to the fact that the trenches would

Fig. 9.3 "Freshwater lake, Eucalyptus 4 years old". Near Famagusta. As with Fig. 9.2, it is likely that the figure in the photo is A. H. Unwin himself. Also from the Oakden commemorative album (© UK National Archives, CO 1069/697)

often contain stagnant water of a type favoured by mosquitoes, and perhaps fewer larval predators. Grove and Rackham (2001) suggest that the expectation of governments that mosquitoes would not breed in drained fenlands was "naïve", and note that some of the worst malaria seen by the historian of mosquito-human interactions, W. H. S. Jones, was in the Lake Kopáis area of Greece after it was drained (Jones 1907). Indeed, the fact that Waterer was still constructing major projects to combat malaria in 1940, and that final eradication seems to have been

Fig. 9.4 Photos 31 and 32 from R. R. Waterer's application for Colonial Office development funds, 1941. The left hand photo (31) shows the salt meadows at Akrotiri with the "thriving plantation established in background". The right hand photo (32) shows the "[m]arshland adjoining that shown in 30 above [the same location as photo 31] planted with fast growing Eucalyptus about 4 years old" (© UK National Archives, CO 1069/698)

dependent on DDT (Aziz 1947),[11] perhaps indicates the limits of drainage as a strategy in low-lying areas.

9.4 A Forest on the Move? Impacts on Native Plant Communities

The growth of the Akrotiri forest continued through the middle of the 20th century, with Port Jackson wattle (*A. saligna*), river red gum (*Eucalyptus camaldulensis*), tuart (*E. gomphocephala*) and casuarina (*Casuarina cunninghamiana*) as the main components (Michaelides 1961; Christodoulou 2003). It is not clear exactly at which point the impacts of these species on native plant communities began to be of concern to ecologists on Cyprus. Carp (1980), for example, does not mention the issue for 'Limassol Salt Lake' under the 'Threats' section in his *Directory of*

[11]Following successful implementation of the strategy described by Aziz (1947), Cyprus was declared officially malaria free in 1950, with Mehmet Aziz styled as the 'Great Liberator' in the London press (Morgan 2010).

Wetlands of International Importance.[12] Hadjikyriakou and Hadjisterkotis (2002) highlighted the potential problems with wattle and casuarina, and note that Georgiades (1994) had already classified a number of non-native species as invasive on Cyprus, including wattle. The thorough study of Christodoulou (2003) appears to be the first field-based work to attempt a quantification of the impacts of a non-native species on Cyprus, focusing on the spread of wattle into the salt marshes of Akrotiri; this was followed by the study of Hadjichambis (2005), who noted the impacts of the same species in dune habitats. Our observations on the peninsula follow the approach of Christodoulou (2003) in combining GPS tablet-based mapping of stands of woody non-native species around the lake, with plot recording within forest stands to the north. Christodoulou (2003) used a 'gradient transect' method of recording plots through stands of wattle, thereby intending to capture the response of plant communities to the presence of the invader. Our observations and methodological development were limited to the two-week period of our visit in October 2015 (OLP, JMP, MO, and JOM), and therefore we opted to record the plant communities of stands of non-natives throughout and around the Akrotiri forest in order to understand better the nature of the communities persisting under the plantation and in surrounding stands. Such observations should complement the more restricted studies of Christodoulou (2003) and Hadjichambis (2005); the results of our GPS mapping and plot-based recording will be made available through an open data repository (Peyton et al. in prep.)

We mapped over 102 ha of woody non-natives on the Akrotiri peninsula (Fig. 9.5), our study area being circumscribed by the Akrotiri-Kolossi road to the west, the northern boundary of the forest to the north, the sea in the east, and the Akrotiri-Lady's Mile Beach road to the south. Eucalyptus species and casuarina were largely restricted to the forest north of the lake (Fig. 9.5), whilst wattle was present throughout the forest, along the road in the south, and in restricted locations on the Lady's Mile dunes in the east, where evidence of recent wattle control was also observed (along with the presence of the Cyprus RDB species, sea daffodil, *Pancratium maritimum*). As noted above, given the extensive use of Port Jackson wattle throughout the island, particularly to stabilise sand drifts (Bovill 1915; Christodoulou 2003), it can be challenging to separate natural spread from plantings; however, it was normally clear from the distribution of individuals that wattle and casuarina had spread outwards from their plantings in the forest and along roads. Although eucalypt saplings were occasionally observed within the Akrotiri forest, these were never abundant and were not observed outside the forest, or seen encroaching onto the salt marshes. Casuarina is most abundant in the south-western corner of the Akrotiri forest, where evidence of spread into the adjoining wetlands

[12]The information included was derived from a report written by the Conservator of Forests, L. Leontiades, so this absence could reflect a forester's bias away from considering plantations as a threat, as much as a lack of spreading non-natives.

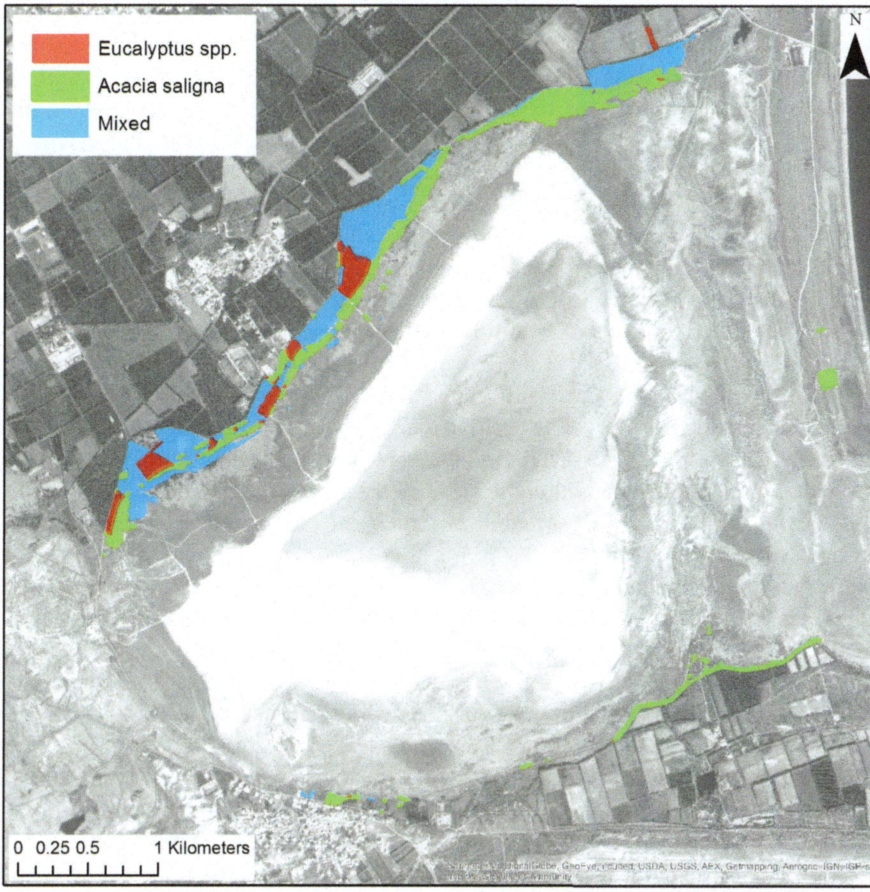

Fig. 9.5 Stands of woody non-natives recorded in October 2015. Eucalypts include *E. camaldulensis* and *E. gomphocephala*. The "Mixed" category contains both stands of casuarina and intimate mixtures of eucalypts and wattle that could not be separated at the scale of the mapping. Field notes associated with these polygons provide more detail on their composition (Peyton et al. in prep.). Note that the polygons in this image have been given thick borders to ensure that smaller stands along the southern lake shore are visible at this scale, therefore all polygons appear slightly larger than their actual mapped areas

appeared clear from the presence of young trees some distance from the main stands (also noted by Hadjikyriakou and Hadjisterkotis 2002).

At the time of our visit, the leading edge of casuarina had been burnt in the recent past, apparently as a means of controlling the advance of the species (G. Hadjikyriakou, retired Cypriot forester, pers. comm.) The spread of wattle into the salt marshes of the north-eastern corner of the wetlands, observed by Christodoulou (2003), was still ongoing at the time of our visit (Peyton and Mountford 2015), and it was clear that a previous ditching operation in the area (clearly seen in Google Earth satellite photography) was facilitating its spread, presumably through the provision of a drier micro-habitat for seedling establishment, as promoted by Hutchins some one hundred years earlier. The ditching in the area was reportedly intended to support the establishment of another non-native species, the desert fan palm (*Washingtonia filifera*), although this initiative was not successful (G. Hadjikyriakou pers. comm.).

The plant communities found within the mature forest do not generally correspond well with (semi-)natural types, but rather represent relict assemblages modified by the shade of wattle, casuarina and eucalyptus. Where the water-table is close to the surface and the shade is less dense, as along access tracks and in clearings, there are patches of common reed (*Phragmites australis*) and saw sedge (*Cladium mariscus*), related to the *Phragmition* reed-swamp, and this vegetation becomes prevalent along the south edge of the forest toward the lake. Although sometimes ascribed (as by AP Marine/Atlantis 2012) to the Annex 1 habitat type 7210 'calcareous fens with *Cladium*', the saw sedge vegetation at Akrotiri is in reality a modified reedbed, though with RDB species such as sea flax (*Linum maritimum*) and water mint (*Mentha aquatica*) in places. These rarities are also present where the tree-cover is absent (or almost so), and where cutting or grazing exclude common reed and saw sedge, to be replaced by a mixed rush-community of the *Juncion maritimi* (Annex 1 habitat 1410 Mediterranean salt-meadows). Very locally where the growth of reeds and rushes is suppressed by traffic or trampling but where the shade of non-native trees may be dense, a wet sward has developed which shows a distant floristic relationship to Mediterranean humid grasslands of the *Molinio-Holoschoenion* (Annex 1 habitat 6420). The grassland is dominated by genera such as *Aeluropus* with patches of forbs such as water germander (*Teucrium scordium* ssp. *scordioides*) and common fleabane (*Pulicaria dysenterica* ssp. *uliginosa*); this habitat is also the home of the RDB species crypsis (*Crypsis factorovskyi*) and frog fruit (*Phyla nodiflora*). However, the community found on the wet Akrotiri tracks is probably best considered a ruderal type rather than a semi-natural grassland.

Elsewhere, ruderal communities are also prevalent where the soil conditions are drier toward the northern edge of the forest. These ruderal grasslands are almost certainly referable to the *Bromo-Oryzopsion miliaceae* alliance and, as well as the planted trees, support populations of other invasive species such as African boxthorn (*Lycium ferocissimum*) and Bermuda buttercup (*Oxalis pes-caprae*). A different vegetation type exists in ruderal situations where the soil is compacted and the drainage impeded as, for example, along the well-illuminated track that divides the main forest block from the reedbeds by the lake. Here, sticky fleabane (*Dittrichia viscosa*) and couch-grasses (*Elymus elongatus* etc.) grow together in moist saline conditions. Adjacent to the forest tracks this fleabane-couch community harbours strong populations of non-native composites such as fleabane (*Conyza*) species and saltmarsh aster (*Symphyotrichum squamatum*).

Further away from the main forest blocks, Christodoulou (2003) found that Port Jackson wattle displaced the most sensitive salt marsh and salt meadow species, and reduced the abundance of the remaining species. This effect was particularly pronounced in heavily invaded plots. This mirrors the impacts of wattle in other Mediterranean-type and arid environments (Holmes and Cowling 1997; El-Bana 2008). Our observations suggested a similar pattern under wattle in halophytic environments, and provide an updated data source against which to assess spread (Peyton and Mountford 2015; Peyton et al. in prep.). We also sought to describe the habitats throughout the plantation as a whole, and this broader approach indicated that much of value and interest persists beneath the canopy of Australian trees and shrubs, albeit in a much modified form from the marshlands observed by British foresters in the late 19th and early 20th centuries. The close proximity of the (semi-) natural Fassouri marshes, west of the Akrotiri-Kolossi road, provides an instructive contrast to the afforested area (see Tsintides et al. 2002 for a photograph of these marshes), and also highlights how the gradual establishment of the State Forest must have affected, not just local biodiversity, but also livelihoods, and a people's sense of place.

9.5 Ecosystem Services, but Which, and for Whom?

The historical and biological evidence presented above documents changing values and practices against a background of ecological fact (as far as it is known), but the response(s) of the Cypriot is glaringly absent: were the 'ecosystem services' promoted by the British and provided by non-native species useful to local people? Some reports from foresters suggest that they (eventually) were (e.g. see Couppis 1954 on the 'Village Fuel Areas' established throughout Cyprus), but this is of course one side of the story, and written for the imperial forestry community at that. Our researches to date have not managed to provide much insight into the historic, human element of our story, as Harris (2007) managed to do for various, broader strands within the domain of Cypriot forestry.

Nonetheless, documents in the Cyprus State Archives attest to the fact that local Cypriots valued marshland enough to seek to appropriate it for personal use (Land Registry and Surveys: Minute Papers LRS1/19947, 1919) or to reclaim it from government ownership (LRS1/25601, 1932). The marshlands were clearly valued for grazing by locals, although, as noted above, forestry personnel considered the land unproductive, and as "*only* support[ing] rough grazing and reed production" (our emphasis; caption to photo number 30, CO 67/310/16). Despite this, we have not so far located information relating to the loss of access to the Akrotiri marshlands as a result of afforestation (or indeed through the natural spread of non-natives), but a thorough exploration of the Cyprus State Archives could yield records of complaints lodged by locals, especially since the area became a Main State Forest, and as such is likely to be well documented (Harris 2007).

Today, the Akrotiri forest north of the salt lake is used by locals and visitors for various purposes, including the collection of wood fuel, apiculture, basket weaving, shooting (although the forest is within a general game reserve in which hunting is prohibited), photography and natural history, and other recreational purposes. Apiculture in particular may be the most profitable use, with locals supplementing their incomes with hives located in many places throughout the forest (the eucalypts river red gum and tuart are highly valued for this purpose; Doughty 2000). The harvesting of wattle from the forest for wood fuel has also been encouraged by the Department of Forests in recent years by charging a small fee per truckload. Although uptake has been considerably less than initial interest in the scheme suggested (G. Bullivant and K. Martinou, JSHU, pers. comm.), we have observed such harvesting in the forest. However, this minor use is unlikely to contribute much to the reduction of the impacts of wattle on ecological communities (and does not of course affect casuarina), and we do not know whether wood collectors consider this aspect of their activity.

The biodiversity value of the area has already been discussed in detail above, and this value is frequently cited as a key feature of the area in many documents, including the management plan for the area (Sovereign Base Areas Administration 2012), a local community consultation on this plan (Sovereign Base Areas Administration 2006), and other reports (e.g. Perdiou 2010; Hadjikyriakou 2011). More broadly, Cypriot interest in and support for the environment appears to be strong, based on both qualitative and quantitative social survey work (Michaelidou and Decker 2003), some of which has focused on the Akrotiri peninsula (Birol et al. 2008). However, it is not clear that national surveys (Michaelidou and Decker 2003), or surveys conducted in terms of 'Willingness To Pay' thought-experiments (Birol et al. 2008), will necessarily capture local responses to changes to management or land use in the area, or the amount to which the non-native forest, or local native habitats such as the salt marshes, are valued, habitats which may be less emotive to Cypriots than the mountain forests (cf. Michaelidou and Decker 2003).

Development of the surrounding area is welcomed in some quarters (e.g. see Christou 2016), but there is also concern about the impacts such developments will have on the rare and protected species and habitats of the peninsula (P. Charilaou, Akrotiri Environment and Education Centre, pers. comm.). Developments that have begun are already having small impacts: for example, a new drainage pipe that flows into the salt lake (34°38′29.57″ N, 32°59′24.44″ E) has resulted in an area of fresher marsh and increased the cover of common reed (*Phragmites australis*) in this location (OLP and JMP, pers. obs. March 2017) at the likely expense of the Annex 1-listed salt meadow communities. The drainage pipe is associated with a new golf course which has been approved for construction on former citrus plantation land. Whether such developments will lead to environmental changes that facilitate non-native species invasions in the area remains to be seen, but, if other development plans for the western Limassol area come to fruition (Christou 2016), then it is possible that the increased demand on fresh water, and the possibility of eutrophication, may interact to promote further invasion, particularly as both wattle and casuarina are able to establish in the fresher, drier areas of the marsh.

9.6 Conclusion

We have noted earlier the complex interplay between ecological fact and the changing systems of valuation brought to bear on the world over time, whether through science, administration, or the over-arching effects of cultural time and place. We have attempted to describe an ongoing, multifaceted, biological invasion in a set of threatened Mediterranean habitats, documenting the ecological and historical processes that have interacted to produce the current forest on the peninsula at Akrotiri and its overspills. The environmental historian Stephen Pyne (2005) has emphasized the necessity of utilising an interdisciplinary approach, linking science and the humanities, in order to illuminate and solve environmental problems. Although we make no claim to have solved any problems, we hope that this chapter, a collaboration between ecologists and an environmental historian, has at least described the elements of a problem clearly, and to has illuminated some of its ecological and historical elements and foundations.

The non-native-driven habitat conversion engineered by British foresters, and the associated invasions, have been driven both by ideological narratives and the practical process of finding species that could thrive in the Cypriot environment, to provide, at least in the eyes of the planters, multiple benefits to the island. We are by now well into an era that values nativeness, and that legislates for the protection of rare and endangered native species and habitats; whether these, often top-down (Michaelidou and Decker 2003), norms will lead to the protection of local diversity from invasive non-natives, or other drivers of environmental change at Akrotiri, is a question for the future. The outcome will provide a fascinating insight into the environmental values of modern-day Cypriots.

Acknowledgements We thank C. S. Christodoulou, Simon Pooley, Chris Preston, Ana Isabel Queiroz and Ioannis Vogiatzakis for useful comments on the text. JOM and JMP thank COST Action TD1209 Alien Challenge for funding their Short Term Scientific Mission to Cyprus; we also thank this COST Action for funding image reproduction and licensing. OLP thanks NERC CEH Wallingford for a Learning & Development award to fund his participation, and A.-M. Catterall (Druce-Fielding Herbarium, University of Oxford) and H. Alexander (UK National Archives) for their assistance. We would also like to thank all of those who helped us at Akrotiri, including Capt. G. Bullivant (JSHU), P. Charilaou (AEEC), G. Hadjikyriakou, and N. Andreou (JSHU).

References

Allen TFH, Hoekstra TW (2015) Towards a unified ecology, 2nd edn. Columbia University Press, New York

AP Marine/Atlantis (2012) Hydrological study & further studies to be incorporated in the Akrotiri Peninsula Management Plan. Final report. AP Marine Environmental Consultancy Ltd., Nicosia

Aziz M (1947) A Brief Account of the *Anopheles* (Malaria) Eradication in Karpas-Cyprus, 1946. J R Sanitary Inst 67:498–509

Barrios V, Carrizosa S, Darwall WRT, Freyhof J, Numa C, Smith K (2014) Freshwater key biodiversity areas in the Mediterranean basin hotspot: Informing species conservation and development planning in freshwater ecosystems. IUCN, Cambridge and Malaga
Barton GA (2002) Empire forestry and the origins of environmentalism. Cambridge University Press, Cambridge
Bergmeier E, Abrahamczyk S (2008) Current and historical diversity and new records of wetland plants in Crete, Greece. Willdenowia 38:433–453
Birol E, Koundouri P, Kountouris Y (2008) Integrating wetland management into sustainable water resources allocation: the case of Akrotiri wetland in Cyprus. J Environ Planning Manage 51:37–53
Bovill AK (1902) Memorandum on plantation work in Cyprus. Government Printing Office, Nicosia
Bovill AK (1915) Report on plantation works in Cyprus 1879 to 1914. Government Printing Office, Nicosia
Bovill AK (1920) Report on the natural resources & present development of the forests of Cyprus. Government Printing Office, Nicosia
Britton RH, Crivelli AJ (1993) Wetlands of southern Europe and North Africa: Mediterranean wetlands. In: Lieth H (ed) Wetlands of the world I: inventory, ecology and management. Springer, Netherlands, pp 129–194
Carp E (ed) (1980) A directory of wetlands of international importance in the Western Palearctic. IUCN, Gland
Castree N (2014) Making sense of nature. Routledge Abingdon, Oxon
Christodoulou CS (2003) The impact of *Acacia saligna* invasion on the autochthonous communities of the Akrotiri salt marshes. B.Sc. thesis, Department of Forestry, University of Central Lancashire
Christou A (2016) Requirements and opportunities in the making. Cyprus Weekly Project Limassol Supp 76–77
Couppis TA (1954) Village fuel areas, Cyprus. Empire Forestry Rev 33:252–255
Davis DK (2016) The arid lands: history, power, knowledge. The MIT Press, Cambridge
de Mas Latrie L (1870) The island of Cyprus: its present situation and the remains from the middle ages. Firmin-Didot, Paris
Delipetrou P, Makhzoumi P, Dimopoulos P, Georghiou K (2008) Cyprus. In: Vogiatzakis IN, Pungetti G, Mannion AM (eds) Mediterranean Island Landscapes. Springer, Dordrecht, pp 171–203
Doughty R (2000) The eucalyptus: A natural and commercial history of the gum tree. The John Hopkins University Press, Baltimore
ECDC (2005–2017) http://ecdc.europa.eu/en/healthtopics/vectors/mosquitoes/pages/anopheles-sacharovi-factsheet.aspx. Accessed 20 Dec 2016
El-Bana MI (2008) Effect of invasion by exotic *Acacia saligna* (Labill.) H. Wendl. on native species diversity across an aridity gradient along the coastal mediterranean dunes of Sinai Peninsula. Catrina 3:41–48
Georgiades C (1994) The adventive flora of Cyprus, taxonomic, floristic, phytogeographic, ecophysiological study. Ph.D. thesis, Athens University (in Greek)
Grove RH (1995) Green imperialism. Cambridge University Press, Cambridge
Grove AT, Rackham O (2001) The nature of Mediterranean Europe: An ecological history. Yale University Press, New Haven and London
Gucel S, Kadis C, Ozden O, Charalambidou I, Linstead C, Fuller W, Kounnamas C, Ozturk M (2012) Assessment of biodiversity differences between natural and artificial wetlands in Cyprus. Pak J Bot 44:213–224
Guillemard FHH (1888) VIII.-Ornithological notes of a tour in Cyprus in 1887. Ibis 30:94–124
Hadjichambis ACh (2005) Conservation biology of threatened coastal habitats of Cyprus: Flora, vegetation, ecology and management. Ph.D. thesis, Department of Biology, University of Athens, Athens (in Greek)

Hadjikyriakou T (2011) A bird-habitat association study at Akrotiri Peninsula Cyprus. M.Sc. thesis, University of the West of England
Hadjikyriakou G, Hadjisterkotis E (2002) The adventive plants of Cyprus with new records of invasive species. Zeitschrift für Jagdwissenschaft 48(Supplement):59–71
Hand R, Hadjikyriakou GN, Christodoulou CS (eds) (2011–2017) Flora of Cyprus—a dynamic checklist. http://www.flora-of-cyprus.eu. Accessed 7 Jan 2017
Harris SE (2007) Colonial forestry and environmental history: British policies in Cyprus, 1878–1960. PhD. thesis, University of Texas at Austin
Harris SE (2012) Cyprus as a degraded landscape or resilient environment in the wake of colonial intrusion. Proc Natl Acad Sci 109:3670–3675
Hobohm C, Bruchmann I (2011) Are there endemic vascular plants in wet habitats of Europe? Transylv Rev Syst Ecol Res 12:1–14
Holmboe J (1914) Studies on the vegetation of Cyprus, based upon researches during the Spring and Summer 1905. Grieg, Bergen
Holmes PM, Cowling RM (1997) The effects of invasion by *Acacia saligna* on the guild structure and regeneration capabilities of South African fynbos shrublands. J Appl Ecol 35:317–332
House of Commons (1880) Report by Her Majesty's High Commissioner for the year 1879. Cyprus. No. 2 (1880). Harrison and Sons, London
House of Commons (1881) Correspondence respecting the affairs of Cyprus. June 1881. Eyre and Spottiswoode, London
House of Commons (1896) Annual reports for the year 1894–85. Cyprus. Eyre and Spottiswoode, London
Hulme PE, Pauchard A, Pyšek P, Vilà M, Alba C, Blackburn TM, Bullock JM, Chytrý M, Dawson W, Dunn AM, Essl F, Genovesi P, Maskell LC, Meyerson LA, Nuñez MA, Pergl J, Pescott OL, Pocock MJO, Richardson DM, Roy HE, Smart SM, Štajerová K, Stohlgren T, van Kleunen M, Winter M (2015a) Challenging the view that invasive non-native plants are not a significant threat to the floristic diversity of Great Britain. Proc Natl Acad Sci 112:E2988–E2989
Hulme PE, Bernard-Verdier M, Bufford JL, Godsoe W (2015b) Rapid Anthropocene speciation reveals pull of the recent: a response to Thomas. Trends Ecol Evol 30:635–636
Hutchins DE (1909) Report on Cyprus forestry. Waterlow & Sons Ltd., London
JNCC (2008) Information sheet on Ramsar wetlands: Lake Akrotiri. JNCC, Peterborough
Jones WHS (1907) Malaria: A neglected factor in the history of Greece and Rome. Macmillan and Bowes, Cambridge
Joosten H (1999) Human impacts. Maltby E, Barker T (eds) The wetlands handbook. Wiley-Blackwell, Chichester, PP 689–718
Kent M (2012) Vegetation description and data analysis: a practical approach, 2nd edn. Wiley-Blackwell, Chichester, West Sussex
Lockwood JL, Cassey P, Blackburn TM (2009) The more you introduce the more you get: the role of colonization pressure and propagule pressure in invasion ecology. Divers Distrib 15(5): 904–910
Madon PG (1881) The replanting of the island of Cyprus, 1880. Forest conservancy of the island of Cyprus. Presented to both Houses of Parliament. 1881 (Reprinted 1930. Cyprus Government, Nicosia)
Markogianni V, Tzirkalli E, Gücel S, Dimitriou E, Zogaris S (2014) Remote sensing application for identifying wetland sites on Cyprus: problems and prospects. Hadjimitsis DG, Themistocleous K, Michaelides S, Papadavid G (eds) Second international conference on remote sensing and geoinformation of the environment (RSCy2014). Proc. of SPIE Vol. 9229, 1U1–9
Meikle RD (1977) Flora of Cyprus, vol 1. The Bentham-Moxon Trust, Royal Botanic Gardens, Kew
Meikle RD (1985) Flora of Cyprus, vol 2. The Bentham-Moxon Trust, Royal Botanic Gardens, Kew

Michaelides ED (1961) Brief notes on the reclamation of marshy and saline areas in Cyprus. Forest Department, Ministry of Agriculture and Natural Resources, Nicosia

Michaelidou M, Decker DJ (2003) European Union policy and local perspectives: Nature conservation and rural communities in Cyprus. Cyprus Rev 15:121–145

Morgan T (2010) Sweet and bitter island: A history of the British in Cyprus. IB Tauris & Co. Ltd, New York

Mucina L, Bültmann H, Dierßen K, Theurillat JP, Raus T, Čarni A, Šumberová K, Willner W, Dengler J, García RG, Chytrý M (2016) Vegetation of Europe: hierarchical floristic classification system of vascular plant, bryophyte, lichen, and algal communities. Appl Veg Sci 19(S1):3–264

MWO (2012) Mediterranean wetlands: Outlook. First Mediterranean wetlands observatory report—technical report—2012. Tour du Valat

MWO (2014) Land cover—spatial dynamics in Mediterranean coastal wetlands from 1975 to 2005. Thematic collection, issue #2. Tour du Valat

Noirfalise A (1987) Map of the natural vegetation of the member countries of the European Community and the Council of Europe. Office for Official Publications of the European Community, Luxembourg

Papacostas T (2001) The economy of late antique Cyprus. In: Decker M, Kingsley SA (eds) Economy and exchange in the East Mediterranean during late antiquity: proceedings of a conference at Somerville College, Oxford, 29th May, 1999. Oxbow, Oxford, pp 107–128

Pearce F (2015) The new wild. Icon Books Ltd., London

Perdiou A (2010) Designation and management within the Sovereign Base Areas in Cyprus. Sanctuary 39:56–57

Perennou C, Beltrame C, Guelmami A, Tomas Vives P, Caessteker P (2012) Existing areas and past changes of wetland extent in the Mediterranean region: an overview. Ecol Mediterr 38:53–66

Peyton JM, Mountford JO (2015) COST action short term scientific mission—Lake Akrotiri, Cyprus. Centre for Ecology and Hydrology, Wallingford

Pyne SJ (1982) Fire in America: A cultural history of wildland and rural fire. Princeton University Press, Princeton

Pyne SJ (2005) Environmental history without historians. Environ Hist 10:72–74

Rackham O (1990) Ancient Landscapes. In: Murray O, Price S (eds) The Greek city: from Homer to Alexander. Clarendon, Oxford, pp 85–117

Rackham O (2017) Landscape history of Cyprus: A preliminary account. Pungetti G (ed) Island landscapes: An expression of European culture. Routledge, Abingdon

Richardson DM, Pyšek P, Rejmanek M, Barbour MG, Panetta FD, West CJ (2000) Naturalization and invasion of alien plants: concepts and definitions. Divers Distrib 6:93–107

Savile AR (1878) Cyprus. Government Printing Office, Nicosia

Silvertown J (2015) Have ecosystem services been oversold? Trends Ecol Evol 30:641–648

Sinka ME, Bangs MJ, Manguin S, Coetzee M, Mbogo CM, Hemingway J, Patil AP, Temperley W, Gething PW, Kabaria CW, Okara RM, Van Boeckel T, Godfray HCJ, Harbach RE, Hay SI (2010) The dominant *Anopheles* vectors of human malaria in Africa, Europe and the Middle East: occurrence data, distribution maps and bionomic précis. Parasites & Vectors 3:117. https://doi.org/10.1186/1756-3305-3-117

Smith A (1887) Through Cyprus. Hurst & Blackett, London

Smith KG, Barrios V, Darwall WR, Numa C (eds) (2014) The status and distribution of freshwater biodiversity in the Eastern Mediterranean. IUCN, Gland

Sovereign Base Areas Administration (2006) Akrotiri Peninsula Environmental Management Plan: Results of the consultation workshop held at Akrotiri Village on 11th May 2006. http://www.sbaadministration.org/images/AEEIC/consultations/Mgt_Plan_consultation_results.pdf. Accessed 10 June 2017

Sovereign Base Areas Administration (2012) Akrotiri Peninsula Environmental Management Plan. Version 2.0. http://www.sbaadministration.org/images/AEEIC/publications/20121003_EMP_Intro.pdf. Accessed 10 June 2017

Strayer DL, Dudgeon D (2010) Freshwater biodiversity conservation: recent progress and future challenges. J North Am Benthological Soc 29:344–358

Tererai F, Gaertner M, Jacobs SM, Richardson DM (2013) *Eucalyptus* invasions in riparian forests: effects on native vegetation community diversity, stand structure and composition. For Ecol Manage 297:84–93

Thirgood JV (1987) Cyprus, a chronicle of its forests, land, and people. University of British Columbia Press, Vancouver

Thomas CD (2015) Rapid acceleration of plant speciation during the Anthropocene. Trends Ecol Evol 30:448–455

Thomas CD, Palmer G (2015) Non-native plants add to the British flora without negative consequences for native diversity. Proc Natl Acad Sci 112:4387–4392

Tsintides CT, Hadjikyriakou NG, Christodoulou SC (2002) Trees and shrubs in Cyprus. Foundation A.G. Leventis and Cyprus Forest Association, Lefkosia

Tsintides CT, Christodoulou CS, Delipetrou P, Georghiou K (2007) The Red Data Book of the flora of Cyprus. Cyprus Forestry Association, Lefkosia

Wild AE (1879) Report on the forests in the south and west of the island of Cyprus. Presented to both houses of Parliament of Her Majesty. Cyprus. No. 10 (C-2427). Harrison and Sons, London

Chapter 10
The Long and Entangled History of Humans and Invasive Introduced Plants on South Africa's Cape Peninsula

Simon Pooley

Abstract This chapter provides an environmental history of plant introductions and fire—and how their unintended consequences have been framed and managed—at the Cape of Good Hope, South Africa. The chapter explains why the plants which have proved invasive were introduced to the region, examines the effects of urbanisation on attitudes to introduced tree plantations, and describes the development of concern over the effects of fires and introduced plants on the indigenous fynbos vegetation. The chapter recounts the complex history of environmental management on the Peninsula, discussing the advantages and limitations of the powerful narrative linking invasive introduced plant control with fires and water supplies, and recent controversies between invasion biologists and commercial forestry managers.

Keywords Wildfire · Invasive plants · Cape of Good Hope · Environmental management · South Africa

10.1 Introduction

This is a history of fire, water and plant introductions, and the unintended consequences of how these have been framed and managed at the Cape of Good Hope, South Africa. In this chapter I explore why the plants which have proved invasive were introduced to the region, the effects of urbanisation on attitudes to introduced tree plantations, and the development of concern over the effects of fires and introduced plants on the indigenous fynbos vegetation. The chapter shows how a series of major fires ignited controversy over introduced plants and fire control, and recounts the complex history of environmental management on the Peninsula, drawing attention to economic, social, cultural and political as well as environmental influences on policy and management practise. Finally, there is a discussion of the advantages and limitations of the powerful narrative linking invasive intro-

S. Pooley (✉)
Birkbeck University of London, London, UK
e-mail: s.pooley@bbk.ac.uk

A. I. Queiroz and S. Pooley (eds.), *Histories of Bioinvasions in the Mediterranean*, Environmental History 8, https://doi.org/10.1007/978-3-319-74986-0_10

duced plant control with fires and water supplies, which developed after the unprecedentedly large fires of January 2000. Parallels—and differences—with the fire situation in the Mediterranean basin region, where fires have played an important role in shaping the biota for millennia, are explored in a panel 'Fires in Mediterranean-type ecosystems.'

10.2 Plant Introductions

Plants have been introduced to the Cape Peninsula at least since Europeans settled there in 1652. The first Dutch commander, Jan van Riebeeck, and his fellow Dutch East India Company (VOC) employees set out to establish a refreshment station for Dutch ships sailing between the Netherlands and Batavia, the capital of the VOC's territories in the Far East. They diligently investigated the possible uses and commercial potential of the local fauna and flora, introduced some early legislation in an effort to preserve scarce timber and firewood resources on the Peninsula, and introduced a wide variety of fruit trees, vines, ornamental trees, cereal crops and vegetables (Pooley 2009).

Over the ensuing 154 years of Dutch colonial rule at the Cape, the local landscape was transformed through wood cutting, the planting of numerous introduced plants, and sedentary farming including livestock raising (the Khoikhoi had used the Peninsula for seasonal grazing). There is no evidence to suggest that any of these introduced plants proved invasive, however. The only early records of problematic invasive plants are the scientific explorer Anders Sparrman's observations from the mid-1770s on the spread of the indigenous but unpalatable shrub renosterbos (*Elytropappus rhinocerotis*), on land used to graze livestock. This, he observed, was exacerbated by farmers' attempts to burn it out (Pooley 2014, p. 25). Most of the plants that ultimately proved to be invasive of the indigenous fynbos and renosterveld vegetation were introduced during the period of British colonial rule (1806–1910), from the mid-1800s. As the northern Peninsula developed from the mid-1800s, the sandy flats surrounding the Table Mountain Chain came to be seen as a major inconvenience and obstruction to development. A suite of hardy shrubs able to handle the harsh conditions in the nutrient-poor sandy soils, and adapted to the occasional fires which swept the undeveloped Flats and mountain slopes, were imported from Australia. Wattles (*Acacia* species), hakeas and casuarinas were imported from the 1840s to stabilise driftsands (Shaughnessy 1980; Bennett 2010; Pooley 2014).

The lack of firewood and timber had always been a problem for European settlers at the Cape, becoming ever more so as the town grew and railways were built from the 1860s. Private individuals established the first timber plantations on the lower slopes of Table Mountain; for example Van Breda had a full grown pine forest on his property Oranjezicht by 1861. In 1876, a Department of Forests and Plantations was established, run by a professional forester, Joseph Storr Lister. His father had overseen tree-planting and driftsands reclamation along hard roads on the Cape Flats, and after a spell in India and England, Joseph returned to work as Superintendent of Drift Sands Plantations on the Flats. He supervised the planting of *Acacia saligna* and various eucalypts and pines across the Peninsula. In 1876 he established the first large plantation of trees in the Cape Colony, planting blue gum (*Eucalyptus globulus*) to drain swampy ground believed to be a source of fevers in the inland town of Worcester. The plantation was a success and repurposed as a commercial plantation, the first rotation being sold to the De Beers Consolidated Mining Company mining diamonds at Kimberley (Pooley 2014, pp. 35–36; Lister 1957).

State foresters subsequently set out to establish commercial plantations of imported trees on the Peninsula, efforts which gained significant impetus with the appointment in 1880 of a Superintendent of Woods and Forests, the French forester Count Médéric de Vasselot de Regné, who had been trained at Nancy. Cecilia, Tokai and Uitvlugt plantations were established on the Peninsula in 1884, and a nursery and an arboretum were established at Tokai by 1886. State forestry plantations were also established at Devil's Peak (1893) and Rifle Range/Ottery (1897). State foresters encouraged municipalities and private individuals to establish plantations, and in this period, German immigrants established wattle plantations on the Cape Flats. The City of Cape Town sponsored competitions to reward the planting of wattle plantations from the 1880s. In 1894 it established its first plantation at Kloof Nek, below Lion's Head, and then a large plantation at Newlands (Pooley 2014, p. 36; Bennett 2010; Sim 1907).

Pine trees were planted for timber (and initially to help encourage a moister local climate), and the main species (many more were introduced) included the Mediterranean species the Aleppo pine *Pinus halepensis*, the cluster pine *P. pinaster*, the stone pine *P. pinea*, the Monterey pine *P. radiata* which is native to California and Mexico (called *P. insignis* in the earlier literature), and the Scots pine *P. sylvestris* from Eurasia. Eucalypts were introduced from Australia for timber and to stabilise driftsands, the main species including the sugar gum *Eucalyptus corynocalyx* (*cladocalyx* F.Muell.), the karri *E. diversicolor*, and the blue gum *E. globulus*, and were planted in strips to protect pine plantations from wind and fire (particularly blackbutt *E. pilularis* and red mahogany *E. resinifera).* The Australian wattles (*Acacia* species) were grown for timber (especially Port Jackson wattle *A. saligna*, mainly for poles, and Australian blackwood *A. melanoxylon*) and tannin (especially black wattle *A. decurrens* and golden wattle *A. pycnantha*), to stabilise driftsands (especially red-eye wattle *A. Cyclops* and long-leaved wattle *A. longifolia*) and to protect pine plantations from wind and fire (notably *A. decurrens*). Australian Hakea species were introduced to provide shelter, serve as hedges and stabilise driftsands (sweet hakea *H. drupacea*), of which three species have become particularly problematic in the long term (silky hakea *H. sericea,* rock hakea *H. gibbosa* and *H. drupacea*). A few of these plants, and the Australian myrtle (*Leptospermum laevigatum*) were introduced as ornamental plants and have proved invasive (Pooley 2014, p.164 and pp. 168–175; CGH 1900 G.39-1901, p. 12; CGH 1902, p. 17; CGH 1905, pp. 17, 19 and 25).

In brief, although fynbos plants are adapted to certain fire regimes, many of the introduced species are better adapted to exploiting post-fire periods, growing more quickly and taller, establishing dense fire resistant stands which shade out understorey plants and exclude the indigenous shrubs. Although they ignite less easily than fynbos plants, they burn more intensely and are more harmful to indigenous ecosystem processes and structure than fynbos fires. The invasive plants can be grouped by dispersal strategy: the Australian myrtle, hakea and pines are serotinous, releasing winged seeds when their fire-proof cones or follicles are burnt by fires; the wattles drop hard coated seeds, forming seedbanks which are stimulated to germinate by fires (heat and smoke). They spread mainly along waterways (Richardson et al. 1992; Keeley et al. 2012).

10.3 The First Concerns Over Introduced Plant Species

The first major public debates over the desirability of introduced plants arose in connection with the provision of ecosystems services (or disservices) to the growing city. Two major issues for the expanding city and its satellite villages were: the suspected impact of tree plantations on water supplies from Table Mountain; and the inflammability of these trees and the threat this posed to surrounding settlements (and the natural beauty of the Peninsula).

Following the discovery of diamonds and gold in the interior in the 1860s and 1880s respectively, Cape Town grew from an estimated 45,000 residents in 1875 to 171,000 by 1904. By 1900 its satellite villages on the Peninsula were substantial municipalities, and new suburbs were being developed. Some of the forestry department's new plantations on Table Mountain were removed to make way for reservoirs (seven were built from 1896–1907), and Hutchins' claims in the mid-1890s that 'reafforesting' Table Mountain would increase water supplies and improve water quality were disputed. By 1896 tree planting atop Table Mountain was suspended, the City Council having convinced the Minister of Agriculture that tree-planting was adversely affecting the City's water supply. In 1904, the city was still running out of water in summer (Pooley 2014, pp. 43 and 126–27; Wall 2008; CGH 1896 G17-97).

The suspension of tree planting on top of Table Mountain, and residents' opposition to tree planting, were importantly influenced by fires on the mountain. Two big fires on the slopes above Cape Town and its suburb Rondebosch in 1892 and 1893 respectively resulted in fierce objections to plantations on the mountain, both because of the dangers posed by fires, and their disfigurement of the Peninsula's natural beauty. The forestry department lamented that private plantation owners were at fault for allowing their plantations to degenerate into patchy woods interspersed with rank understorey flora, and that properly run plantations characterised by dense planting and maintenance, firepaths and vigilant firefighters would solve the problem. Although the forestry department were allowed to continue to plant trees around the lower slopes of Table Mountain—providing they undertook extensive safety precautions including building lookouts, clearing fire breaks and digging ditches around plantations—introduced trees had become a target of public disapproval (CGH 1892 pp. 27–93, p. 17; CGH1893, pp. 50–94, pp. 30–31) (Fig. 10.1).

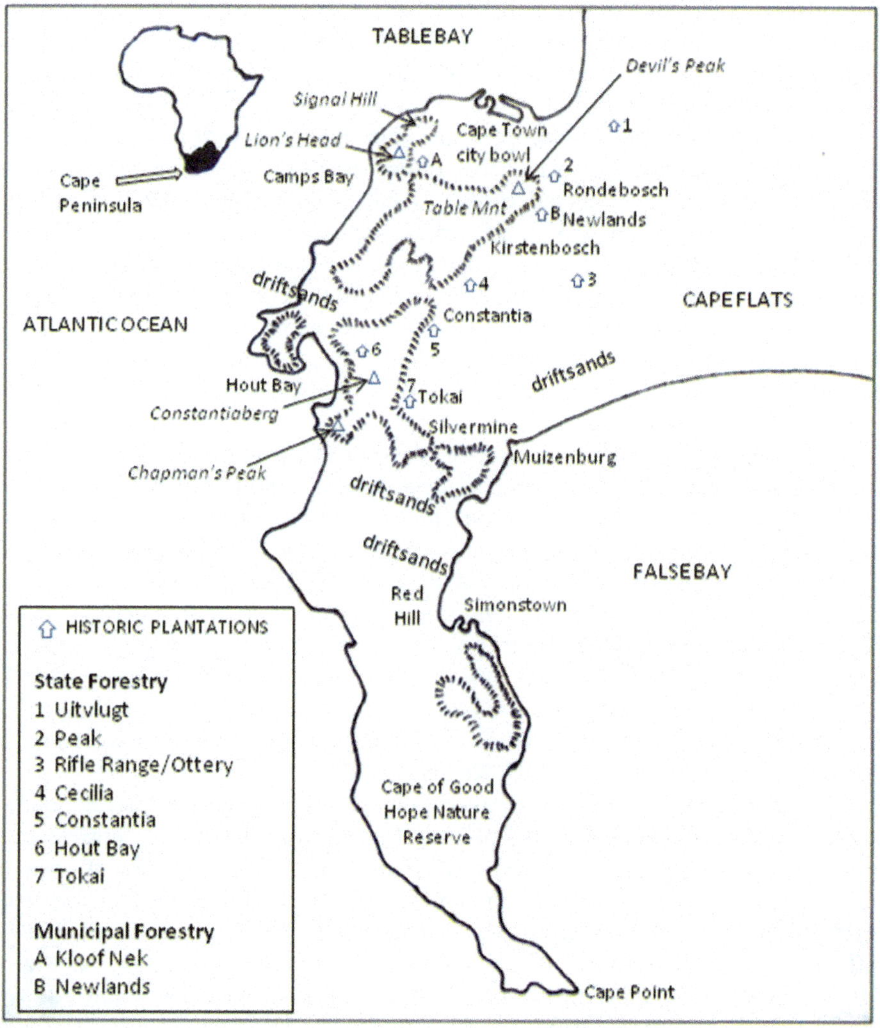

Fig. 10.1 Map of the Cape Peninsula showing historic plantations, sandrifts, key places and mountains. Most of these plantations had disappeared by the 1970s, and the Cape of Good Hope Nature Reserve has been incorporated into the Table Mountain National Park. *Credit* Simon Pooley

10.3.1 Concern for the Indigenous Flora

Concern that introduced plants would invade and overwhelm the indigenous flora of the southwestern Cape was remarked on from time to time from the mid-1800s, for instance in the region of Caledon in 1855, around Bathurst in 1863, in the 1880s by Cape botanist Peter MacOwan and in 1908 by another prominent Cape botanist, Rudolf Marloth (Van Wilgen et al. 2016). It became a sustained narrative in the 1920s, however, expressed prominently and consistently in the *Journal of the Botanical Society of South Africa* (*JBSSA*), which despite its title focused mainly on the Cape flora.

The 'floral nativism' developed at the Cape has been well described by Van Sittert (2003a, b), myself (Pooley 2010), Carruthers (2011) and Bennett (2015), and doesn't require lengthy reiteration here. What is important to recall is that the national botanical gardens at Kirstenbosch on the eastern slopes of Table Mountain was founded in 1913, not long after the creation of the Union of South Africa in 1910, and the chief instigator and first director Harold Pearson had explicitly argued for establishing a national botanical garden as an 'expression of the intellectual and artistic aspirations of the New Nation'. This he believed would 'encourage a proper appreciation for the rare and beautiful flora with which Nature has so lavishly endowed it' (Pearson 1910, pp. 1–4).

In the 1920s, Harold Compton (second director of Kirstenbosch, from 1913) and his fellow Botanical Society members developed a narrative of indigenous flora (they focussed on 'flowers') under dire threat from the ignorance, wilful destructiveness and callousness of some members of the public, including competitors in the many and popular flower shows, and the unchecked greed of commercial flower sellers and those in favour of commercial plantations over indigenous forests. Compton had instigated fire control measures including warning signs and safe barbecuing areas at Kirstenbosch following a large bushfire above the gardens in May 1923. In 1924, Compton (1924, p. 3) wrote that 'an attempt is also being made to get rid of the cluster-pines, hakeas and wattles which have been spreading so rapidly on the open slopes' [above Kirstenbosch].

Compton (1927, p. 3) urged people to garden with indigenous plants, arguing that 'apart from the natural beauty of our indigenous plants … these flowers have their intimate association with scenes of beauty: they carry memories of open-air days, among mountains and vleis and on the rolling veld. Lovely as they are, what associations have roses and sweet-peas that can compare with these?' Further, in the aftermath of the First World War, he argued that 'the other reason why we should grow South African flowers in our gardens is because they *are* South African. When love of country is linked with beautiful things like flowers, instead of with hideous things like war, then we shall have a truer patriotism than we know to-day' (Compton 1927, p. 3).

Compton (1928, p. 5) worried that much of Table Mountain was public land, and the Crown Land was 'administered by the Forest Department, which cares for its plantations but exercises little or no surveillance over the areas without timber'. The

Cape Town municipality managed its catchment area on Table Mountain, but planted introduced trees and '[favoured] the spread of that aggressive weed the Blackwood Wattle in Disa Gorge'. Compton urged the authorities to establish a nature reserve on Table Mountain, to enable the appointment of a ranger 'for fire-protection, for combating the spread of exotics, for the collecting of litter and for checking the depredations of flower pickers'.

In 1929, Compton published an article on 'The Vegetation of the South West Region' by a rising star on the local botanical scene, Robert Adamson, who had arrived in the country in 1923 to take up a professorship at the University of Cape Town. Adamson (1929, pp. 7–12) commented on the widespread invasion of the 'original' indigenous vegetation by the (indigenous) shrub renosterbos (*E. rhinocerotis*), which he attributed to 'interference with' and 'destruction of' the indigenous vegetation through burning and overgrazing. He also bemoaned the 'impoverishment' of the flora in the mountain regions, including Table Mountain, as a result of frequent fires. Drawing on his surveys of Table Mountain's flora (Adamson 1927), he argued (1929, p. 12) that 'the introduction of alien trees has brought about very marked changes…'. He commented that 'pines, and still more Australian species of Acacia have attained complete dominance over the native flora in places. Another Australian genus, Hakea, is locally ousting the indigenous bush'. When Sir Arthur Hill (1929), the Director of the Royal Botanic Gardens, Kew, visited the Cape in 1929, he complained of the domination of the vegetation of the Cape Flats by Eucalypts, pines, venothera (evening primrose, *Oenothera* species, from the Americas) and wattles.

10.3.2 Major Fires and Invasive Introduced Plants

Public opposition to imported tree plantations in particular re-emerged with a vengeance following a very big fire which broke out on the northern slopes of Table Mountain on 25 December, 1935, spreading into Peak Plantation. The foresters struggled to muster fire-fighters as it was a Christmas Day, and the fire raged for two days, finally requiring 700 men to bring it under control. The fire caused £4000 worth of damage to the plantation, and part of the reason it had been so difficult to fight was that it had spread into the pine-tree-invaded slopes of Devil's Peak, above the plantation. The City and the white sands of Milnerton Beach were dusted with gray ash. In the aftermath, there was a major public outcry in the newspapers against planting inflammable trees on the mountain slopes above the city, which dominated the headlines for three weeks after the blaze (Pooley 2014).

While a minority rallied to the defence of these introduced trees, bemoaning the otherwise shadeless windy slopes, and foresters defended the plantation for adding beauty to the naked slopes and for stopping soil erosion, the majority of correspondents to the newspapers focussed on the aesthetically displeasing aftermath of the fires and the threat to the indigenous vegetation. A vocal contingent championed by Lady Florence Phillips, a noted conservationist and wife of the Randlord Sir

Lionel Phillips, argued for fencing off the mountain from vagrants and other undesirables (ARFD 1935/36, pp. 6 and 16; *Cape Argus* 26 December 1935; *Cape Argus* 27 December 1935). Unruly and dangerous intruders, both human and floral, were to be excluded.

In the event, little came of calls to clear introduced species from the mountain. Some desultory efforts were made to clear invasive cluster pines in the immediate aftermath of the fire, and the municipality cleared a few hakea and *Acacia* in 1939 and 1940, but with the development of the Second World War, labour was not available for such work (Lückhoff 1951). Although there were few concerted management efforts to control invasive introduced plants in the first half of the twentieth century, there were important developments in research, conservation thinking and public perception in this period.

The period from 1921–1936 was an extended dry period, and the Great Depression had meant reduced resources for preventive control measures for fire and plant invasions. Fire incidence increased, and public concern increased over both of these interlinked threats to the landscapes and fauna and flora of the Peninsula. Alongside this, the opening up of the Peninsula through road building (in part to provide jobs during the Depression) and rising ownership of private cars was spreading the risk of anthropogenic ignitions, with very high fire incidence in 1938 and 1939. Capetonians worried that the scenic beauty of the Peninsula was under threat from quarrying on the mountain's slopes, the opening up of the Peninsula's mountains to the general public through road development, and the completion of the cable car link to the top of Table Mountain in 1929 (Pooley 2014, pp. 127–128).

For the first time the Peninsula was zoned into 11 fire control zones, and cooperative firefighting arrangements were set up between the Peninsula's municipalities and private estate owners. Propaganda efforts were funded to educate the public about the risk of veld fires and competitions held in schools (MM1940, CLFB, p. 41; Lückhoff 1951, p. 138). The Cape Town Municipality's Chief Forester Officer Hugo Brunt was praised for his efforts at overseeing a fire suppression campaign on municipal lands, and amid escalating fire incidence in the wider region, it was argued in the press that his fire-exclusion strategy should be expanded to the Western Cape's mountain regions (*Cape Times* 1942). This call provoked a memorandum to the Royal Society of South Africa from a retired forester John Henkel (1943), who worried that any such fire exclusion strategy would be impossible to achieve, dangerous, and ecologically harmful.

Henkel's memorandum came at a time of heightened sensibilities over national identity and protecting the country's natural heritage, as Prime Minister Jan Smuts led a divided country to war in support of the Allies (many Afrikaner nationalists supported the Nazis against their old adversaries the British). Smuts had appointed a committee to form a National Veld Trust in 1943 (*Cape Times*, 7 January), and land management issues had been militarised in the public discourse, with careless or deliberate fire starters being termed traitors in the newspapers. 'War' was declared on soil erosion, and on the Peninsula there was concern over invasive Argentine ants undermining the economy (*Cape Times* 1943, 20 November and 4 December). Smuts (1942) among others saw the country's natural heritage as a neutral and hence valuable source of common identity for the country's divided European population, and thus its conservation took on nation-building political overtones (Van Sittert 2003b). There was also a sense (as there had been following the First World War) that Nature would provide a means of healing troubled spirits in the aftermath of war (Wicht 1945, pp. 7–9 and 53).

Amidst this period of political and environmental ferment, and in response to Henkel's letter, the Royal Society of South Africa appointed a committee to investigate the protection of the montane vegetation of the southwestern Cape. Chaired by the forestry researcher Christiaan Wicht, who was directing hydrological experiments at the Department of forestry research station at Jonkershoek near Stellenbosch, it published its report in 1945 (largely written by Wicht, with a chapter on 'characteristics of the vegetation' by Robert Adamson). This report (Wicht 1945) is now regarded as a landmark publication in the conservation of the region's flora (Van Wilgen et al. 2016).

Although the members of the Royal Society committee adhered to the received wisdom of the time that fire retards or interrupts the succession of vegetation to its 'natural' climax state, and can impact negatively on water supplies, they concluded that fire does have an important ecological role in fynbos and does not *necessarily* result in soil erosion or affect water supplies adversely. Poor burning practises—notably too-frequent burning over many years, followed by grazing—could result in such negative impacts, as could fires in areas with shallow sandstone soils on mountain slopes in areas with low rainfall (Adamson 1945, pp. 26–31, 38). From a management perspective, Wicht (1945, pp. 39–45) observed that attempts to exclude fire—ongoing since 1652—had all failed. He concluded that controlled burning in non-sensitive areas is preferable, supplemented by the creation of fire-belts, access paths, lookouts and labour for firefighting during the fire season.

Of the six processes identified in the report as major causes of vegetation deterioration, including conversion of the veld to other uses, erosion, flower gathering, pasturing, spread of undesirable species, and veld burning (Adamson 1945, pp. 22–23), the commissioners identified 'the spread of vigorous exotic plant species' as 'one of the greatest, if not the greatest, threats' to the fynbos. They noted the important role of fire in facilitating these biological invasions and advocated fire suppression as a key management strategy to control them. The Wicht Committee was sceptical that private landowners could handle dense stands of invasive species, and called on the government to take the initiative (Van Wilgen et al. 2016, pp. 32–33).

In the short term, nothing came of these expert proposals. Although the Department of Forestry formulated a policy for prescribed burning of fynbos vegetation in 1948, it was not implemented in the face of strong opposition to burning from the public and (largely) Department of Agriculture officials involved in soil conservation. There was also judged to be insufficient guidance on when to burn (which season, and how often) (Pooley 2014, pp. 83–86; Van Wilgen et al. 2016, p. 69).

On the Cape Peninsula, it is perhaps unsurprising that calls for prescribed burning (in part to prevent large dangerous fires resulting from a build-up of fuel loads in the vegetation) came to naught. The war years had seen high fire incidence, partly due to a temporary population boom and increased ignitions, and the decline in fire prevention measures and available fire control personnel resulting from recruitment into the armed forces. A voluntary firefighters force had helped, but this was disbanded in 1945, and Hugo Brunt died in the same year. Amid an extended dry period from 1946–1949, fire incidence continued to increase steeply. The considerable anti-fire propaganda schemes had not had the desired impact, and in 1947 the Cape Peninsula Fire Protection Area was declared. The focus was on clearing inflammable vegetation, prohibiting veld burning and maintaining firebreaks.

10.4 Post-war Environmental Concern Over Invasive Plants

Concern over the situation on the Peninsula came together at a public meeting organised by the South African Association of Arts in 1948, where a resolution was passed requesting the appointment of a government commission to assess the situation (Hey 1978). Despite a wide range of measures from additional manpower to legislation in the period to 1949, fire incidence continued to rise, and in January 1950 a fire broke out on Table Mountain above the City and burned uncontrollably for four days. In another incident, two navy personnel were killed fighting a fire above Simonstown on the southern Peninsula. Following a public meeting, the Department of Agriculture donated funds to enable the Cape Peninsula Fire Protection Committee to greatly increase fire control work. This included efforts to control inflammable introduced plants (Pooley 2014, pp. 210–214) (Fig. 10.2).

Letters published in the *Cape Times* showed a similar range of opinions on introduced plants as evidenced in the aftermath of the Christmas Day 1935 fire. On the one hand some argued that indigenous vegetation was more effective at protecting the soil and used less water than pines. On the other hand, a Mr. R. Pothier asserted that the indigenous fynbos was highly inflammable, fires were easier to fight in plantations, and walking through shady pines was preferable to walking in 'sun-scorched, snake-ridden bush'. Singling out pines for removal because they were introduced species, he pointed out, should logically extend to removing oaks, gums, poplars and many other attractive introduced trees (*Cape Times* 1950, p. 8).

The called-for commission on the state of the mountain (the Van Zyl Commission) reported to the Minister of Lands in 1951. The report recommended the prevention and control of fires on the mountain, and 'eradication of noxious weeds and alien vegetation threatening the indigenous vegetation' to restore the mountain 'to its natural state of beauty' (cited Hey 1978, p. 17). The Commission motivated for the establishment of a Table Mountain Preservation Board to oversee this, which was constituted in 1952. Table Mountain was proclaimed a national monument five years later.

Common sight in Cape Town during dry summer. Veld and mountain fires anywhere in the Peninsula are a threat to public safety, to life and property, to the rich indigenous fauna and flora, to the conservation of the soil, and veld and water supplies.

CAPE CAMPAIGN AGAINST "FIRE BUGS"

Peninsula Inhabitants Draw Up Five-year Plan to Avert
An Annual Threat of Disaster to Mountains and Forests

Fig. 10.2 Feature story (1954) in the journal of *The National Veldtrust* (15(4): 6–7) on the five-year plan devised in 1953 'to make widespread forest and veld fires a thing of the past'

Carl Lückhoff, a well-known Capetonian plant collector and conservationist and author of *Table Mountain: our national heritage after three hundred years* (1951, p. 116), explained in this book that it was only after the January 1950 fire that the Forestry Department began clearing inflammable invasive species from the mountain above the Cape Town city bowl. These 'exotic' flora including pines, hakeas and wattles were, he argued, relentlessly advancing on the lower slopes and summit plateaus of Table Mountain, and the Table Mountain chain south to the Constantiaberg and Chapman's Peak.

Although well publicised in Cape Town since the 1920s, and following big fires in 1935 and 1950, it was only in the late 1950s that the issue of invasive introduced plants was formally tackled by a conservation body.[1] The Kirstenbosch Botanical Gardens' Wild Flower Protection Society Committee set up a Control of Alien Vegetation Committee in 1958, which published its findings as *The Green Cancers*

[1]The Wild Flower Protection Ordinance, No. 21 of 1957, set out to protect 'any plant indigenous to the Republic of South Africa, except noxious weeds' (cited in Hey 1963, p. 68).

of South Africa: the menace of alien vegetation (1959), which despite its title focussed on the Cape. The Committee's findings were amplified by the findings of University of Cape Town botanist Anthony Hall, who in 1959/1960 surveyed invasions along the Peninsula's mountains from Muizenberg Mountain in the south to Table Mountain. He recorded four *Acacia* species, three pines, and three Hakea, which he argued were shading out fynbos plants (Moll and Trinder-Smith 1992, pp. 135–137).

In 1962, an illustrated booklet *Bokkie the Grysbuck* (Bagnall and Louw 1962) was published promoting fire and invasive plant awareness. It narrates a heartrending tale of a small antelope orphaned by a bush fire set by careless picnickers. Rescued by firefighters, Bokkie is nursed back to health and released in a nature reserve, where he will be safe from 'the fires which ravage the mountain and destroy so many animals'. However, this Edenic landscape is threatened by hakeas and pines 'which are strangers and intruders in South Africa'. The denouement shows happy scenes of children chopping out invasive plants (Bagnall and Louw 1962, pp. 33 and 37). The iconic 'prevent bush fires' poster based on Bokkie was used throughout South Africa in ensuing decades, and was resurrected on the Cape Peninsula following the big fires of January 2000.

Thus, by the beginning of the 1960s, there was widespread concern over the impacts of alien invasive vegetation on the indigenous flora, with specific legislation in place to protect the indigenous flora (except for 'toxic weeds'). The position was summarised by the Director of the Department of Nature Conservation in the Cape Province, Douglas Hey, in an editorial in which he singled out several Hakea and Acacia species, *Pinus pinaster* and Australian myrtle as 'green cancers' and 'the greatest enemy of our natural flora' (Hey 1963, p. 69).

10.4.1 Management of Invasives on the Peninsula, c.1959–1978

Despite this public and expert concern over the state of the natural environment on the Peninsula, and despite education measures like the *Bokkie* book (sponsored by the provincial Department of Nature Conservation and the Cape Peninsula Fire Protection Committee), efforts to manage invasions, and research on invasive species, were slow to materialise.

The Table Mountain Preservation Board had managed to purchase and incorporate private land into the protected area, consolidating management to some degree, but lacked funds for conservation work and the authority to impose conservation measures on private landowners (Hey 1978). Colin Grohl, Director of the City Council's Parks and Forests department (also instrumental in the *Bokkie* pamphlet), discussed this situation with the Provincial Secretary of the Cape, W. J. B. Slater, and in June 1963 a meeting of all the major NGOs and provincial and municipal authorities was convened by the Administrator of the Cape, Dr. Nico

Malan, to propose the establishment of a large nature reserve on Table Mountain. Key issues included eradicating undesirable introduced vegetation and controlling veld fires and soil erosion. The Table Mountain Nature Reserve was duly proclaimed in November 1963, on City Council land (Hey 1978).

Annual reports show that the Corporation of the City of Cape Town's Department of Parks and Forests was clearing invasive plants (only Hakea are listed) from forest stations from 1959 onward, but only in 1964 was the programme of clearing and reclearing described as a 'vigorous campaign' (ARCE1979, Annual Report of the Parks and Forests Branch (ARPFB), p. 19). This mostly reflected expanded effort directed at clearing around its dams at the Wemmershoek and later (from 1967) Steenbras plantations, both distant from the Peninsula. *Pinus pinaster* was added to the list of cleared plants from 1967, and efforts at clearing 'alien vegetation' from nature reserves (Silvermine and Table Mountain, both established in the 1960s) were recorded from 1967–1970. From 1968, Port Jackson (*Acacia saligna*), *Pinus pinaster* and *Albizia* were listed as invasive introduced plants subject to control measures (ARCE 1958–79).

Invasive plant control was first recognised by the state Forestry Department as a discrete management goal in 1960, and department members attended a conference focussed on invasive Hakea management convened by the Department of Agricultural Technical Services in Stellenbosch in 1961 (ARFD RP20/1961, p. 12; RP80/1964, pp. 8–9). An interdepartmental action committee was formed which subsequently surveyed invaded areas and undertook awareness campaigns, and some gains were made in controlling Hakea (Van Wilgen et al. 2016, p. 71).

By the early 1970s (the years of the OPEC oil crisis), the Forestry Department acknowledged that invader control programmes were being scaled back as 'whenever funds have to be cut back these operations are rated a lower priority than others' (ARFD1972/73 RP38/1974, p. 9). This was particularly the case on the Cape Peninsula as the department had virtually ceased forestry activities: only Tokai, Devil's Peak and Cecilia plantations remained by the late 1960s, and Peak was phased out as a state plantation by 1971. Following big fires in 1973 and 1974 and post-fire clearing, the Municipality phased out commercial timber growing on the Peninsula in 1975 (ARFD and ARCFO c.1970–76; Shaughnessy 1980, pp. 260, 267 and 310). All of this resulted in reduced fire protection and invasive plant clearing activities on the Peninsula. The City Council was concentrating its efforts on the catchment areas of its Wemmershoek and Steenbras dams (ARCE 1963–76).

In 1974, a report on the status of Table Mountain published by the Cape Town Section of the Mountain Club of South Africa called for urgent measures to prevent the deterioration of Table Mountain, noting fire and alien vegetation control as priorities. In the same period, the Cape Divisional Council was under pressure to approve numerous applications for private development on the southern Peninsula, and a series of meetings were held to discuss comprehensive environmental planning on the Peninsula's mountains, with reports published in 1972 and 1975. An important report was prepared by University of Cape Town botanists Moll and Campbell (1976), which received the support of the Table Mountain Preservation Board. However, the authors' call for total eradication of alien vegetation was judged unrealistic and too expensive, and the recommendation was rather to target sparsely infested areas of fynbos and protect them from future invasions, and aim for containment of other invaded areas (Hey 1978).

In 1976, McLachlan et al. resurveyed Hall's sample points from 1959/1960 and while they found a reduction in the distribution of some species (*Pinus pinaster* and *Acacia Cyclops*), they found increased distributions of *Hakea gibbosa*, *H. sericea* and *Acacia saligna* (Moll and Trinder-Smith 1992, pp. 137–139). There was clearly still cause for concern. The municipality's policy on invasive introduced plants at this stage was to clear all such plants from above the 300 m (984ft) contour line on the Peninsula's mountains. Below 300 m, non-invasive introduced species were to be left in place to provide shade for 'high usage nature-orientated recreation' (ARCE 1979, p. 19). In the opinion of Douglas Hey, Director of a one-man commission on the 'future control and management of Table Mountain and the southern Peninsula mountain chain' (1978, p. 62): 'despite all the good work on eradication which has been done over the past 15 years, the threat of aliens on the natural vegetation has not even been checked'. He identified impenetrable stands of hakea on private lands as a particular problem, and 'a great fire hazard,' also singling out *Acacia cyanophylla*,[2] *A. Cyclops*, *A. melanoxylon*, stinkbean (*Albizia futens*) and cluster pines as particularly problematic invaders (1978, p. 63).

10.5 Research on Invasive Plants, 1974–1989

With the proclamation of the Mountain Catchment Areas Act 63 of 1970, foresters were put in charge of the country's catchments. In recognition that plantations of introduced trees could have significant impacts on water supplies in catchment areas, the Forest Amendment Act, Act 40 of 1972, put restrictions on where afforestation could occur, controlled through a permit system (Van Wilgen and Richardson 2012a). The Department belatedly instituted a programme of prescribed burning. To work out how best to burn fynbos in these areas, and manage invasive plants, Gerrit Wagner and then John Fenn developed a system of coordinated working between researchers, planners and managers, which was implemented in many catchments until the terminal decline of state forestry as an active management institution in the late 1980s (ARFD RP18/1972, p. 9; Marais 2008; Kruger pers. comm. 2017).

In this period when management of invasive plants on the Peninsula was being called for, but scaled back for economic reasons, research began in earnest. Excluded from the International Biological Project (IBP) because of the country's Apartheid policies, South African researchers set up a programme of biome projects coordinated by The Cooperative Scientific Programmes secretariat of the Centre for Scientific and Industrial Research (CSIR). This included a Fynbos Biome Project,

[2]The blue-leaved wattle is no longer considered a problematic invasive in South Africa.

initiated in 1977, which included work on introduced invasive plants as a threat (Huntley 1992; Kruger 1978). The programme built on a scientific research programme initiated by state foresters in 1974 and led by Frederick Kruger at Jonkershoek, which drew in researchers who went on to become key figures in fynbos research and management including William Bond, David Le Maitre, David Richardson, Brian van Wilgen, and others.

An important international association which remained open for South African researchers was the MEDECOS network, a group of ecologists who had begun making comparative studies of Mediterranean-type climate regions in the late 1960s. The third MEDECOS meeting was hosted in Stellenbosch in 1980, where delegates were forcibly struck by the invasions of undisturbed fynbos ecosystems by introduced plants (disturbed ecosystems were supposed to be vulnerable; undisturbed ones not). The delegates motivated for an international programme to research and inform management of invasive species. The resulting Scientific Committee on Problems of the Environment (SCOPE) programme was launched in 1982, and the South African programme a year later, both resulting in books synthesising the available knowledge (Macdonald et al. 1986). This work included important discoveries on how such invasions impacted on fire regimes in fynbos, and on how key species invaded the fynbos (Van Wilgen et al. 2016, p. 33; Pooley 2014, pp. 98–101).

The control efforts of the 1950s and 1960s had proven relatively ineffective because little was known about the reproductive strategies of the invading plants. During the 1970s studies by Anthony Hall, Hugh Taylor, Eugene Moll and others resulted in more effective control measures, in particular, repeat treatments after initial clearing for invaded areas. A systematic invasive plant management plan was adopted in the Cape of Good Hope Nature Reserve in 1981, and by 1996, *Pinus pinaster* and *Acacia saligna* were judged no longer a threat, with only *A. cyclops* remaining an issue (Privett et al. 2001; Van Wilgen 2009, pp. 339–340). However, the fractured nature of land ownership outside of nature reserves made management very difficult on the Peninsula, as cleared areas could be easily invaded from adjacent privately owned properties which had not been cleared. It was realised that burning was a key control method, but prescribed burning was opposed by the public (Ibid.; *Cape Times* 1973).

10.6 Repercussions of the Collapse of Apartheid, c.1986–1994

The period 1978–1989 can be regarded as a golden era in research into fire and plant invasions in the fynbos, and certainly by the mid-1980s management was benefitting from these discoveries. However, as a result of international economic sanctions, internal strife and a war in Angola, by the late 1980s the Apartheid State was in a state of collapse. In 1986, the government set in motion a process of devolution of state forestry which in 1990 resulted in the transfer of plantations to a parastatal, the South African Forestry Company Limited (SAFCOL), and placed the management of mountain catchments into the hands of provincial administrations ill equipped to manage them. Shortfalls in funding, experienced managers and labour, slashed budgets and (for prescribed burning) the fear of litigation from the expanding wildflower industry, contributed to a widespread abandonment of prescribed burning and invasive plant control. The Fynbos Biome Project terminated in 1989, and State Forestry ceased to exist as a powerful research, planning and management entity, with all research projects terminated by 1995 (Pooley 2014, pp. 110–112).

By the early 1990s, fire management on the Cape Peninsula was reduced to mainly firefighting, and invasive plant clearing to a minimum. At the same time, the City was booming, and informal settlements were springing up on the Peninsula. With the creation of the 'New South Africa' in 1994 international tourism thrived, especially with the hosting of the rugby World Cup in 1995. There were also important changes in the protected areas on the Peninsula: on 29 May 1998, President Nelson Mandela proclaimed the Cape Peninsula National Park (now Table Mountain National Park), in part the result of a sustained public campaign to better protect the Peninsula's flora and fauna stimulated by very large fires in 1986 and 1991. Importantly, this unified the management of most of the Peninsula's nature reserves under a single authority, South African National Parks (SANParks). Mandela (1998) looked to the striking natural landscapes and unique flora of the Peninsula for a symbol of national unity, as had Smuts and Pearson before him.

What hadn't been anticipated in the long-awaited creation of a National Park was that many of the experienced staff employed by the Municipality and Forestry Department on the Peninsula elected to take severance packages and not transfer to SANParks. Thus a great deal of experience of working to prevent and fight fires on the complex typography of the Peninsula was lost (Prins 2007; Pooley 2014).

Finally, the 1990s (with the exception of 1996) were to be a very dry period, and it is therefore unsurprising that there were major fires recorded in every year except 1996 and 1997. Forsyth and Van Wilgen (2008) found a significant increase in areas burning frequently in the period 1985–2007, in comparison with 1970–1990. Explicit links were made between fires and invasive introduced plants, as environmental threats. In February 1991, a big fire burned on Devil's Peak and in May, storms washed 400 m^3 (14,126 ft^3) of fire debris, mud and rocks down Tin Mine Ravine, flooding streets in the city below and costing millions of rands to clean up. Researchers concluded that the flood was caused when stormwater rushed over soil hardened by the intense fires resulting from the burning of introduced invasive plants (ARCE1991/92, pp. 7–8). Invaded sites carry as much as eight times more fuel than typical fynbos shrublands (Van Wilgen and Richardson 1985).

In this period of political turmoil and transition, state funds for research into invasive plants and fire had dried up, as had resources for control measures. In November 1993, former forestry researchers, conservation managers, university academics and others resolved to motivate for continued funding through focussing on ecosystem services. The management threat they focused on was the invasion of catchment areas by introduced plants, and in 1994 they argued that unchecked plant invasions could reduce water supplies to the city of Cape Town by 30%. Guy Preston helped develop a proposal to combine invasive plant control with poverty alleviation through employing and training the poorest of the poor, and this was presented to the new Minister of Water Affairs and Forestry, Kader Asmal, in June 1995. The plan chimed well both with the political agenda of the new ANC-led government in South Africa which took power in 1994, and international developments in conservation and development thinking summarised in the Rio Declaration on Environment and Development in 1992. The result was Working for Water, a national water conservation campaign which included as a central component the removal of invasive introduced plants, creating jobs for the poor in the process (Preston 2007; Le Maitre et al. 1996; Van Wilgen et al. 1996, pp. 184 and 189).

The Working for Water programme was a great success, proving to be the only one of the government's new Reconstruction and Development (RDP) programmes to spend their budget. The pace at which the initiative grew raised challenges for management, and there were some important oversights in the early years. Working for Water workers were active on the Cape Peninsula, notably on the heavily invaded areas on the southern Peninsula where plants had escaped plantings intended to stabilise driftsands. These woody plants were cut and stacked on the hillsides, where they dried out, providing high concentrations of inflammable fuel (Preston 2007; Kruger et al. 2000).

10.7 The Great Fires of January 2000, and After

On Sunday 16 January 2000, two fires broke out on the southern Peninsula, one near the access road to the picnic site in Silvermine Nature Reserve, and one near the informal settlement at Red Hill. Weather conditions were extremely favourable to fire spread, with high winds and temperatures, and low humidity. The Silvermine fire spread quickly through fynbos invaded by introduced woody plants, and the Red Hill fire torched piles of cut invasive plants stacked by Working for Water workers, before tearing through mature fynbos. Continuous high winds and changes in wind direction made fire fighting even more difficult, and the fires were upgraded from a Code Red to a provincial emergency, with a Joint Operating Centre established to coordinate firefighting forces from four municipalities, South African National Parks, Heyns Helicopters, the forestry company SAFCOL (which had taken over Cecilia and Tokai plantations) and the South African Air Force and Navy. It took six days of firefighting to contain the fires, which together burned 8370 ha (20,682 acres) of the Peninsula, destroying eight buildings and damaging 51. The operational costs together with estimated costs of rehabilitation were in excess of R5 million, and the fires resulted in insurance claims of R40 million (Aupiais and Glenn 2000; Kruger et al. 2000).

It was the most serious fire event yet recorded on the Peninsula, and briefly made international headlines with footage of flames racing across the slopes above the capital city's suburbs. However, the ecological damage was slight, chiefly soil repellency caused by intense flames where invasive plants had burned (Kruger et al. 2000). In fact, the fires were arguably beneficial to areas of fynbos which had not been burned for decades, largely as a result of public antipathy to burning. Some fynbos species (notably the rare silver tree *Leucadendron argenteum*) may become senescent, i.e. show reduced vigour and reproductive capacity, if not burned for intervals of more than 35 years (Rebelo et al. 2006).

In the aftermath of these fires there was an outpouring of public support for improved fire control (as there had been following previous large fires), manifested as the Ukuvuka Operation Firestop Campaign. This campaign was importantly shaped by the goals and influence of Working for Water, and the first of its ten objectives was to 'secure full control over invading alien plants along the Table Mountain chain and adjacent areas' (Ukuvuka 2000). The campaign was heavily skewed towards controlling fire through controlling 'alien invasive plants'. Department of Water Affairs and Forestry billboards featured a dramatic scene of flaming fynbos towering over a house, proclaiming that 'every house that burned in the wild fires along Table Mountain in January 2000 was surrounded by invasive alien plants'. Thus for many members of the public, alien plants were held responsible for these huge fires. This was despite the official report on the fires being careful to state that these plants did not cause the fires; they exacerbated them in places where dense infestations existed (Kruger et al. 2000, p. 84).

Neely (2010, p. 880) argues that local scientists remained silent on this incorrect attribution of blame for the big fires to alien plant invasions, as from c.2003 their funding was derived mostly from Working for Water. Working for Water and Ukuvuka both advocated sustainable management programmes based on sound ecological science which were also ideally suited to the political climate of the new South Africa, linking biodiversity conservation and poverty relief. Another way of looking at this is to recognise a conjunction between a political imperative and a particular 'strong' strand within invasion biology, well represented at the Cape (discussed below).[3] Either way, the popular perception linking the fires with invasives was not an entirely helpful message on a Peninsula containing both the country's parliamentary capital, its suburbs and informal settlements, and a national park designed to protect the inflammable fynbos. It did not incline Capetonians to favour prescribed burning.

From the point of view of scientific research, this proved a fertile period, however. Once again concerns over introduced plants, fire and water had come

[3]For evidence of the vigorous debate on the impacts of bioinvasions within invasion biology see Russell and Blackburn's article, and the many responses, in *Trends in Ecology & Evolution*, January 2017, 32(1).

together in a powerful narrative about environmental degradation, and the need for research and scientifically informed management. An important symposium series sponsored by Working for Water was launched in 2003, generating a synthesis of scientific knowledge informing invasive plant management in the country, published in a special issue in the *South African Journal of Science* (Vol. 100, issue 1–2, 2004). In 2004, the Centre for Invasion Biology was established at the University of Stellenbosch, one of six Centres of Excellence created by the country's Department of Science and Technology. Under the initial direction of Prof. Steven Chown, and since of Professor David Richardson, the Centre has provided a focus for the many research efforts going on in different organisations, and has served as a catalyst for further research into bioinvasions (Van Wilgen et al. 2016, pp. 33–35; Neely 2010, pp. 884–885).

But what of invasive introduced plant management in the wider Cape Floral Region? An assessment of control measures in the Cape Floral Region (Van Wilgen et al. 2012) concluded that although there had been some progress with the control of some invasive introduced tree species, notably *Acacia cyclops* and *A. saligna*, and *Hakea* species had declined, significant control efforts and expenditure had failed to reduce the extent of invasions by *Acacia mearnsii* or *Pinus* species relative to 1996.

The 'invasive alien species strategy: greater Cape Floristic Region' initiated in 2007 was regarded as a useful exercise but a failure in that the recommended governance models had not been adopted, and the goals were neither measurable nor time-bound. Developed by the Cape Action for People and the Environment (CAPE) programme, the strategy was developed with funding from the Global Environment Facility and the World Bank, and in retrospect it was concluded that the external funding and origins of the strategy, meant senior South African officials decided to ignore it (Van Wilgen et al. 2016, p. 35).

10.7.1 Radical Narratives

Jean and John Comaroff (2001) have argued that the rhetoric of the anti 'alien' species campaign in South Africa is bound up with xenophobia in the new South Africa (prominent at that time). I have argued (Pooley 2010) that there are much stronger historical continuities in these debates over nature, nationalism and belonging than the Comaroffs allow, but their argument draws attention to the attempt to create a single voice, or normative position, on biological invasions. Certainly, the publicity campaign (8–15 October 2000) stimulated by the Department of Agriculture's regulations on invasive species proposed in 2000, entitled 'AlienBusters,' attempted this. It drew on images of evil extraterrestrials threatening planet earth to mobilise the public's fear (and fascination with) the 'other' to promote the eradication of invasive species (Lidström et al. 2015). The cohesiveness between political, scientific and management goals which Neely praises has provided a strong framework to motivate for programs to manage plant invasions in the region's unique fynbos vegetation.

Invasion biologists' narrative about the need to resist 'biotic homogenization' or the 'McDonaldisation of nature' has been remarkably effective internationally in recent decades. This narrative has succeeded in catalysing action against what Rob Nixon (2011) has called the 'slow violence' of complex environmental problems like biological invasions. These kinds of complex problems unfold slowly over decades or longer, and are the result of a multitude of ecological and anthropological causes. They are thus difficult to narrate as dramatic stories demanding immediate action (Lidström et al. 2015, p. 3).

A complication for such powerful narrative frameworks on the Cape Peninsula (as elsewhere) is that in fact there remains a significant diversity in attitudes to different species of invasive introduced plants and how they should be managed. The powerful narrative which has been developed at the Cape, linking scientific research, environmental management and social justice, tends to obscure and dominate the real diversity of views on what should be done (an ethical, not a scientific, question).

Environmental histories remind us that the Cape Peninsula has been inhabited by humans for millennia, and its fire regimes and vegetation have been transformed to varying degrees for at least 2000 years, beginning with the visits of Khoikhoi (or KhoeKhoe) herders and their vast herds of cattle and sheep. Transformations accelerated with European settlement from 1652. Modern vegetation maps do not show what actually grows on the Peninsula, but rather what their compilers' think potentially would grow on the Peninsula were it not for urbanisation and other human transformations and introductions. The underlying assumptions are that climate and fire regimes have remained essentially unchanged—despite indisputable evidence of anthropogenic climate change and changes in the temporal and spatial clustering of fires. In short, contemporary ecological conditions on the Peninsula tend to be measured against an idealised precolonial past (Pooley 2014).

In the strong versions of this narrative, supported for instance by SANParks on the Peninsula, introduced species are aliens and should be removed from all protected areas. All invasive introduced plants should be removed from the Peninsula. Ecologists in the region have contributed to this polarising narrative which tends to obscure the actual spectrum of introduced plants, suggesting that all are harmful whereas there are many examples which are not proven to be so (including some invasive ones).

Since the declaration of Table Mountain National Park in 1998, many of the Peninsula's smaller stands of pines and eucalypts have been cut down or burned and not replaced. Small local activist groups have emerged to protest against the removal of shade-giving trees in picnic areas. This was capitalised upon by Thomas Pakenham, when he launched his book *In Search of Remarkable Trees* (2007) in Cape Town. He accused white 'eco-fascists' of being a 'tree-Taliban' trying to ethnically cleanse the country of introduced trees (*Cape Times* 2007, p. 3; *Sunday Argus* 2007, p. 11). This was a misrepresentation: the campaign was focused on invasive species, carried out by poor non-whites, and funded by the ANC-led government (Pooley 2014).

The powerful narrative presenting invasive alien plant control efforts as a 'war on aliens' obscures a long and complex history of interactions between introduced and local species, and the complexity of human histories which have shaped the

Peninsula's landscapes for hundreds of years. As Lidström et al. (2015, p. 26) put it, this complex history is 'flattened' to present a simpler narrative about saving a 'threatened "now" defined against a static and idealised "then".'

More specifically, and problematically, this narrative pits conservationists and invasion biologists against commercial foresters, who since the winding down of state forestry in the 1990s have been major players in controlling (and some argue causing) the spread of the region's most important invasive plants. This is exacerbated by the rocky relationship that developed between the private forestry companies, and Guy Preston and others in the formative years of Working for Water (Preston 2007). In 2011, Preston took on responsibility for implementing the National Environmental Management: Biodiversity Act regulations on alien and invasive species.

The animosity between commercial forestry and conservation researchers surfaced publically in an ill-tempered exchange in 2012, following the online publication of a summary of a paper entitled 'Three centuries of managing introduced conifers in South Africa' by two prominent local conservation biologists (Van Wilgen and Richardson 2012a). Their paper argued that both in terms of the Conservation of Agricultural Resources Act (43 of 1983, amended 2001), and Forest Stewardship Council (FSC) certification, commercial forestry operators in South Africa were failing to fulfil their obligation to prevent the spread of invasive pines. These pines were negatively impacting on water resources, causing more intense fires and resultant soil erosion, and crowding out native species.

In a fierce rebuttal, the President of the South African Institute of Forestry, Ben du Toit (2012) disputed claims about water use by plantations, denied that plantations were resulting in most invasions (arguing good management, and the introduction of less invasive hybrids, were reducing invasions), and denied that plantations were having serious impacts on the biodiversity of the fynbos. He accused Van Wilgen, Richardson and the press release summarising their work of bias, pseudoscience, and perpetrating a socially irresponsible attack on an economically and socially important rural industry.

Van Wilgen and Richardson (2012b) responded with a point-by-point rebuttal, standing by their claims, most of which are reiterated in a recent paper (Van Wilgen et al. 2016). Commercial foresters in turn remain aggrieved that their efforts at what they call 'integrated weed management,' including their phasing out of highly invasive *Pinus pinaster*, planting of less invasive species like *P. elliottii* around the perimeters of plantations of more problematic species like *P. radiata*, and the use of biological control agents to combat plants like *Acacia mearnsii*, are not acknowledged by conservationists (du Preez 2017).

10.8 Conclusion

The Cape Peninsula harbours a wide variety of introduced plants including commercially important ones such as grapes. Among them are a remarkably persistent suite of invasive plants introduced by Europeans for a range of purposes: timber, driftsand stabilisation and horticulture. These are valued differently by a range of interest groups. Well-to-do Capetonians living around the mountain slopes like the remaining pine plantations where they walk their dogs. They dislike the 'scruffy' wattles and hakea. The poor use these wattles and hakea for fuel for warmth and cooking, as building materials, and a means of earning an income (selling them to the well-to-do for their barbecues). For some conservationists, all of these plants detract from the ecological and aesthetic integrity of the Peninsula's landscapes. For some members of the public, stands of introduced trees bring welcome relief from the sun and wind in the otherwise treeless fynbos. Further, some trees have become engrained in the culture of the place, including the iconic stone pines below Devil's Peak and Lion's Head, the eucalypts on the flanks of Signal Hill, the palms of Camps Bay, and the camphor trees and oaks of Constantia (Pooley 2010).

In a long review and meditation on the history of scientific research and management of the Cape Floral Region from 1945–2015, some of the country's foremost ecologists, environmental historians and managers note that a strong emergent theme in invasion science has been the comparison of biological invasions in different regions and how they have been managed (Van Wilgen et al. 2016, p. 37). Although interdisciplinary in aspiration, comparative studies still tend to focus on ecological factors and quantified histories of management impacts, despite increasing acknowledgement of the importance of understanding the environmental histories of these social-ecological systems, and despite the realisation that environmental policy and management are influenced by much more than scientific knowledge and technical expertise.

As the example of the Cape shows so well, as important as control efforts are, they are unlikely to eradicate all introduced invasive species or to exclude all of the fires which favour them. By admitting more voices into the discussion of what to do about these plants, and toning down dominating and divisive narratives about fighting aliens, it may be possible to create space for a more nuanced discussion of the policies which should shape environmental management in the region. In the absence of a powerful state management agency with the political clout and managerial resources to enforce and implement sound environmental policies, collaboration between conservation and commercial forestry managers, scientific researchers and regulatory agencies, private landowners and the public—and the inclusive processes which this will require—seems a sound way forward.

It may be that what is required are management plans for hybrid rather than pure ecosystems, or a patchwork of the two. This will require formulating legislative and management responses which are implementable, and target actually harmful species identified in socially inclusive ways. The history recounted in this chapter suggests it may be advisable to take a triage approach, and prioritise some landscapes for protection, and some plants for management, rather than strive to restore unattainable idealised landscapes which not everyone aspires to inhabit (Fig. 10.3).

Fig. 10.3 Post-fire lilies spring up among charred protea shrubs after a fire on the slopes of Lion's Head, Cape Town. In the background are stands of non-invasive stone pines, on the slopes of Table Mountain above the city bowl. *Credit* Simon Pooley

Fires in Mediterranean-Type Ecosystems

Simon Pooley

Invasive grasses, shrubs and trees can profoundly reshape local plant communities and vegetation structure. The invasion of Mediterranean ecosystems by fire-adapted plants can be said to introduce a further, often very significant invasive 'beast': fire. Fire acts like a generalist herbivore, consuming all woody biomass whether alive or dead, toxic or protein-deficient, so competing with biotic consumers for resources as well as facilitating vegetation change.

Fire regimes
Fire regimes are particular, prevailing combinations of frequency and seasonal timing of burning, fuel types, fire intensity and spatial distribution of individual fires. Plants adapt to particular fire regimes, developing strategies for coping with the prevailing rates of variation in fire regime parameters. Mediterranean climates, with wet and mild winters promoting vegetation growth, followed by long, hot, dry and windy summers, provide ideal fire conditions. However, natural ignitions (usually caused by lightning) are rare. Thus while Mediterranean plants are fire prone and well adapted to specific fire regimes, it is difficult to talk about 'natural fire regimes' (i.e. excluding human influence).

In the Mediterranean basin, fire regimes have developed in cultural landscapes where humans have deliberately (and accidentally) influenced the frequency and seasonality of burning. In turn, these anthropogenic fire regimes have selected for fire adapted plants. Further, over millennia, humans have brought about major transformations of the fuel, for example breaking up fuel bodies, clearing and protecting fields, or creating plantations of inflammable trees (Pyne 1997; Pausas et al. 2008; Keeley et al. 2012).

In some of the world's Mediterranean type regions, anthropogenic transformations of the landscape and ignition incidence have created fire regimes more favourable to introduced species than to the indigenous fire-adapted flora. For example, more frequent and more intense fires can favour introduced plants like quick-recovering, fire-resistant stand-producing wattles, or flammable grasses. Globally, Mediterranean-type ecosystems are spatially the most degraded (93%) in terms of having fire regimes falling outside of natural rates of variation. This is unsurprising considering their attractiveness to human development and their ecological fire dependence (Shlisky et al. 2008).

Fires are usually high-intensity crown fires which burn all the vegetation above ground. The plants of these regions are adapted to fire regimes of moderate frequency and variability of intensity, and have evolved an extraordinary array of strategies to cope with fire. The two major strategies are resprouting, and germination from a previously dormant seedbank, and individual species have evolved numerous approaches to surviving fires through these methods.

Plant responses to fires
Fires are followed almost immediately by an ephemeral flora which survives for only a year or two. They exploit the nutrient flush following fires, including the open space and sunlight produced when the usually dominant shrubs are consumed by fire. Geophytes like certain lilies and species of onion resprout from underground bulbs and corms after fire, flower profusely to form beautiful fields of colourful blooms, and set seeds which germinate when the winter rains arrive.

Suffrutescent species survive in dormant seed banks until stimulated by the heat of fires and germinate in the first rains after the fire.

The shrub flora also emerge following fires, but only begin to dominate after several years as they begin to form an overstorey, shading out the smaller plants, finally forming a closed canopy over them. The shrubs also regenerate either by resprouting after fire or from seedbanks, or in some cases, use both strategies. These woody evergreen shrubs form continuous bodies of well aerated fine vegetation which burns readily. Many Mediterranean pines have serotinous cones, where hot fires melt the waxy seals triggering seed release. Some fire-adapted species may experience local extinctions after very long periods without fire (senescence), but too-frequent fires are much more of a hazard. In most areas humans have so increased fire frequency that long fire-free intervals are unlikely (Pyne 1997; Keeley et al. 2012). In sum, the high biodiversity of this flora has been shaped, rather than threatened, by fires.

Fire history of the Mediterranean basin

The major difference between the fire histories of the Mediterranean basin, and those of the other Mediterranean type climate regions, is the antiquity of widespread and intensive land use in the former region. The first evidence of fire management in the Mediterranean Basin dates from the spread of crop farming and domestic grazing in the Neolithic period ($\sim$10,000–4500 BP), and during the Bronze Age fire was very widely used to improve pastures, dispose of waste and recycle nutrients in fallow fields, and clear forests and scrubland. By the late Bronze Age, extensive areas had been terraced and irrigated, and the economies of the Roman and Muslim empires in particular resulted in major land use changes, including logging to build large navies. Landscapes were cleared and opened, progressively from east to west. Anthropogenic activities created and maintained landscapes with low and fragmented fuel levels, through close cultivation and grazing, and fire regimes were fuel-limited (Pyne 1997; Keeley et al. 2012).

Pyne (1997) describes the immemorial practice of free-burning fire of nomadic (or transhumant) herders of livestock, and the tension between their seasonal burning habits and settled agriculturalists striving for closed ecological cycles in which labourers took over the tasks of unruly fire. There were naturally significant regional variations, encapsulated for Pyne in the divergent styles of pastoralism exhibited on the peninsulas of Iberia, Italy and Greece. In the 1700s, Enlightenment thinkers participated in an agricultural revolution informed by the ideals of classical agronomists, interpreting the region's physical and cultural landscapes as degraded ones ruined by imprudent land use, notably the destructive foraging and burning habits of shepherds. Pyne (1997) describes the attempts of French administrators to regulate and control fire policy nationally, legislated into the Code Colbert of

1669, and refined through the application of scientific expertise to guide national and later colonial forestry policy and management.

In the nineteenth century, industrialisation and urbanisation began to transform the primarily agricultural landscapes of the Mediterranean, emptying the countryside, leaving abandoned fields and vineyards. State foresters came to rival agronomists as shapers of these landscapes, and French forestry in particular, drawing on its domestic and colonial experiences on Mediterranean islands and in North Africa, implemented a military approach to fire control which proved widely influential and has had a lasting legacy in the region (Pyne 1997).

From the mid-twentieth century, rural depopulation and land abandonment along with reduced livestock grazing pressure has resulted in a shift in fire regimes along the northern rim of the Mediterranean basin. Old fields have been abandoned, become more shrubby, or been planted up with pines and eucalypts. Fuel beds have become large and continuous. Along the coastal areas development in response to tourism has increased the wildland-urban interface, resulting in increased fire ignitions. Together with hotter drier climatic conditions in recent decades, this has resulted in more frequent, and larger fires. While until recently continuing intensive rural land use in the southern and eastern Mediterranean rim has maintained fuels at low levels, resulting in relatively smaller areas burned annually, rural depopulation resulting from industrialisation and political instability is driving a similar shift in fire regimes to those occurring in the northern rim countries (Pyne 1997; Pausas et al. 2008; Keeley et al. 2012).

Fire and bioinvasions: the Mediterranean basin and other MTCs

For European botanists educated in the ecological orthodoxy of the early twentieth century, which regarded climate and soil as the chief determinants of vegetation, the fire-prone vegetation of the world's Mediterranean type climate regions was a puzzle. (Some plants, for instance juniper, rosemary and species of pine, even produce volatile substances which make them more inflammable.) It did not fit their linear conception of vegetation succession, where the vegetation progresses from pioneer communities colonising bare soil, through successive stages of increasing complexity, to reach a stable climax stage in equilibrium with prevailing climatic conditions. Botanical and forestry experts interpreted the dominant tree and shrub communities of Californian chaparral, Chilean matarral, Australian kwongan or mallee, Mediterranean-basin maquis or macchia, and South African fynbos and renosterveld, as a stage en route to mature forest—and not a climax community.

What confused botanists, was that this Mediterranean-type vegetation is autosuccessional, meaning the *climax* species are present in the *pioneer* communities. The structure and species mix of particular communities are shaped by time since the previous fire, and the nature of that particular fire. However, fire was interpreted as a disturbance to the 'natural' progression (or equilibrium state) of the vegetation. It was often seen as destructive of biodiversity, whereas fire is a naturally recurring phenomenon in these plant communities, and is actually an important driver of their biodiversity. Into the 1970s, attempts were still being made by land managers to 'correct' this anomaly, either through excluding fire in the hope that the scrub would succeed to forest, or by planting trees (usually non-native species). This was especially the case in mountain catchment areas where trees were believed to best protect soils and water supplies.

Plant invasions by introduced species are a major problem in Mediterranean-type ecosystems, where altered fire regimes favour non-native plants including grasses and (introduced) shrubs and trees. Successful invaders usually exploit the post-fire invasion window more successfully than the indigenous flora, regenerating and maturing more quickly, and establishing dense fire-resistant stands which persist through subsequent fires. Although such stands of introduced plants (for examples of eucalypts or wattles) may be less easily ignited, once alight, they burn more intensely than the indigenous flora they have invaded, and can damage the soil and cause excessive runoff and erosion. Studies suggest that post-fire ecological damage of this kind usually occurs in transformed landscapes, notably pine plantations established on old fields (Pausas et al. 2008).

Keeley et al. (2012) argue that it is the long coevolution of Mediterranean basin plants with heavy human land use[4] (notably fire use) that has made them such successful invaders of other Mediterranean type climate ecosystems. These plants were introduced by European settlers in contexts of large-scale anthropogenic ecosystems disturbances, where settlers were clearing forests, cultivating crops, grazing livestock and deploying fires to aid them in these activities. Nearly two-thirds of invasive plants in California, Chile, South Africa and southern Australia come from the Mediterranean basin. Within the Mediterranean basin, on the other hand, the shrubs seem resistant to both local disturbances and invasions by introduced plants. An exception is areas with acidic soils and (relatively) high rainfall, where Australian Acacia and Hakea species are becoming invasive, notably in Portugal.

South Africa shares a number of characteristics with the more recent fire histories and current situations of the Mediterranean Basin countries, especially the 'fire club' comprising Portugal, Spain, France, Italy, and Greece. First, the depopulation of rural areas, particularly with accelerating industrialisation and use of fossil fuels following the Second World War, resulted in the decline of traditional biomass usage, the neglect of previously enacted

[4]See Pyne (1997) on the Mediterranean as an anthropogenic landscape, its biota sculpted by anthropogenic stresses, notably fire, browsing and grazing.

control measures, biological invasions by fuel-heavy inflammable plants, and a lack of labour for fire fighting.

Rapid urbanisation especially from the 1970s contributed to an extended wildland-urban interface, making fire control more difficult, and concentrating resources on fire fighting to defend dwellings and infrastructure rather than undertake fire prevention measures. Urban-oriented fire brigades have taken over fire control from forestry departments in wildland urban interface regions, and often lack the skills to prevent or adequately combat wild fires. Prescribed block burning along the wildland urban interface, or intermix, remains controversial and sporadically applied. The fractured nature of land ownership in such areas makes applying uniform or coordinated fire control measures difficult if not impossible. A shift in emphasis from plantation forestry to recreation and nature conservation in the last third of the century also resulted in a decline (or collapse) of fire control measures particularly in mountain regions (Anon 2007; Pooley 2014; Keeley et al. 2012; Doblas-Miranda et al. 2017).

European Mediterranean fires, c.1990–present

The perception is that in Europe, there was an upward trend in fire incidence and spread during the 1970s and 1980s, and some have argued this continues to the present.[5] This is despite the heavy investment in fire fighting capacity in the EU since the 1990s, which has managed to maintain the status quo in average years, but proved inadequate to contain megafires in years of difficult conditions with extreme fire weather. Despite the warnings of fire experts in the early 1990s, policy makers were slow to recognise the socio-economic determinants of this deteriorating fire situation. Major fires burned in southern Europe in 2000, with numerous fire disasters over the next decade, for example in Portugal in 2003 and 2005, Spain in 2006, Greece in 2007 and 2009, and Israel in 2010 (Xanthopolous 2009).

From 1998–2009, 35 major forest fires killed 307 people and caused losses of €6917 billion across Europe. On average, per annum, 70% of these fires occurred in the Mediterranean region, accounting for 85% of burned area. While 95% of ignitions were attributed to human causes, fire spread was attributed primarily to fuel levels and weather conditions. There was no overall trend either in area burned or number of fires, but it is asserted that fires are becoming more intense and difficult to control. They are also occurring at higher altitudes and in northern regions of the Mediterranean basin where they were uncommon in the past. The combination of increasing climate-driven fire risk, land abandonment and loss of grazers resulting in homogenous landscapes of fuel-heavy vegetation, and fire suppression rather than fire control policies led by urban-style fire services resulting in the accumulation of fuel, are shifting the fire regime towards large, frequent, intense fires (Anon 2010; Keeley et al. 2012).

[5]Data is incomplete and not always commensurable, and Turco et al. (2016) find an overall decline in area burned from 1985–2011, with the exception of Portugal.

At the time of writing, in 2017, forest fires had caused havoc in Portugal, Spain and Italy. Perhaps it is time, as Keeley et al. (2012) maintain, to move on from a fire management paradigm of fire suppression and a preference for dense forests, to one focused on sustaining natural ecosystems—which necessarily includes attempting to maintain the fire regimes the local biota have evolved to thrive in. In the case of the Mediterranean basin, this could mean restoring traditional agricultural land use practises. This approach would involve maintaining particular favourable fire regimes in landscapes, rather than trying to suppress all fires and treating those that do break out as natural disasters to be controlled through emergency responses. It would also involve controlling the spread of invasive, inflammable introduced trees and shrubs, and greater caution in where these are deliberately planted, avoiding close proximity to roads and homesteads.

Acknowledgements
My thanks to Stephen J. Pyne for his comments on this text. The views expressed are mine, as are any errors.

References

Anon (2007) Recommendations of session C: Europe, Southeast Europe, Mediterranean North Africa and Caucasus. In: Fourth international wildland fire conference. International Forest Fire News, vol 36, pp 1–3

Anon (2010) Mapping the impacts of natural hazards and technological accidents in Europe: an overview of the last decade. European Environment Agency Technical report No. 13/2010. EEA, Copenhagen

Doblas-Miranda E, Alonso R, Arnan X et al (2017) A review of the combination among global change factors in forests, shrublands and pastures of the Mediterranean region: beyond drought effects. Global Planet Change 148:42–54

Keeley JE, Bond WJ, Bradstock RA, Pausas JG, Rundel PW (2012) Fire in Mediterranean ecosystems: ecology, evolution and management. Cambridge University Press, Cambridge

Pausas JG, Llovet J, Rodrigo A, Vallejo R (2008) Are wildfires a disaster in the Mediterranean basin? A review. Int J Wildland Fire 17:713–723

Pooley S (2014) Burning Table Mountain: an environmental history of fire on the Cape Peninsula. Palgrave, London

Pyne S (1997) Vestal Fire: an environmental history, told through fire, of Europe and Europe's encounter with the world. University of Washington Press, Seattle

Shlisky A, Meyer R Waugh J, Blankenship K (2008) Fire, nature, and humans: global challenges for conservation. Fire Manage Today 68(4):36–42

Turco M, Bedia J, Di Liberto F et al (2016) Decreasing fires in Mediterranean Europe. PLoS ONE 11(3):e0150663

Xanthopolous G (2009) Wildland fires: Mediterranean. Crisis Response 5 (3):50–51

References

Adamson RS (1927) The plant communities of Table Mountain: preliminary account. J Ecol 15:278–309

Adamson RS (1929) The vegetation of the South West Region. J Bot Soc S Afr 15:7–12

Adamson RS (1945) Characteristics of the vegetation. In: Wicht WL (ed) Preservation of the vegetation of the South Western Cape. Report for The Royal Society of South Africa. Royal Society of South Africa, Cape Town

Aupiais L, Glenn I (eds) (2000) The Cape of flames: the great fire of January 2000. Inyati Publishing, Cape Town

Bagnall DG, Louw H (1962) Bokkie the grysbuck. Department of Nature Conservation & Cape Peninsula Fire Protection Committee, Cape Town

Bennett BM (2010) A global history of Australian trees. J Hist Biol 44(1):125–145

Bennett BM (2015) Margaret Levyns and the Decline of Ecological Liberalism in the Southwest Cape, 1890–1975. S Afr Hist J 67(1):64–84

Carruthers J (2011) Trouble in the garden: South African botanical politics ca. 1870–1950. S Afr J Bot 77:258–267

Comaroff J, Comaroff JL (2001) Naturing the nation: aliens, apocalypse and the postcolonial state. J South Afr Stud 27(3):627–651

Compton RH (1924) News and notes. J Bot Soc S Afr (JBSSA) 10:3

Compton RH (1927) News and notes. JBSSA 13:3

Compton RH (1928) News and notes. JBSSA 14:5

Du Preez B (2017) pers. comm. From Braam du Preez, Forestry Risk Manager for MTO Cape, 12 March

Du Toit B (2012) "No" to biased and pseudo-scientific reporting, says Forest Scientist. Available at: http://blogs.sun.ac.za/news/files/2012/05/Forestry-in-Fynbos-reply-BdT.pdf. Accessed on 9 June 2017

Forsyth GG, Van Wilgen BW (2008) The recent fire history of the Table Mountain National Park and its implications for fire management. Koedoe 50(1):3–9

Henkel JS (1943) Preservation of mountain vegetation of the South-West Districts of the Cape Province. Memorandum to the Royal Society of South Africa

Hey D (1963) The conservation of flora in the province of the Cape of Good Hope. Lantern 13:68–69

Hey D (1978) Report on the future control and management of Table Mountain and the Southern Peninsula Mountain Chain. Repub S Afr Pretoria, Government Printer

Hill (Sir) Arthur (1929) Personal diary. Kew Gardens Library and Archive, AWH/1/8 (1929)

Huntley B (1992) The Fynbos Biome Project. In: Cowling RM (ed) The ecology of fynbos: nutrients, fire and diversity. Oxford University Press, Cape Town, pp 1–5

Keeley JE, Bond WJ, Bradstock RA, Pausas JG, Rundel PW (2012) Fire in Mediterranean ecosystems: ecology, evolution and management. Cambridge University Press, Cambridge

Kruger FJ (1978) Description of the Fynbos Biome Project. National Scientific Programs Unit: CSIR, South African National Scientific Programs Report 28

Kruger FJ, Reid P, Mayet M, Alberts W, Goldammer JG, Tolhurst K (2000) A Review of the Veld Fires in the Western Cape during 15 to 25 January 2000. Department of Water Affairs and Forestry Report, South Africa, 145 pages

Le Maitre DC, Van Wilgen BW, Chapman RA, Kelly DH (1996) Invasive plants and water resources in the Western Cape Province, South Africa; modelling the consequences of a lack of management. J Appl Ecol 33:161–172

Lidström S, West S, Katzschner T, Pérez-Ramos MI, Twidle H (2015) Invasive narratives and the inverse of slow violence: alien species in science and society. Environ Humanit 7:1–40

Lister MH (1957) Joseph Storr Lister, the first chief conservator of the South African Department of Forestry. Journal of the South African Forestry Association 29:10–18

Lückhoff CA (1951) Table Mountain: our national heritage after three hundred years. AA Balkema, Cape Town
Macdonald IAW, Kruger FJ, Ferrar AA (1986) The ecology and management of biological invasions in southern Africa. Oxford University Press, Cape Town
Mandela RN (1998) Message from President Nelson Mandela on the occasion of World Environment Day. 5 June
Marais C (2008) Interview with the author. Cape Town, 3 October
Moll EJ, Trinder-Smith T (1992) Invasion and control of alien woody plants on the Cape Peninsula Mountains, South Africa 30 years on. Biol Cons 60:135–143
Neely AH (2010) 'Blame it on the weeds': politics, poverty, and ecology in the new South Africa. J South Afr Stud 36:869–887
Nixon R (2011) Slow violence and the environmentalism of the poor. Harvard University Press, Cambridge
Pakenham T (2007) In search of remarkable trees: on safari in Southern Africa. Weidenfeld & Nicolson, London
Pearson H (1910) Text of speech sent to Col. Pierie, Director of Kew. Miscellaneous Reports: South Africa; National Botanic Garden, 1910–1928, Kew Gardens Archive, 302, R6 (4 November 1910), 1–13
Pooley S (2009) Jan Van Riebeeck as pioneering explorer and conservator of natural resources at the Cape of Good Hope (1652–1662). Environ Hist 15:3–33
Pooley S (2010) Pressed flowers: notions of indigenous and alien vegetation in South Africa's Western Cape, c. 1902–1945. J South Afr Stud 36(3):559–618
Pooley S (2014) Burning Table Mountain: an environmental history of fire on the Cape Peninsula. Palgrave, London
Preston G (2007) Interview with the author. Pearson Hall, Kirstenbosch, Cape Town, 19 October
Prins P (2007) Interview with the author. Newlands Fire Base, Cape Town, 19 October
Privett SDJ, Cowling RM, Taylor HC (2001) Thirty years of change in the fynbos vegetation of the Cape of Good Hope Nature Reserve. Bothalia 31:99–115
Rebelo A, Boucher C, Helme N, Mucina L, Rutherford MC (2006) Fynbos biome. In Mucina L, Rutherford MC (eds) The vegetation of South Africa, Lesotho and Swaziland. Strelitzia 19. South African National Biodiversity Institute, Pretoria, pp 53–219
Richardson DM, Holmes PM, Cowling RM (1992) Plant and animal invasions. In: Cowling RM (ed) Ecology of fynbos: nutrients, fire and diversity. Oxford University Press, Cape Town, pp 271–308
Shaughnessy GL (1980) Historical ecology of alien woody plants in the vicinity of Cape Town, South Africa. PhD thesis, University of Cape Town
Sim TR (1907) Forests and forest flora of the Colony of the Cape of Good Hope. Taylor & Henderson, Aberdeen
Smuts JC (1942) The Spirit of the Mountain. In Plans for a better world: speeches of field-marshal the right honourable J. C. Smuts. Hodder and Stoughton, London
Ukuvuka (2000) The Santam/Cape Argus Ukuvuka Operation Firestop Campaign draft business plan: 2000–2004. Ukuvuka, Cape Town
Van Sittert L (2003a) Making the Cape Floral Kingdom: the discovery and defence of indigenous flora at the Cape ca. 1890–1939. Landscape Res 28(1):113–129
Van Sittert L (2003b) The bourgeois eye aloft: Table Mountain in the Anglo urban middle class imagination, c.1891–1952. Kronos 29:161–190
Van Wilgen BW (2009) The evolution of fire and invasive plant management practice in fynbos. S Afr J Sci 105:335–342
Van Wilgen BW, Richardson DM (1985) The effect of alien shrub invasions on vegetation structure and fire behaviour in South African fynbos shrublands: a simulation study. J Appl Ecol 22:955–966
Van Wilgen BW, Richardson DM (2012a) Three centuries of managing introduced conifers in South Africa: benefits, impacts, changing perceptions and conflict resolution. J Environ Manage 106:56–68

Van Wilgen BW, Richardson DM (2012b) Fynbos forestry debate response. Available at: http://blogs.sun.ac.za/news/files/2012/05/Fynbos-Forestry-debate-response.pdf. Accessed on 9 June 2017
Van Wilgen BW, Cowling RM, Burgers CJ (1996) Valuation of ecosystem services. BioScience 46:184–189
Van Wilgen BW, Forsyth GG, Le Maitre DC, Wannenburgh A, Kotzé JDF, Van Den Berg E, Henderson L (2012) An assessment of the effectiveness of a large, national-scale invasive alien plant control strategy in South Africa. Biol Cons 148:28–38
Van Wilgen BW, Carruthers J, Cowling RM et al (2016) Ecological research and conservation management in the Cape Floristic Region between 1945 and 2015: History, current understanding and future challenges. Trans R Soc S Afr 71(3):207–303
Wall K (2008) Water supply: reshaper of Cape Town's local government a century ago. In: van Zyl JE, Ilemonbade AA, Jacobs HH (eds) Proceedings of the 10th annual water distribution systems analysis conference, WDSA2008, Kruger National Park, South Africa
Wicht CL (1945) Preservation of the vegetation of the South Western Cape. Report for The Royal Society of South Africa. Royal Society of South Africa, Cape Town

Archives

State Forestry Department Annual Reports

Series reconstructed from personal copies and libraries as follows:

Rhodes House Library, University of Oxford

Republic of South Africa: Department of Forestry: Annual report 1951–67, Shelf No.610.44s. 5

Plant Sciences Library, University of Oxford

BN/South Africa/Department of Environment Affairs: Report, Annual–Forestry Branch; and on Microfilm, Bulletins, P3, BN/Rumania, Forestry and Environmental Conservation Branch Annual Report

(This library now rehoused in the Radcliffe Science Library)

Reports series (with abbreviations used in text) as follows:

Cape of Good Hope (CGH): Department of Agriculture, Reports of the conservators of forests, Cape Colony (to 1909)

Government of the Union of South Africa (UG): Forest Department: Report of the Chief CONSERVATOR of forests (ARFD), 1910–58/59

Republic of South Africa (RP): Annual reports of the Department of Forestry (ARFD), 1959/60–78/79

RP: Annual report of the Department of Water Affairs and the Department of Forestry and Environmental Conservation (ARFD), 1979/80–83/84

RP: Annual report of the Department of Environment Affairs (ARFD), 1984/85–94

ARFD (Annual Report of the Forestry Department). UG53-1936; RP20/1961; RP80/1964; 1972/73 RP38/1974; RP18/1972

CGH (Cape of Good Hope): 1900 G.39-1901; CGH 1902, 17; CGH 1905; CGH 1892 27-93; CGH1893 G.50-94; CGH 1896 G17-97

Municipal Sources

Chancellor Oppenheimer Library, University of Cape Town, Government Publications Department

Minute of His Worship the Mayor, Cape Town (MM): 1900–1953

Shelf Ref.: G682 VC2 s.2247

Incorporating Annual Reports of the:

Chief Officer of the Metropolitan Fire Brigade (ARFB)
Superintendent of public gardens and tree planting (ARPGTP), 1900–09
Superintendents of public gardens, and tree planting (separately), 1910–28
Director of Parks and Gardens (ARDPG), c.1928–53
City Lands and Forests Branch (CLFB), c.1939–53
The Corporation of the City of Cape Town: Annual Report of the City Engineer (ARCE), Shelf Ref. G 682 VC ED(ENGI), including
Incorporating the Annual Reports of:
Forests, Parks and Gardens (ARFPG) c.1954–60
Parks and Forests Branch (ARPFB) 1961–94/95

Newspapers

Cape Times and Cape Argus newspaper reports accessed on microfilm at the African Studies Library, University of Cape Town, Ref. BZA 76/72, MP 1025, and Rhodes House Library, Oxford, call number 620.18 t. 2
Cape Argus (1935) £250,000 havoc in great fire. 26 December
Cape Argus (1935) To safeguard mountain. 27 December
Cape Times (1942) Stop those mountain fires. Editorial, 27 November
Cape Times (1943) National Veld Trust. Editorial, 7 January
Cape Times (1943) Start of war on soil erosion. 20 November
Cape Times (1943) The enemy within. Editorial, 4 December
Cape Times (1950) Letters to the Editor. 1 February
Cape Times (1960) £1.2 million new science drives round Cape. 29 January, front page
Cape Times (1973) Fire could ruin flora. 18 April: 2
Cape Times (2007) 'Tree Taliban' ignores please to save green heritage. 3 February: 3
Sunday Argus (2007) Author blames 'eco-fascists' for tree-culling. 7 October, 11

Index

A

Acacia saligna, 198, 221, 231–233
Acacia species, 220, 221, 230
Acclimatisation, 15
Adamson, Robert, 225, 227
Adaptation, 58, 115, 153, 156, 164
Aesthetic dimensions, 4, 187
Africa, 13, 16, 22, 26, 59, 107, 136, 149–151, 153, 156, 173, 175
African box thorn, 210
African clawed frog, 11, 133, 134, 136, 137, 141, 144, 145
African National Congress (ANC), 235, 238
Agriculture, Department of (S. Africa), 91, 227, 228, 237
Akrotiri Peninsula, 12, 16, 195–198, 201, 208, 212
Algeria, 11, 12, 53, 174
Algerian Psammodromus, 112
Alien reptiles, 106
Alien species, 1, 2, 26, 32, 39, 40, 51, 53–56, 62, 63, 65, 66, 100, 109, 110, 124–126, 145, 182, 183, 237
Alytes muletensis, 110
Alytes obstetricans, 112
Amathia verticillata, 26
Amphibians, 13, 54, 105, 110, 113, 133, 134, 137
Angola, 11, 153, 233
Animal personality, 157, 158
Anopheles sacharovi, 201
Anopheles superpictus, 200
Anthropocene, 2, 8
Anthropochorous, 16
Anthropogenic dimensions, 2
Apartheid, 232, 233
Aquaculture, 9, 29, 30, 33, 39, 54–57, 59, 60, 62, 66
Aquarium trade, 9, 25, 55, 61
Argentine ant, 10, 87–100
Arsenical baits, 93
Asmal, Kader, 235
Astacus leptodactylus, 53
Atherinomorus lacunosus, 31
Atlantic Ocean, 9, 22, 61
Australia, 12, 13, 16, 25, 172, 185, 198, 220, 221, 245
Australian myrtle, 221, 222, 230
Australian wattle, 221

B

Balearic Archipelago, 106, 108, 109, 111, 115
Balearic green toad, 109–111
Balearic Islands, 10, 61, 105, 107–110, 112, 113, 119, 125, 126, 134
Balearic shearwater, 120
Balearic Warbler, 120
Ballast water, 9, 23, 38, 40, 51, 57, 62
Barcelona Convention, 38, 40, 53
Barn owl, 119, 120
Baseline, 9, 13, 23, 30, 66
Behavioural interference, 114
Bermuda buttercup, 210
Biddulph, Sir Ralph, 201
Biocontrol, 2, 134
Biodiversity, 1, 4, 9, 10, 22, 26, 40, 53–55, 60, 63, 87, 100, 106, 107, 109, 126, 150, 156, 171, 179, 180, 186, 197, 199, 201, 211, 212, 236, 239, 243, 245

A. I. Queiroz and S. Pooley (eds.), *Histories of Bioinvasions in the Mediterranean*, Environmental History 8, https://doi.org/10.1007/978-3-319-74986-0

Bioinvasions (Biological invasions), 1–3, 5–10, 13, 14, 16, 17, 40, 196
Biosafety, 11, 146
Birds, 11–14, 108, 112, 113, 116, 119, 120, 149, 150, 152, 158–163, 179, 180, 187
Bird trade, 152
Bokkie the Grysbuck, 230
Bovill, A. K., 202
Braudel, Fernand, 14
British Empire, 202
Bronze Age, 243
Brunt, Hugo, 226, 228
Bufotes balearicus, 109
Bycatch, 9, 51

C
Cabrera, 107, 110
California, 16, 88, 98, 100, 114, 141, 175, 185, 221, 245
Callinectes sapidus, 27, 56, 61, 62
Calonectris diomedea, 120
Canaanite, 116
Cape Colony, 221
Cape Divisional Council, 231
Cape Floral Region, 237, 240
Cape of Good Hope, 219, 223, 233
Cape Peninsula, 88, 220, 223, 227, 228, 230, 231, 234, 235, 238, 239
Cape Town, 135, 221, 222, 225, 229–231, 235, 238, 241
Cape Town Municipality, 225, 226
Casuarina cunninghamiana, 198, 207
Casuarinas, 12, 220
Caulerpa cylindracea, 33
Caulerpa taxifolia, 9, 25
Caulerpa taxifolia var. *distichophylla*, 33
Centre for Invasion Biology (CIB), 237
Centre for Scientific and Industrial Research (CSIR), 232
Chamaleo chameleon, 112
Charadrius dubius, 119
Charybdis helleri, 63
Charybdis longicollis, 63
Chestnut blight, 2
Chile, 16, 88, 175, 245
Chronology, 12, 14, 176
Chrysemys picta, 112
Civilizations, 109, 120, 172
Climate, 1, 15, 22, 23, 107, 108, 123, 136, 157–159, 162, 179, 204, 221, 236, 238, 244–246
Climate change, 2, 6, 22, 63, 66, 113, 114, 123, 156, 182, 183, 186, 187, 238
Colonial Office (Britain), 200, 205
Colonisation (by species), 3, 6, 16, 155
Columbretes Archipelago, 115
Common chameleon, 112, 113
Common midwife toad, 112, 113
Common quail, 119
Common waxbill, 11, 149–153, 156, 163
Conservation /ists, 7, 15, 53, 65, 100, 113, 115, 119, 121, 125, 137, 139, 141, 149, 150, 163, 172, 177, 181–183, 186, 197–199, 226, 227, 229, 230, 234–236, 239, 240, 246
Containment, 2, 87, 99, 232
Control, 2, 4, 8, 10–12, 15, 38–40, 65, 66, 87, 91–96, 99, 100, 106, 112, 121, 123–125, 133, 134, 138, 141–145, 177, 179, 181, 184, 185, 187, 208, 219, 224–229, 231–237, 239, 240, 243, 244, 246
Convergent evolution, 6
Cordgrass, 171, 173, 175, 176, 181, 186
Cory's shearwater, 120
Côte d'Azur, 10
Coturnix coturnix, 119
Crabs, 9, 13, 27, 51–54, 60–64
Crayfish, 9, 10, 13, 51–54, 56–58, 64, 65
Crops, 14, 15, 90, 98, 99, 220, 245
Cultur* (culture/cultural) dimensions, 7, 89
Cyprus, 12, 16, 63, 195–205, 207, 208, 211, 213
Cyprus State Archives, 211

D
Dalmatian wall lizard, 114, 115
DDT (insecticide), 207
Decapods, 9, 10, 51, 52, 55, 56, 64, 66
Desert fan palm, 210
Diadema setosum, 36
Disease, 28, 30, 35, 36, 39, 59, 155
Distribution/ distribution mapping, 1, 2, 11, 27, 40, 56, 60, 65, 88, 136, 141, 142, 145, 153, 154, 156, 160, 161, 163, 121, 123, 124, 171, 173, 174, 176, 183, 196–198, 208, 232, 242
Domesticated animals, 14
Driftsands, 4, 12, 220, 221, 235
Dussumieria elopsoides, 34
Dutch East India Company, 220
du Toit, Ben, 239
Dyer, William Thiselton, 202
Dyspanopeus sayi, 62

E
Economic impacts, 2, 10, 21, 27, 63
Ecosystem functions, 62, 199

Ecosystem services, 7, 30, 40, 173, 180, 181, 183, 184, 186, 187, 199, 201, 211, 234
Education, 106, 180, 187, 212, 230
Elytropappus rhinocerotis (renosterbos), 220
Emys orbicularis, 110, 111
Endangered, 118, 120, 163, 174, 213
Endemic, 22, 32, 53–55, 106, 107, 115, 118–121, 124, 125, 171, 197
Environmental history, 3, 12, 13, 15, 199, 219
Eradication, 2, 10, 40, 52, 61, 65, 87, 94, 95, 98, 124, 125, 133, 134, 141–143, 145, 179, 184, 185, 206, 228, 231, 232, 237
Erinaceus europaeus, 121
Eriocheir sinensis, 10, 53, 56
Erosion, 16, 182, 187, 200, 227, 245
Erythraean bioinvasion, 32
Estrilda astrild, 11, 149–151, 153
Eucalyptus camaldulensis, 198, 207
Eucalyptus globulus, 201, 221
Eucalyptus gomphocephala, 198, 207, 209
Eucalyptus species, 202, 208
Europe, 10–12, 14, 16, 22, 23, 26, 29, 39, 55, 57, 64, 65, 87, 88, 91, 100, 106, 107, 112, 139, 152, 153, 171, 173, 176, 182, 185, 197, 246
European pond turtle, 110–112
European serin, 162
European turtle dove, 119
Exotic birds, 161

F

False smooth snake, 111, 112, 114
Famagusta, 200, 201, 204, 206
Feral, 16
Ficopomatus enigmaticus, 27
Fire
  aesthetic dimension, 4
  anthropogenic, 4, 6, 226, 238, 242
  burning practises, 227
  ecology, 245
  fire fighting, 235, 246
  harmful impacts, 64
  incidence, 13, 226, 228, 246
  major fires, 219, 234, 246
  regimes, 5, 16, 221, 233, 238, 242–244, 247
Fishes, 34, 35, 53, 59, 116, 181, 187
Fishing, 4, 25, 29, 33–35, 57, 61, 62, 142, 181
Fistularia commersonii, 35
Fleabane, 210
Flood, 11, 133, 138, 139, 145, 181, 185, 234
Forest, 107, 177, 196, 198, 200–203, 205, 207, 208, 210–213, 220, 224, 229, 231, 232, 239, 244–247
Foresters/ Forestry, 6, 12, 13, 15, 16, 141, 199–202, 204, 211, 213, 219, 221, 222, 225, 227, 229, 231–236, 239, 240, 244, 246
Forestry, municipal (Cape), 225, 226
Formentera, 106, 108, 110, 113, 115, 117–121
France, 10, 12, 16, 23, 27–29, 35, 38, 53, 57, 64, 87–92, 95–97, 99, 100, 112, 113, 174, 178, 245
Freshwater, 10, 12, 13, 22, 51–55, 64, 65, 139, 196, 197, 204, 206
Frogs, 10, 11, 13, 105, 111, 133–145
Fucus spiralis, 25
Fynbos, 17, 219–222, 227, 228, 230, 232–237, 239–240, 244
Fynbos Biome Project, 232, 234

G

Gammaesano, 93
Gecko, 10, 111
Genet, 120
Genetta genetta, 120
Geolog* (geological/geology), 5, 6, 22, 106
Goats, 109
Granivorous birds, 11, 161, 162
Grapevines, 2
Grasses, 13, 161, 163, 176, 210, 241, 242, 245
Grasslands, 11, 210
Grazing, 15, 16, 173, 205, 210, 211, 220, 227, 243–245
Greece, 35, 37, 53, 56, 59, 61, 197, 206, 243, 245, 246
Grohl, Colin, 230
Guillemard, F., 200
Gulf of Mexico, 13, 61
Gymnesic Islands, 106

H

Habitats Directive (EC), 11, 35, 39, 53, 138, 197
Hakea species, 221, 237, 245
Hall, Anthony, 230, 233
Hay, Douglas, 232
Health hazards, 36
Hedgehogs, 121
Hemidactylus turcicus, 110
Hemorrhois hippocrepis, 112, 122, 123
Henkel, John, 226
Herders/ herding, 5, 109, 238, 243
Hermann's tortoise, 110, 111
Herpetofauna, 10, 105, 106, 110, 112
Hill, Sir Arthur, 225
Historical analysis, 7
Historical contingencies, 133

Holmboe Jens, 204
Hooker, Sir Joseph, 202
Horseshoe whip snake, 112–114, 118, 119, 122, 123
House of Commons (Britain), 199, 200, 202, 203
Hull fouling, 21
Human-mediated, 26, 125
Human values, 196
Humid grassland, 210
Hunting, 4, 5, 17, 180, 212
Hutchins, E. E., 199, 200, 204, 205, 222
Hybrid ecosystems, 7
Hybridization, 114, 174, 186
Hyla meridionalis, 111
Hyperpredation process, 119

I
Iberia, 11, 61, 149, 150, 153–157, 160–163, 177, 243
Iberian Peninsula, 12, 61, 63, 64, 112–114, 118, 134, 149, 150, 153, 163, 171, 174–176, 178
Ibiza, 11, 106–108, 110, 113, 115–125
Ibiza wall lizard, 110, 112, 118–120, 124
Iceria purchasii, 95
India, 27, 201, 220
Indian Ocean, 9, 22, 51, 59, 63
Indigen* (indigenous/ indigeneity), 3, 4, 7, 9, 11, 15, 16, 21, 26, 29, 57, 185, 219, 220, 222, 224, 225, 228, 230, 245
Industrialisation, 16, 244, 245
Insecticide, 95, 100
Interdisciplinary research, 11
Introduced species, 4, 6, 7, 10, 21, 29, 110, 114, 221, 226, 228, 232, 238, 242, 245
Introductions, 3, 4, 6, 7, 9, 10, 12–14, 16, 21, 23, 25, 27, 29, 30, 32, 36, 39, 40, 55, 59, 66, 99, 110, 112, 113, 115, 118, 125, 134, 146, 150, 154, 171, 176, 219, 238
Invasional meltdown, 119
Invasion biology, 2, 157, 236
Invasion ecology. *See* invasion biology
Invasive species, 2–4, 7–10, 13, 14, 39, 55–57, 64, 98–100, 109, 118, 124, 125, 134, 142, 145, 149, 150, 156, 161, 163, 171, 181, 210, 227, 229, 230, 233, 237, 239, 240
Islands, 3, 5, 10, 13, 16, 61, 105–110, 112, 114–121, 125, 126, 134, 153, 172, 244
Israel, 32, 37, 53, 56, 60, 63, 246
Italian wall lizard, 111, 112, 114, 115, 125
Italy, 10, 12, 16, 23, 26, 30, 35–38, 53, 54, 56–58, 60, 61, 64, 65, 87–90, 92, 94, 96–99, 174, 177, 243, 245, 247

J
Jones, W.H.S., 206
Jonkershoek, 137, 227, 233

K
Kentish plover, 119
Kew, Royal Botanic Garden, 202
Khoikhoi, 220, 238
Kirstenbosch National Botanical Garden, 224, 229
Kruger, Frederick J., 232, 236

L
Laboratory, 54, 93, 94, 135, 145
Ladder snake, 110, 112, 113, 118, 119, 122
Lagocephalus sceleratus, 35
Land abandonment, 244, 246
Lanius senator, 119
Larnaca, 200, 201
Latrie, L. De Mas, 200
Legislation, 9, 10, 39, 94–96, 99, 100, 172, 220, 228, 230
Le Maitre, David, 233
Leptospermum laevigatum, 221
Leucadendron argenteum, 236
Levant, 16, 30, 32, 34, 37, 53, 54, 60, 201
Levant Sea/Levantine Sea, 32, 61
Lilford's wall lizard, 110, 115
Limassol, 200, 202–205, 207, 212
Linepithema humilis, 10, 87, 88
Lisbon metropolitan area, 133
Lister, Joseph Storr, 220
Little ringed plover, 119
Littorina saxatilis, 26
Livestock, 15, 220, 243–245
Lizards, 10, 11, 105, 110–116, 118, 120, 121, 125, 126, 134
Lückhoff, Carl, 229
Lycium ferocissimum, 210

M
MacOwan, Peter, 224
Macroprotodon mauritanicus, 111
Macrorhynchia philippina, 37
Madon, P. G., 199
Malan, Nico, 231
Malaria, 14, 195, 197, 200, 201, 205, 206
Mallorca, 106, 107, 110, 112, 113, 117

Mallorcan midwife toad, 110, 112
Malpolon monspessulanus, 112
Mandela, Nelson, 234
Maquis, 198, 201, 244
Mariculture, 21, 23, 27–29
Marine Strategy Framework Directive, 39, 66
Maritime trade, 112
Marloth, Rudolf, 224
Marshes, 12, 51, 57, 58, 107, 154, 174, 176, 178–187, 196, 197, 200–202, 211
Marsupenaeus japonicus, 56, 59
Mathematical modelling, 11
Mauremys leprosa, 112
MEDECOS network, 233
Mediterranean basin, 6, 10, 13–16, 51–53, 105, 106, 108, 134, 150, 156, 176, 220, 242–247
Mediterranean climate region, 15, 136
Mediterranean house gecko, 110, 116
Mediterranean region, 3, 4, 6, 12–14, 17, 53, 54, 99, 111, 150, 171, 246
Mediterranean Sea, 4, 9, 10, 12, 14, 21–26, 28–30, 32, 35–40, 52, 53, 55, 56, 59–63, 66, 106, 171
Mediterranean tree frog, 111, 112
Mediterranean turtle, 112
Mediterranean-type climate, 4, 133, 145, 233
Mediterranean type climate regions, 15, 243, 244
Megafauna, 16
Melicertus hathor, 59
Menaethius monoceros, 63
Menorca, 11, 106, 107, 110, 112, 113, 115, 117
Meridionalisation, 52
Messinian salinity crisis, 22, 106
Metapenaeus monoceros, 59
Metapenaeus stebbingi, 59
Michell, Roland L.N., 202
Microevolution, 156, 157, 164
Mining, 221
Moll, Eugene, 233
Molluscs, 13, 30, 31, 63, 182
Montpellier snake, 112–114, 118
Moorish gecko, 110, 116
Moroccan rock lizard, 111
Mouflon, 16
Mountain Club of South Africa, 231
Mountains, 12, 14, 199, 223, 224, 226, 230–232
Mozambique, 11, 153
Mullus barbatus, 34
Myicola ostreae, 29
Mytilicola orientalis, 29
Mytilus edulis, 29

N
Narratives (about invasions), 2
Nationalism, 237
National Veld Trust, 226
Native species, 10, 23, 30, 35, 52, 54, 56, 59, 61, 64, 66, 98, 109, 111, 112, 120, 125, 143, 150, 153, 156, 162, 163, 176, 177, 181, 184, 185, 195, 198–201, 208, 210–213, 239, 245
Natrix maura, 111
Naturalised species, 3
Naturalization, 116
Natural resources, 4
Natural spread (of alien species), 201, 208
Nature, 2, 4, 5, 7, 8, 13, 28, 100, 137, 141, 149, 159, 176, 199, 208, 223–225, 227, 230–235, 237, 238, 244, 246
Neolithic Age, 13
Newspapers, 225, 226
Non-Indigenous Species, 9, 23, 24, 26, 28, 30, 39, 55, 171, 180, 182
North Africa, 16, 57, 106, 114, 171, 244
North America, 2, 11, 13, 14, 55, 57, 63, 134, 171
Novel ecosystems, 162
Nursery trade, 110, 113

O
Occelated lizard, 112
Olive trees, 11, 113, 118, 125
Orange groves, 90, 95, 98
Orconectes limosus, 53, 56
Orconectes virilis, 53
Origin, 1, 2, 7, 8, 22, 35, 54, 60, 64, 91, 112, 113, 171
Ornamental plants, 11, 106, 114, 221
Ostrea edulis, 29
Ostreopsis cf. *ovata*, 38
Ottoman empire, 16, 200
Oxalis pes-caprae, 210

P
Pacifastacus leniusculus, 53, 56
Painted turtle, 112
Pakenham, Thomas, 238
Passive transport, 112
Pastoralism, 6, 243
Pathway, 30, 55, 60, 66, 111–114
Pearson, Harold, 224
Pelophylax perezi, 111
Penaeus aztecus, 29
Penaeus japonicus, 29

Penaeus kerathurus, 59
Penaeus merguiensis, 29, 30
Penaeus pulchricaudatus, 30
Penaeus semisulcatus, 59, 60
Percnon gibbesi, 61
Perez's frog, 112
Pests, 89, 95, 96, 99
Pets/pet trade, 11, 106, 109, 110, 113, 134, 149
Phillips, Lady Florence and Sir Lionel, 225
Phylloxera, 2, 95
Phytopathology, 92
Pinus (pine tree) species, 237
Pityusic Islands, 106, 107, 110, 115–121, 124, 125
Plantations (timber), Cape Peninsula, 88, 220, 230, 231, 234
Plantations, 15, 16, 200–202, 204, 219–225, 228, 231, 232, 234, 235, 239, 240, 242, 245
Plant communities, 196, 207, 208, 210, 241, 245
Plant pests, 15, 92, 94, 96
(Devil's) Peak
  Cecilia, 231
  Rifle Range/Ottery, 221
  Tokai, 221, 231
  Uitvlugt, 221
Plants (Cape Peninsula)
  and fire, 12, 219, 234
  dispersal strategies, 222
  indigenous, 11
  introduced, 13
  invasive, 3, 12, 15, 222, 230, 231, 233–236, 239, 240
Plants, invasive, 3, 12, 15, 220, 222, 228, 230–236, 239, 240, 245
Plasmodium vivax, 201
Pleistocene, 105–107, 110, 115
Plotosus lineatus, 36
Podarcis lilfordi, 107
Podarcis melisellensis, 114
Podarcis pityusensis, 112
Podarcis sicula, 111
Policy, 2, 7, 8, 10, 12–14, 39, 53, 66, 92, 96, 126, 138, 150, 195, 196, 199, 201, 219, 227, 232, 240, 243, 244, 246
Pond slider, 112, 113
Port Jackson wattle, 198, 204, 207, 208, 210, 221
Ports, 12, 23, 27, 171, 177
Portugal, 10–12, 14–16, 28, 53, 57, 64, 87–90, 94, 96–100, 113, 133, 134, 137, 138, 140, 142, 145, 146, 153–155, 157, 174, 245–247
Portunus segnis, 34, 60
Posidonia oceanica, 32
Post-nature, 8
Prawns, 9, 29, 30, 51, 53, 54, 59
Predation, 106, 113, 114, 116, 126
Preston, Guy, 235, 239
Procambarus clarkii, 10, 53, 57
Provisioning service (ecosystem service type), 181, 183, 222
Psammodromus algirus, 112
Pseudo-indigenous species, 27
Pterois miles, 37
Public awareness/ perceptions, 4, 21, 115, 121, 180
Public participation, 124
Puffinus mauritanicus, 120

R

Rackham, Oliver, 14, 197, 199, 200
Railways, 220
Rainfall, 5, 15, 16, 123, 138, 172, 227, 245
Ramsar, 12, 65, 108, 172, 197
Range expansion, 62, 149, 154–156, 159, 164
Rarity, 65, 198
Red Data Book (RDB), 198, 208, 210
Regné, Count Médéric de Vasselot de, 221
Regulating service (ecosystem service type), 199
Regulations, 10, 87, 90, 96–99, 237, 239
Reptiles, 11, 13, 105, 106, 111, 113, 116, 119, 123, 134
Research, 3, 8, 11, 14, 26, 31, 32, 38, 88, 93, 135, 137, 138, 145, 146, 149, 150, 153, 156, 158, 159, 163, 164, 195, 226, 227, 230, 232–234, 237
Rhopilema nomadica, 35, 37
Richardson, David, 233, 237
River red gum, 198, 202, 204, 207, 212
River restoration, 138
Royal Botanic Gardens (Kew), 202, 225
Royal Society of South Africa, 226, 227
Ruditapes philippinarum, 29
Ruined landscapes, 12

S

Salt lake, 197, 198, 200, 201, 203–205, 207, 211, 212
Salt marshes, 12, 13, 51, 171–174, 178–183, 186, 208, 209, 212
Salt meadows, 197, 205, 207

Sand dunes, 108, 174, 208
Sanitation, 195, 199, 202
Santa Eulària des Riu, 124
Sant Joan de Labritja, 124
Saurida lessepsianus, 34
Savile, A. R., 200
Scelarcis perspicillata, 111
Scientific Committee on Problems of the Environment (SCOPE), 233
Scientific research, 9, 13, 26, 136, 233, 236, 238, 240
Second World War, 27, 226, 245
Serinus serinus, 161
Sexual ornaments, 159
Sexual selection, 149, 159–161
Shipping, 9, 13, 21, 23, 26, 27, 34, 39, 55, 61, 66
Shrubs, 12, 15, 16, 108, 174, 183, 211, 220, 222, 241–243, 245, 247
Siganus luridus, 33
Siganus rivulatus, 33, 34
Silver tree, 236
Slater, W. J. B., 230
Smith, A., 197
Smuts, Jan, 226
Snakes, 10, 11, 105, 106, 109–114, 116–119, 121–125, 134
Soil erosion, 12, 225–227, 231, 239
Soils, 15, 176, 202, 204, 220, 227, 245
South Africa, 11–13, 16, 17, 88, 133, 136, 137, 145, 173, 185, 219, 224, 230, 234–237, 239, 245
South African Forestry Company Limited (SAFCOL), 234, 235
South African Institute of Forestry, 239
South African National Parks (SANParks), 234
Southern blue gum, 201
Spain, 10–12, 16, 23, 28, 37, 53, 54, 56, 57, 61, 64, 87–89, 95, 100, 105, 113, 118, 141, 154, 156, 174, 178, 179, 181, 245–247
Sparrman, Andreas, 220
Spartina densiflora, 12, 171, 175, 186
Spartina patens, 178, 179
Spartina versicolor, 12, 171, 174
Spatial scale, 3, 196
Species Distribution Models (SDMs), 155, 156
Spur-thighed tortoise, 111, 112, 116
Stellenbosch, 17, 227, 231, 233, 237
Streptopelia turtur, 119
Styela plicata, 27
Suez Canal, 9, 13, 21–23, 26, 30–34, 39, 51, 53, 55, 56, 59, 60, 66
Swamps, 4
Sylvia balearica, 120
Symphyotrichum squamatum /saltmarsh aster, 210
Synanceia verrucosa, 36

T
Table Mountain, 220, 222, 224–226, 228–232, 236, 241
Table Mountain National Park, 223, 234, 238
Tarentola mauritanica, 110
Taxonomy, 15
Taylor, Hugh, 233
Testudo hermanni, 110
Tethys Ocean, 22
Thermophilic species, 30
Threatened, 28, 56, 119, 120, 198, 213, 230, 239, 243
Timon lepidus, 112
Toads, 10, 105, 134
Tortoises, 10, 105, 109, 112
Tourism/tourists, 6, 10, 22, 27, 64, 90, 106, 110, 112, 115, 139, 234, 244
Trachemys scripta, 112
Trade, 3, 6, 10–14, 25, 30, 39, 57, 63, 90, 110, 112, 113, 117, 118, 136, 152, 172, 178
Transport, 6, 23, 27, 38, 56, 106, 116, 133, 134, 138, 178
Trees, 12, 13, 15, 16, 89, 90, 95, 107, 113, 125, 195, 196, 203, 205, 209–211, 220–223, 225, 228, 232, 238, 240–242, 245, 247
Tuart, 198, 207, 212
Turtles, 10, 105, 110–112, 134
Tyto alba, 119, 120

U
UK National Archives, 203, 205–207
Upeneus moluccensis, 34
Urbanisation, 6, 219, 238, 244, 246
Urban pest, 89

V
Vacant niche, 163
Value, 7, 8, 54, 89, 90, 100, 145, 164, 172, 180, 187, 196, 197, 199, 211, 212
Van Riebeeck, Jan, 220
Van Wilgen, Brian, 233
Vegetation succession, 244
Viperine snake, 111, 112

W
Washingtonia filifera, 210
Waterer, R. R., 200, 207

Water supplies, 13, 219, 222, 227, 232, 235, 245
Wetland, 12, 54, 173, 195–198, 200
Wicht, Christiaan, 227
Wild, A. E., 201
Woodchat shrike, 119
Working for Water, 235–237, 239

X
Xenopus laevis, 11, 133–135, 145

Z
(Z)-9-hexadecenal, 100

The manufacturer's authorised representative in the EU is Springer Nature Customer Service Centre GmbH, Europaplatz 3, 69115 Heidelberg, Germany. If you have any concerns regarding our products, please contact ProductSafety@springernature.com

Printed and bound by CPI Group (UK) Ltd, Croydon, CR0 4YY
01/10/2026
02225860-0002